The Lerma-Chapala Watershed

Evaluation and Management

The Lerma-Chapala Watershed

Evaluation and Management

Edited by

Anne M. Hansen and Manfred van Afferden

Mexican Institute of Water Technology
Jiutepec, Mexico

Kluwer Academic / Plenum Publishers
New York, Boston, Dordrecht, London, Moscow

ISBN 0-306-46747-X

233 Spring Street, New York, New York 10013

http://www.wkap.nl/

10 9 8 7 6 5 4 3 2 1

A C.I.P. record for this book is available from the Library of Congress

Printed in the United States of America

Contributors

Ministry of Urban Development, Jalisco (SEDEUR)
National Water Commission (CNA)
Mexican Institute of Water Technology (IMTA)
Centrum for International Migration and Development (CIM)
Asociación Mexicana de Limnología (AML)
Aquatic Ecosystem Health and Management Society (AEHMS)

Front cover:
Traditional fishing with *tumbo* in Lake Chapala South shore at San Luis Soyatlán county. The *tumbo* is an extended stationary fishing art for in-water captures. Photography: Pablo Chávez-Hernández, IMTA

Back cover:
Priest of Tláloc (rain god), Río Blanco-Papaloapan culture, Mexico, 600-900 A.D., terracota, 47 cm, Anthropological Museum, Jalapa, Mexico. The Mesoamerican cultures were aware of groundwater drainage in the mountains. Photography: Carlos Blanco

Hieroglyph :  *atl*, water in Náhuatl

Cover design:
Xóchitl Zúñiga

Acknowledgements

The editors wish to express their deepest gratitude to the following referees for their critical review and recommendations to improve the quality of the chapters in this book:

1. Javier Alcocer, National University of Mexico
2. Javier Aparicio, Mexican Institute of Water Technology
3. Alfonso Banderas, Mexican Institute of Water Technology
4. Luis Angel Barrera-Morteo, National University of Mexico
5. Arturo Chacón, Michoacán San Nicolás de Hidalgo University
6. Sonia Dávila-Poblete, Mexican Institute of Water Technology
7. Benjamín de León, Mexican Institute of Water Technology
8. John Dunbar, Baylor University, USA
9. Carlos Escalante, National University of Mexico
10. Mercedes Escamilla, Oficina de Comunicación del Lago, Mexico
11. Anatoliy Filonov, Guadalajara University, Mexico
12. Tim Ford, Harvard School of Public Health, USA
13. Alberto Guitrón, Mexican Institute of Water Technology
14. Carlos Gutiérrez-Ojeda, Mexican Institute of Water Technology
15. John R. Jones, University of Missouri, USA
16. Jess Kelly, Baylor University, USA
17. Anna Kurtycz, France
18. Owen Lind, Baylor University, USA
19. Polioptro Martínez-Austria, Mexican Institute of Water Technology
20. Manuel Martínez-Morales, Mexican Institute of Water Technology
21. Jorge Martínez-Ruiz, Mexican Institute of Water Technology
22. Rodrigo Moncayo-Estrada, CIIDIR-IPN-MICH, Mexico
23. Harvey Shear, Environment Canada
24. Joseph White, Baylor University, USA
25. Pedro F. Zárate-del Valle, Guadalajara University, Mexico
26. Nayeli Zúñiga, ITESM Campus Monterrey, Mexico

Foreword

James O. Leckie
Environmental Engineering and Science Program, Department of Civil and Environmental Engineering, Stanford University, Stanford, California 94305-4020, USA

Nearly 10 years have passed since the beginning of the systematic studies of the Lerma-Chapala Basin coordinated by the Instituto Mexicano de Tecnología del Agua. Although many public and private institutions, universities and research centers have conducted studies on the Lerma-Chapala Basin over the last two decades, the need for a comprehensive summary of the findings of those studies has become increasingly obvious and important for this critical water resource.

The Lerma-Chapala Basin is located in the central part of Mexico, and partly occupies five states. The watershed comprises the Lerma river and Lake Chapala. With a length of over 700 km, the tributary watershed covers approximately 54,000 km^2. The basin accounts for more than one-third of the country's economic activity, one-fifth of all commerce and one-eighth of the nation's agricultural land. The watershed receives 3% of the country's total rainfall, less than 1% of the runoff, and accounts for 13% of the total groundwater.

Lake Chapala represents the final destination of most of the pollution discharged to the Lerma river watershed, where surface water and agricultural land are being contaminated by domestic and industrial discharges. The average depth of the Lake Chapala varies between 4 and 7 m, depending on the season and the annual rainfall. Wind mixing helps reduce stratification, and the lake is usually well mixed and oxidized at all depths. Lake Chapala is the largest natural water body in Mexico and, as such is of major importance to both the ecologic and socio-economic systems of the state of Jalisco.

The Lerma-Chapala Basin has cultural and historical as well as scientific and economic importance. The presence of many indigenous people's origins within the basin adds archeological and anthropological interest to a complex setting. Although the current book focuses on summarizing the science and water resource management issues, it also set the precedent for the knowledge base of the ecology, social-political and economic aspects of the Lerma-Chapala Basin. Such knowledge bases are necessary for the development of coherent public policy for efficient management and use of this finite resource.

Generally speaking, multidisciplinary approaches are obviously required to solve environmental problems. This is at odds with the traditional content of most university training, which requires that students, during the course of their curricula, become ever more specialized. The price of competence in natural, economic and social sciences is linked to this high degree of specialization. Unfortunately, simply putting together teams of specialists cannot attain reconciliation of this requirement with multidisciplinary approaches that are needed to study environmental systems. In order to succeed, scientists must be able to communicate effectively together at the highest possible level, to allow both a constructive exchange of ideas and a clearer realization of the limits of their own specialty. Creating the necessary context to enable effective communication is one of the future challenges facing those responsible for the management and operation of our water resources. The task is complicated by the competing demands on multi-use water resource systems.

One of the main purposes of this book is to bring together the wide range of knowledge providing the basis for intelligent management and effective allocation of water as a scarce resource. The authors' and the editors' participation in this endeavor is a valued contribution toward the goal of development of a coordinated comprehensive knowledge base for the Lerma-Chapala Basin water resources.

Preface

Álvaro A. Aldama
Instituto Mexicano de Tecnología del Agua, Paseo Cuauhnáhuac 8532, Jiutepec, Mor., Mexico

The environmental challenges we face today include many of the same ones that were faced twenty years ago. In spite of the efforts of environmental engineers, scientists, legislators, and the public, our environmental problems remain. Many of our rivers and lakes still suffer the effects of poorly treated sewage and industrial discharges. The Lerma-Chapala basin is no exception to this situation. The Lerma-Chapala basin is of the utmost importance to the development of Mexico, and the control of water use and pollution should be of the highest priority in this region. Urban, agricultural and industrial developments have caused serious conflicts in the use of water and, due to decreased water quality, have limited water availability.

The Lerma-Chapala-Santiago Watershed Council was the first river-basin council to be established in Mexico. It also exemplifies the advances in water management decisions that have been attained through stakeholder participation. Solutions to water management problems are not simple to achieve in such a complex region, given that there is an increasing water demand by industry, agriculture and a growing population. Problems of pollution of surface water, groundwater and soil are also present in the Lerma-Chapala basin.

The importance of an accurate water balance cannot be overemphasized. Interrelationships between groundwater and surface water and errors in terms of water budgets have been the focus of studies carried out in the Lerma-Chapala watershed over the past decades. Lake Chapala is Mexico's largest and most important lake. Cyclic decreases in water levels of the lake have alarmed the population not only in the immediate surroundings, but also nationwide and abroad. These fluctuations have caused both ecological and economical problems, including those related to the excessive presence of water hyacinth and blue-green algae.

In Lake Chapala, as in other lakes, eutrophication is a natural process taking place over long periods of time and has been greatly accelerated by human activities. With the addition of anthropogenic nutrients, algae growth in the lake is increased. Eventually, a process that would have occurred over geologic time scales is decreased to decades. When not controlled, this process may have some undesirable effects on water quality such as: excessive plant growth (green color, decreased transparency, excessive weeds), hypolimnetic loss of dissolved oxygen (anoxic conditions), loss of species diversity (loss of fishery), and taste and odor problems.

Contaminants from human activities in the watershed emanate from both point and non-point sources. Point sources are usually more easily controlled (by wastewater treatment). After point source reductions are accomplished, non-point sources may become the most significant contaminant loading to the water body. Non-point sources are caused by agricultural runoff, storm water runoff or combined-sewer overflows. The control of non-point source pollution is the subject of current research. Both particulate and dissolved forms of contaminant additions are important because even particulate phases become available to the water and biota after desorption or transformation in the natural ecosystem. Bioavailability of contaminants is the subject of much concern and debate as to their relative contribution of particulate and adsorbed pollutants to rivers and lakes. In order to obtain as much information as possible, the distribution and availability of contaminants must be taken into account as new monitoring strategies are designed and implemented.

During decades, public and private institutions, research centers and universities have carried out studies on the Lerma-Chapala basin. The results of these studies have received limited exposure and distribution, making recovery of the information difficult. Often studies are repeated, resulting in wasted resources and discontinuity in the historical data that could be used to further work in water use efficiency and ecosystem health assessment.

Solutions to environmental problems in general and in particular those that have been found in the Lerma-Chapala basin, must be approached from a multidisciplinary point of view. The present book includes important contributions from several universities and research centers: University of Guadalajara (U de G), Mexican Institute of Water Technology (IMTA), Autonomous University of Guadalajara (UAG), Baylor University, and Environment Canada, to name a few. It is meant to serve as a starting point for future research and decision-making based on this knowledge. The editors of this book are privileged to have such a fine group of authors contributing with chapters on the various aspects on watershed management. Many of the authors are experienced professionals in their respective fields. I am convinced that this book gathers the most important technical and scientific information on the Lerma-Chapala basin and will serve as a means to make this information available to the government, research centers and society at large.

Introduction

Eduardo Mestre-Rodríguez
Independent Senior Consultant on integrated water resources management. Member of the World Bank River Basin Water Management Window Team. Querétaro, Qro., MEXICO

Water is either critically scarce or highly abundant in Latin America. Some of the richest countries in per capita water availability in the world are located in this region (i.e. Paraguay, Surinam, and Venezuela). However, paradoxically, some of the poorest nations with regards to water availability coexist in this region, with the added stress that their socioeconomic development patterns are rapidly increasing water demands and triggering fierce conflicts. This in turn endangers water governance and has direct negative feedback on water availability, both in quantity and quality. Argentina, Chile, Peru and Mexico, are Latin American countries where in specific sub-regions water is normally scarce or has become scarce either because of natural causes, like cyclical climatic phenomena, or aggravated by human activities and faulty policy that exert increasing pressure on existing supply and reduce it furthermore by pollution.

Within Latin America, Mexico is different concerning history and culture. Its cultural archetypes, some of them only slightly understood, and deeply intermixed racial roots, as a result of the marriage of two rich cultures, the Spaniards and the inappropriately called Indians, have produced a unique civilization, located in an attractive natural location in the world, with its potentials, asymmetries and drama – natural and manmade.

With a territory of little less than two million square kilometers[1], Mexico is the tenth economy in the world with an estimated GNP of 600 billion US dollars for 2000 ($5,980 dollars per capita) and the eleventh country in terms of population with over 95 million inhabitants, 80% of which live in urban

[1] Mexico covers 1.97 million km^2; it is divided into 31 autonomous states – each with its own elected government – and a federal district, which includes Mexico City.

settlements. Presently, 60% of the population is either middle or high class, whereas 40% is still poor, among which, 18% live in extreme poverty (however, no famine exists). Mexico's annual exports are above 135 billion dollars[2] (eighth, worldwide) as a result of structural economic changes and the benefits of NAFTA[3], that triggered a positive economic reaction that helped sustain overall annual economic growth rates higher than 5.5% in the period 1996-2000. The Mexican peso has steadily strengthened and is currently the strongest currency in Latin America. Expert financial reviews grant the lowest investment risk levels to Mexico together with Chile. Recent favorable negotiations with the European Union and South American economic blocks are expected to reinforce economic activity and exports. In turn, these events will raise water demands, which may unfortunately contribute to aggravate the already critical situation.

Water is scarce in Mexico. A complex system of mountain ranges creates 310 hydrological basins, with different degrees of water availability, hydraulic development and water pollution. Although mean rainfall is about 740 mm – roughly the same as in England, its spatial and seasonal distributions make a difference in the ways water is available. Total average surface runoff together with aquifers mean annual recharge is approximately 460 cubic kilometers. Its average annual water availability per inhabitant is below 5,000 cubic meters. In certain areas, like in the Valley of Mexico, the Lerma river basin, La Laguna endorheic system and the Bravo river (Río Grande), this figure drops below 1,000 m^3 per year. Needless to say, Mexico is poor in terms of water supply. Water exists where population density is low, like in the Southeast, where in less than 25% of the territory, two-thirds of all rainfall and surface runoff occur. In dramatic contrast, population and economic activity concentrate in the higher plateaus where climate is temperate but water is either naturally scarce or has become scarce by severe pressure on allocation and by non-optimal water management.

In many regions, Mexico currently faces an imbalance between water demand and availability, primarily due to natural water scarcity as well as uneven water quality distribution. Rapid urban and industrial growth, among other economic and social factors, has worsened this scenario. Water needs have raised, water uses are fiercely competing with each other and as a result, conflicts are surging. Water quality has deteriorated as urban and industrial effluents are often discharged with insufficient previous treatment. Furthermore, Mexico's past economic and financial crisis, limited hydraulic infrastructure development and triggered low water services (urban and rural

[2] Oil represents less than 20% of overall exports, whereas manufactured goods exports are close to 60%. In lesser extent, agricultural and fisheries are still relevant.

[3] Mexico has become US second commercial partner and a strategic ally in political and economic matters.

water utilities) in many sectors and locations. Presently, in formal or legal terms, only three countries in the American Hemisphere allow water markets (either national or regional): The United States, Chile, and Mexico, each with interesting singularities with regards to underlying concepts, procedures and results. Water markets are rapidly affecting actual water distribution per use and water productivity throughout the country, with extraordinary effects in more developed regions such as Valley of Mexico, Lerma-Chapala and Rio Grande basins. However, water markets, without appropriate regulatory measures might yield unsatisfactory results, as can be seen both in Chile and Mexico. The latter has yet to rule out speculation and monopolization, as well as establish a public participation with regards to sectoral water transfers from one use (irrigated agriculture) to another (industry, urban water utilities). Water markets with insufficient financial moderating instruments may go out of kilter.

Presently, in Mexico water is gradually reflecting more the overall costs involved in having access to it (capital, operation, management, and opportunity costs), especially in those regions where water is or has become scarce. However, in the past, social factors have led to grant important subsides to certain water uses, mainly irrigated agriculture[4]. Water productivity is extremely low, and this in turn affect other potential uses that could utilize presently allocated water for higher productivity uses, without affecting agricultural activities, should water efficiencies be improved. Here lies one of the most important possibilities for future water management optimization in Mexico. By Federal Law, water users who are legal tenants of federal – issued water rights titles have to pay for water use by volume[5]. These revenues are called *derechos de agua* (water rights) and were originally conceived as basin tariffs, which would reflect water opportunity costs, that on one hand could render relevant revenues for the water sector, and on the other hand, induce higher water efficiencies. Periodically water rights rates per cubic meter are published within the Federal Rights Law dispositions – such rates are subject to debate in the House of Representatives that has to approve them before becoming official. Water rights rates are mainly determined according to use and location. Payments should be quarterly made on a self-declaration basis and are thought to find its way back to the water sector, mostly via the National Water Commission (CNA) budget, although revenues only exceptionally return to the specific region where they have been originally produced. In Mexico, revenues are

4 Irrigated agriculture represents more than 75% of all water being used in Mexico.

5 Only the Federal Government, through the CNA is legally entitled to issue water rights, either for surface runoff or groundwater.

close to one billion dollars a year, although the potential is much larger[6]. Basin tariffs are also paid for sewage discharges to national water bodies. Such tariffs are based on the discharged volume on a specific location together with water quality criteria[7]. Effluent discharge water rights were conceived, similar to water abstraction basin tariffs. The objectives were to raise revenues to support water quality management expenditure throughout the country and foster public and private investment in water treatment facilities and schemes to effectively reduce pollution[8]. Water discharge tariffs are also periodically published as part of the water rights legislation, and must be previously approved by Congress.

All Mexican water resources are national property. However, water allowances – *concesiones* – may be in the domain of private persons or organizations[9]. Maximum annual abstraction volumes are indicated in these *concesiones* as well as their exploitation rate. Water allowances are usually secured for less than 25 years, although no legal recommendations exist.

The Lerma-Chapala basin is of paramount importance to Mexico, from the economic, political and social points of view. Consequently, this basin receives priority attention from all three government levels – federal, state and municipal. Public awareness on water issues has led to active participating roles played by water users, non-governmental organizations, and social institutions with a plethora of interests directly or indirectly linked to the water sector.

The Lerma river – 750 km long – is born in Mexico's central high plateau surpassing 3000 meters above sea level (masl) and ends in Lake Chapala (1510 masl) [10]. The Lerma river basin, a subtropical semiarid region with average temperature of 21 °C, and an area of 54,421 km^2 – less than 3% of Mexico's territory – features an average rainfall of 735 mm per year, mainly concentrated in the summer, from which a mean runoff of 5.76 km^3 is derived. The Santiago river flows westward from Lake Chapala to the

[6] Potential payers are at least three times as many as those who presently pay. Financial – fiscal – coactive tools have been enforced but mainly dedicated to industrial users who pay for more than 90% of all revenues. Average annual per capita revenues derived from such payments, are presently less than US$10, whereas in developed countries with similarly based basin tariffs, annual per capita revenues can be over US$60.

[7] Only two parameters (total suspended solids and biochemical oxygen demand) were agreed to be accounted for as reasonable proxies for other relevant pollutants.

[8] Originally, water discharge tariffs were purposely higher than equivalent treatment costs (capital investment, operation and management) to improve water treatment.

[9] However, water allowances are not real rights in terms of private possession. Furthermore, although such is the case, water allowances are subject to alienation, inheritance and pawning.

[10] Lake Chapala is the largest tropical lake in Mexico – 77 km long and 23 km wide. Its maximum storage capacity is 8.13 km^3 with a global surface around 110,000 ha. The lake is shallow, with an average depth of 7.2 m, and a maximum of 16 m.

Pacific Ocean[11]. The Lerma-Chapala river basin comprises portions of the central states of México, Querétaro, Guanajuato, Michoacán, and Jalisco, all of capital importance in Mexico's development. Current basin population is eleven million – one in nine Mexicans – with an annual growth rate slightly less than the national average. The Upper Lerma aquifers supply water to 2 million inhabitants in Mexico City and Lake Chapala supplies water to an additional 2 million inhabitants in Guadalajara. The population in the basin is distributed in 6,224 localities, 18 of which have more than 50,000 inhabitants. Rural population contributes currently with 27% of the basin's total population.

Regional socioeconomic development has been triggered by water availability. Industrial and agricultural production per capita has surpassed national levels. Currently, 795,000 hectares (one eighths of Mexico's irrigated land) belong to the Lerma-Chapala basin. National farm goods exports rely heavily on the performance of this tiny region, which is of utmost importance for Mexico's high-value agricultural exports. The region boasts 9,200 industries, which generate one third of the industrial GNP (9% of Mexico's GNP). Furthermore, 20% of all national commerce and service activities occur in this basin. With three highly developed economic sectors and a superior transportation network – partially financed by private investors – this is in fact one of the richest regions in Latin America.

All surface runoff in the basin, either natural or effluent, is allocated. Water efficiency in agriculture is low and illegal water abstractions occur in certain seasons. Some 28,000 deep-water wells operate in the basin with very low efficiency rates, high electricity consumption and rather low water yields. Nevertheless, almost 70% of the 38 aquifers in the region are overexploited. Conflicts derived from surface runoff uses (mainly for irrigation and drinking water supplies), combined with generally untreated effluents discharges, have originated serious pollution problems, both locally and regionally. Another frequent conflict over water quality occurs in Lake Chapala, which plays a key role as a main water source for Guadalajara. As efforts from very different sources and interests converge in this region, the Lerma-Chapala basin could become a model in many fields, to guide Mexico as well as some other emerging economies, towards attaining integrated water management.

[11] The Santiago river basin is less developed in terms of population and economic activity, except for Guadalajara, the second largest city in Mexico, whose metropolitan area has more than 3.5 million inhabitants.

Contents

SECTION I

NATURAL RESOURCES AND MANAGEMENT IN THE BASIN

Chapter 1

Hydrology of the Lerma-Chapala Watershed

JAVIER APARICIO
Instituto Mexicano de Tecnología del Agua. Paseo Cuauhnáhuac 8532, Col. Progreso, 62550 Jiutepec, Mor., Mexico. email: aparicio@tlaloc.imta.mx

Key words: Hydrology, climate, precipitation, evaporation, runoff, hydraulic infrastructure

Abstract: Hydrologically speaking, the Lerma-Chapala watershed is a sub-basin of the Lerma-Santiago watershed, with a total area of 137,144 km^2, comprising the Lerma and Santiago rivers and Lake Chapala as an internal lake, which forms the boundary between the two river basins. The Lerma river basin, which is considered to end in Lake Chapala, has an area of 52,545 km^2. The mean annual precipitation over the entire Lerma watershed is about 730 mm. Hurricanes and other cyclonic perturbations regularly affect the Lerma river basin and floods occasionally cause damages in the area. The dominant climate in the watershed is subtropical, ranging from sub-humid to arid. The Lerma river is more than 700 km long and receives water from several tributaries. Some of the more important are the Laja, Turbio, Angulo, and Duero rivers. Besides Lake Chapala, there are several important natural lakes, like Pátzcuaro, Cuitzeo and Yuriria as well as a significant number of dams. The main use of these dams is for irrigation purposes of about half a million hectares. The mean annual runoff of the Lerma river to Lake Chapala is about 1,160 Mm^3, although this figure varies considerably from year to year depending on several factors, among which variations in water extracted from the Lerma river is the most important. About 1,000 Mm^3 are extracted from the numerous aquifers in the basin, among which the Toluca, Ixtlahuaca, Querétaro, León, Pénjamo, Aguascalientes and Cuerámaro aquifers are among the most important. They serve both urban and agricultural water needs. In general, groundwater resources in the basin are overexploited.

The Lerma-Chapala Watershed, Edited by Hansen and van Afferden,
Kluwer Academic/Plenum Publishers, New York, 2001

1. INTRODUCTION

In this chapter, a general description of the Lerma-Chapala climate and hydrology is made. This description is intended to constitute a framework for the rest of the chapters of this book, and is focused on general, regional data. Rather than presenting the results of a specific research project, this description is based on the analysis of existing data. The Lerma-Chapala watershed is one of the most important in Mexico due to its economical importance as related to national production and to the size of the population. It is also one of the most studied and monitored watersheds in Mexico. The analysis of hydrological information has been divided in precipitation, runoff, and evaporation to describe each component of the hydrological cycle. Interrelations between the different components are discussed as well, and the reader will also find a description of the hydraulic infrastructure in the basin.

2. GENERAL CHARACTERISTICS, AND TOPOGRAPHY

The Lerma-Chapala watershed is a sub-basin of the Lerma-Santiago basin (Figure 1). The Lerma-Santiago watershed has an area of 137,144 km^2 and it includes the Lerma and Santiago rivers, as well as Lake Chapala, which can be considered as an interior lake located between the two river basins. The Lerma river basin (Figure 2), comprising Lake Chapala and its own watershed, includes part of the states of Guanajuato, Jalisco, México, Michoacán and Querétaro. The surface of the basin represents approximately 3% of Mexico's national territory and is close to 52,545 km^2. This area includes the closed basins of the Yuriria, Pátzcuaro and Cuitzeo lakes (Secretaría de Recursos Hidráulicos, 1974). The Lerma river is more than 700 km long and receives water from several tributaries. Some of the most important are the Laja, Turbio, Angulo, and Duero rivers. In Figure 2, a sketch of the watershed is shown, including the main rivers and surface water bodies and lakes. Table 1 summarizes the main river basin areas and water bodies in the Lerma basin.

From a topographical perspective, the Lerma watershed is relatively flat. About 72% of its area consists of elevations ranging from 1,500 to 2,100 meters above mean sea level (mamsl) including small mountainous regions with elevations up to 2,700 mamsl, representing only about 6% of the total watershed area (see Figure 3).

Table 1 Main river and lake catchment areas

Region	Catchment	Partial area	Region area	Total area
Upper Lerma	Upstream Solís Dam	8538	15843	52545
	Lake Pátzcuaro	1096		
	Lake Cuitzeo	3769		
	San Luis de la Paz closed watersheds	2440		
Middle Lerma	Lake Yuriria	1240	26375	
	Laja river	9629		
	Turbio river	4489		
	Angulo river	2079		
	Others	8938		
Lower Lerma	Huáscato river	457	10327	
	Duero river	2690		
	Sahuayo river	907		
	Zula river	2098		
	Others	4175		

Areas in km^2

For analytical purposes, the Lerma watershed is usually divided in three zones (Figure 2): the *higher Lerma* region, including the area from the origin of the Lerma river at the Almoloya lagoon to the Solís Dam. Its area is of 15,843 km^2 and includes the basins of Lakes Pátzcuaro and Cuitzeo, as well as some closed sub-basins in the San Luis de la Paz area (Secretaría de Agricultura y Recursos Hidráulicos, 1973).

The Middle Lerma includes the area from the Solís Dam outlet to the Yurécuaro gauging station and has an area of 26,375 km^2 (Secretaría de Agricultura y Recursos Hidráulicos, 1979). The Lower Lerma begins at the Yurécuaro station and ends at the Lake Chapala outlet to the Santiago river. It includes the Duero and Huascato rivers and Lake Chapala with its own watershed, and has an area of 10,327 km^2.

3. CLIMATE AND HYDROLOGY

3.1 Climate

According to the Köppen classification system, as modified by García (1964), the dominant climate in the Lerma-Chapala watershed is subtropical, ranging from sub-humid to arid, and a few small regions have a template sub-humid climate.

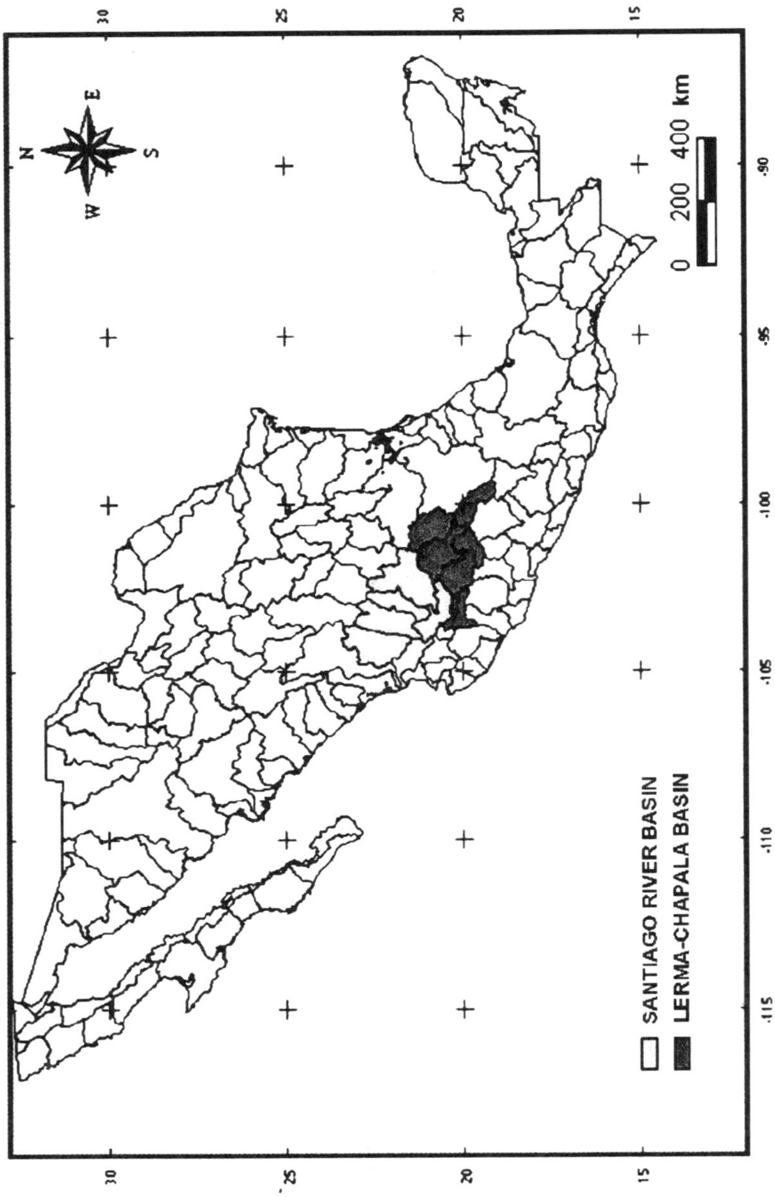

Figure 1 Mexican river sub-basins and location of the Lerma-Santiago region

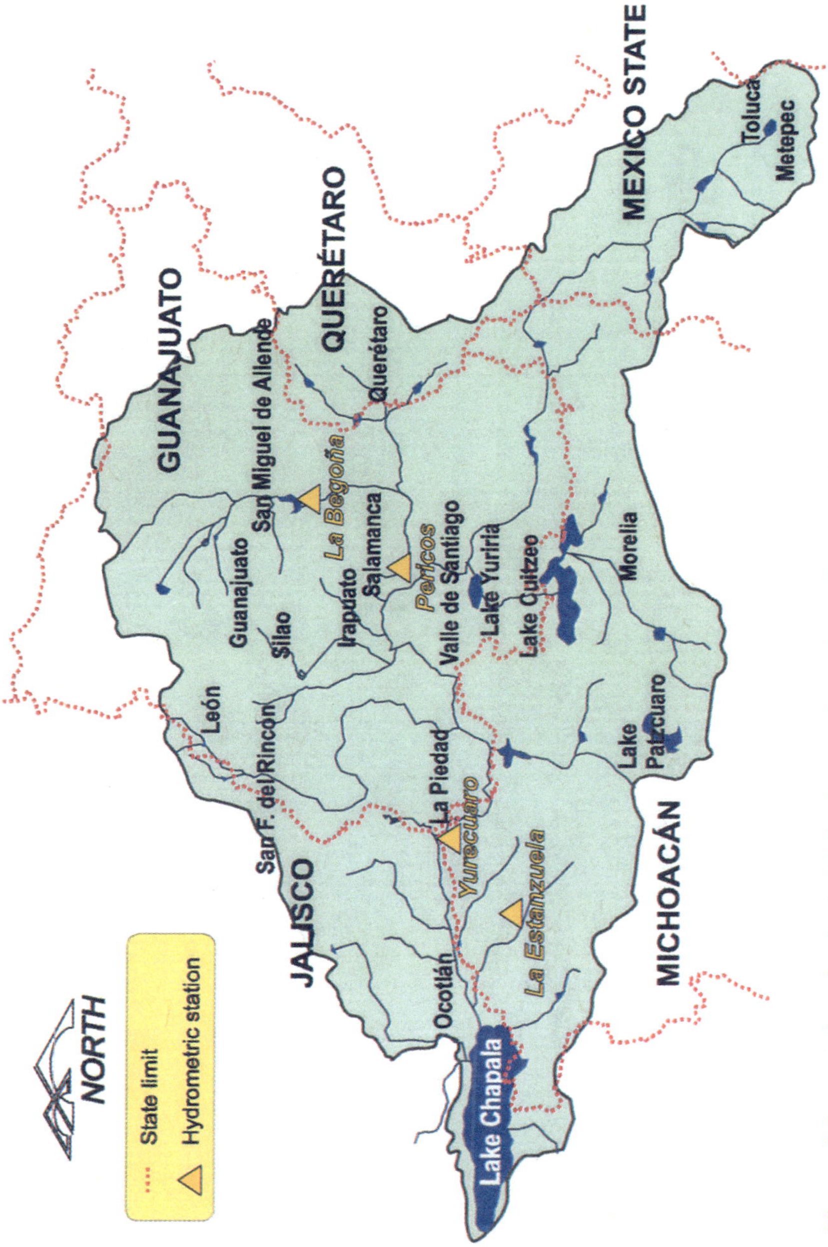

Figure 2 Lerma river basin (with hydrometric stations discussed in this chapter)

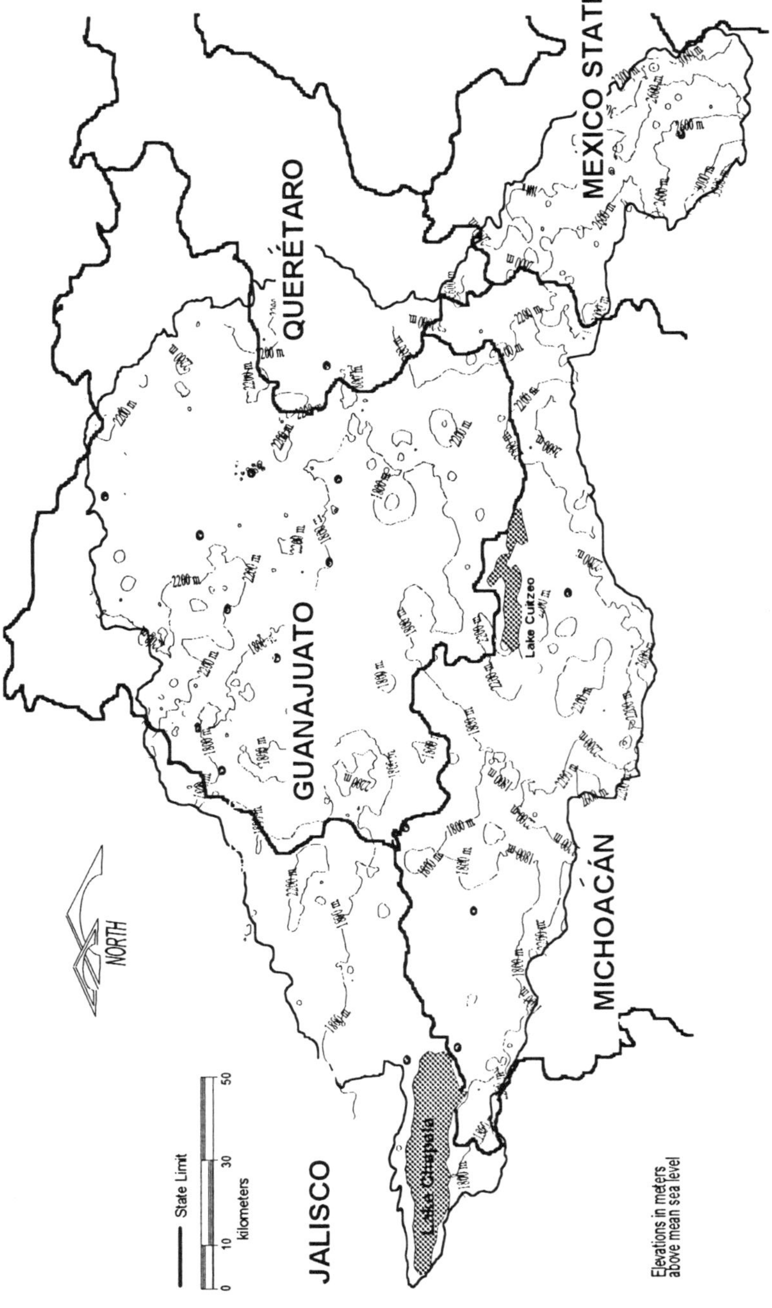

Figure 3 Elevations of the Lerma river watershed

The main climate sub-regions that can be identified by state within the Lerma-Chapala watershed are the following (Medina *et al.,* 1998):

Mexico State. Subtropical template sub humid climate prevails in almost all the state within the Lerma river basin. Only a small part of the area of influence of this river, in the region surrounding the city of Toluca, is classified as cold, template sub humid.

State of Guanajuato. A major part of this state is located within the watershed boundaries. Climates include: subtropical warm arid in the region of Apaseo el Alto, subtropical template arid around San Diego de la Unión, San Luis de la Paz and the limits with the state of San Luis Potosí; subtropical semi-warm semi-arid in the vast region where the cities of La Piedad Cabadas (in the state of Michoacán, near the border with the state of Guanajuato), Irapuato, León and Salamanca are located; subtropical template semi-arid in the region of the cities of Guanajuato, Celaya and San Miguel de Allende and near Cuerámaro; and subtropical template sub-humid in a small area around Coroneo, in the southeastern part of the state.

State of Querétaro. Only a small part of the state is within the watershed boundaries, around the city of Querétaro. The climate in this region is classified as subtropical template arid and subtropical template semi-arid.

State of Michoacán. Four climates are present in the portion of the watershed corresponding to the state of Michoacán, all of them fall within the subtropical classification: semi-warm semi-arid near the Lerma river left bank; template semi-arid near the Lake Chapala Southeastern bank; semi-warm sub-humid in a relatively small area around the town of Ixtlán de los Hervores; and template sub-humid in a more extensive area, consisting of the cities of Pátzcuaro (and its homonymous lake), Morelia, Lake Cuitzeo and its limits with the Balsas river basin.

State of Jalisco. Three climates are present in the Jalisco portion of the watershed: semi-warm, semi-arid near the city of León (in the state of Guanajuato); semi-warm, sub-humid around Atotonilco el Alto; and template, sub-humid in the region surrounding Lake Chapala.

Figures 4a to 4e show annual temperature variations in some selected climatologic stations in the watershed. Also, Figure 5 shows mean seasonal isotherms both for cold and warm seasons. It can be observed that, generally speaking, temperature is relatively uniform both in time and in space in the watershed. Maximum monthly mean temperature rarely exceeds 32 °C and is almost everywhere above 24 °C, except in places like Toluca and the surroundings of the Trinidad Favela Dam, upstream of the Solís Dam in the Upper Lerma, where maximum temperatures are slightly lower. Minimum temperatures are also relatively uniform, between 5 and 15 °C all around the watershed, except in the Upper Lerma, where minimum monthly mean temperatures drop to 1 °C in January.

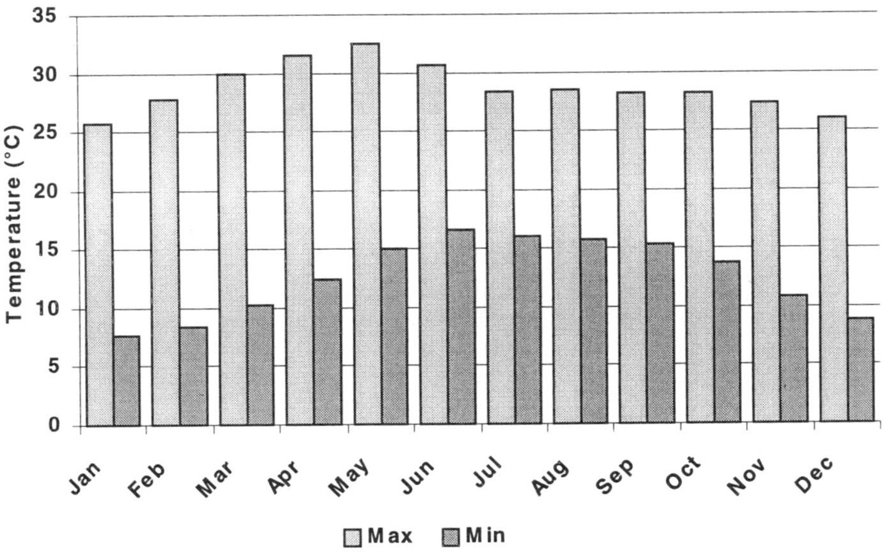

Figure 4a Annual temperature variations in Yurécuaro

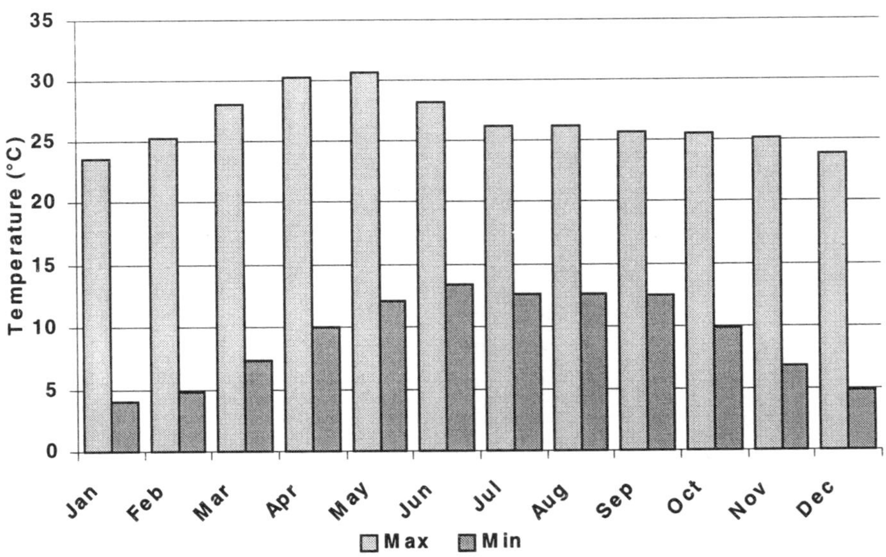

Figure 4b Annual temperature variations in Morelia

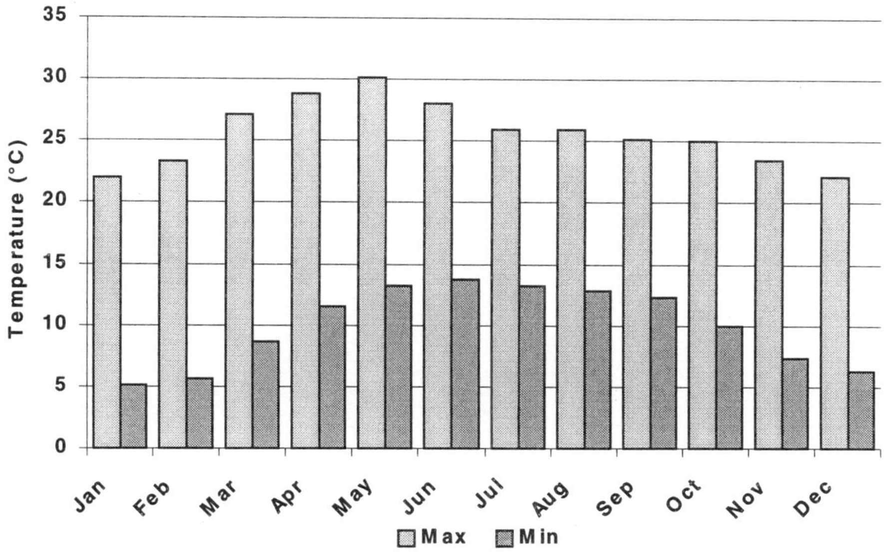

Figure 4c Annual temperature variations in San Miguel de Allende

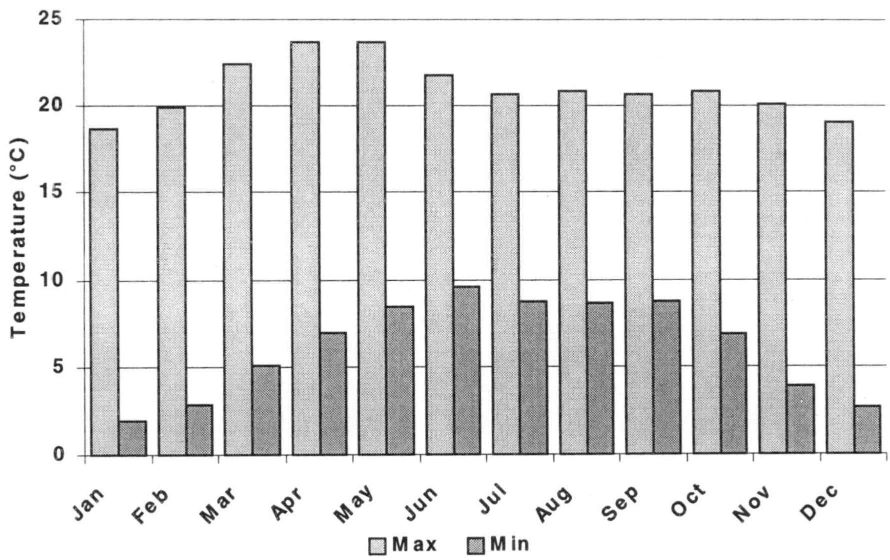

Figure 4d Annual temperature variations in Toluca

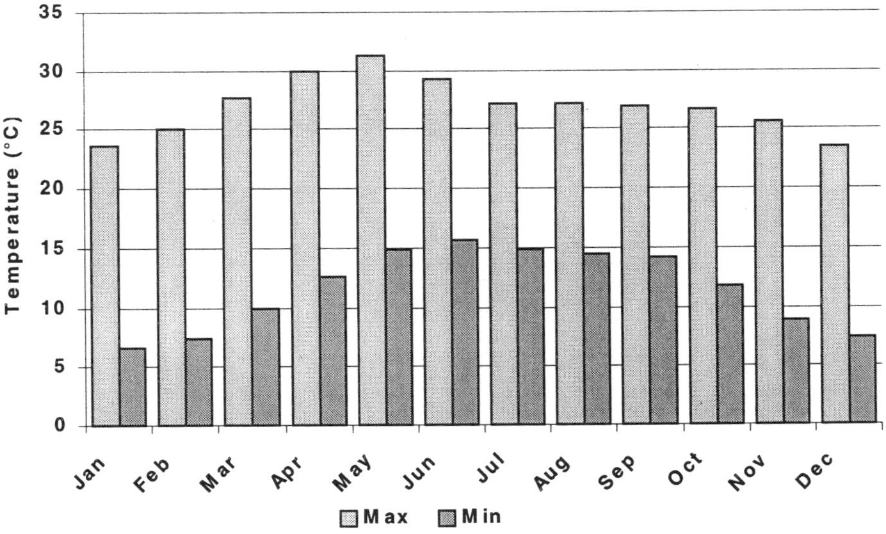

Figure 4e Annual temperature variations in León

3.2 Precipitation

The Lerma-Chapala watershed receives about 3% of Mexico's total precipitation volume. There is a considerable variation of mean annual precipitations in the watershed, ranging from 300 to more than 1000 mm. In this sense, the major part of the watershed (about 46,800 km^2 corresponding to 89% of the total area) can be considered as semi-, which range from 600 to 1300 mm, with a mean value of 730 mm. The arid (300 to 500 mm per year) and semi-humid (1000 mm per year or more) regions correspond to much lower percentages of the total basin area (6% and 5%, respectively). Figure 5 shows the annual mean seasonal isotherms in the Lerma-Chapala basin. In Table 2 the mean annual precipitations per state as well as those of the whole watershed have been included.

As can be observed in Table 2, the topographically more elevated portion of the watershed is where the highest precipitation is recorded, whereas the driest is the northeastern region. In Figures 6a to 6h, the mean annual precipitation distribution along the year in some representative gauging stations is presented. From these graphs, it is worth noting that there is a well-defined rainy season, from June to October, and the dry season takes place during the rest of the year, with occasional small precipitations during December and January. This cycle controls agricultural activities in the basin and partially governs the behavior of water levels in Lake Chapala.

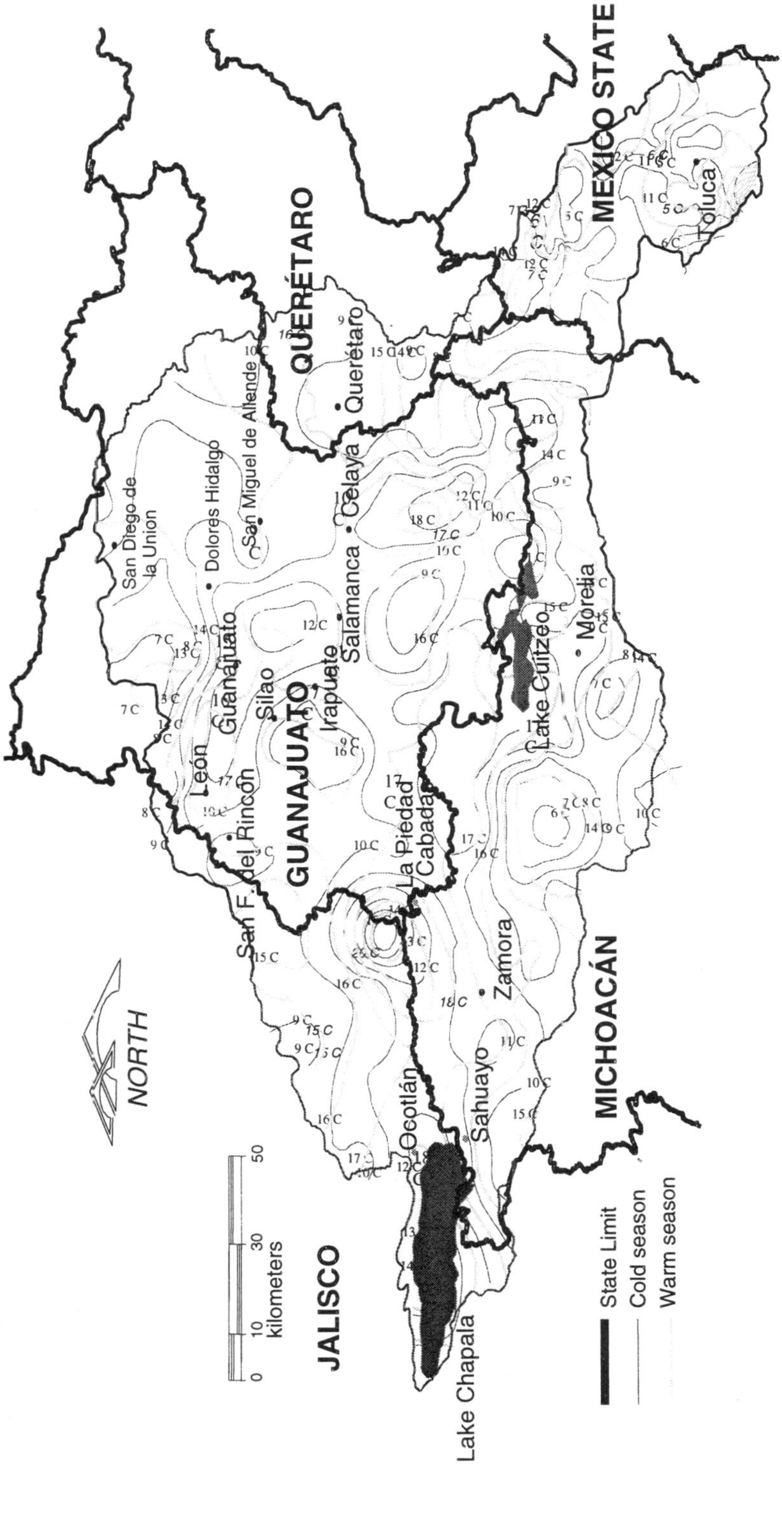

Figure 5 Mean seasonal isotherms

Table 2 Mean annual precipitations

Area	Mean annual precipitation (mm)
Guanajuato	650
Jalisco	800
Mexico State	880
Michoacán	800
Querétaro	600
Lerma-Chapala watershed	730

A fairly good coverage of climate information is available in the watershed. Data from about 340 "traditional" climatologic stations, which keep track of precipitation, maximum, minimum and current temperatures and evaporation are available.

3.3 Evaporation

In Figure 6, mean monthly evaporation data are shown for some selected stations. It is observed that latent evaporation surpasses precipitation during most of the year in all of the stations. It is also noticed that, due to the uniformity of temperatures throughout the year, potential evaporation rates are almost constant.

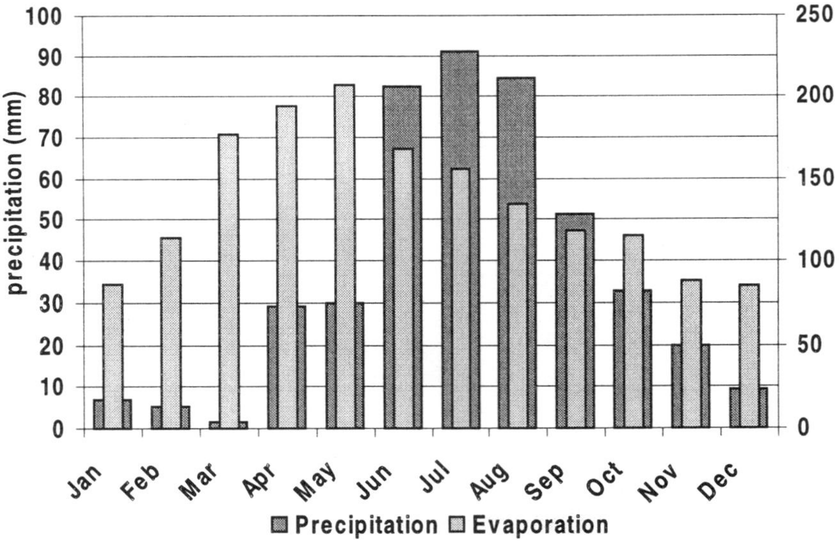

Figure 6a Mean monthly precipitation and evaporation in Dolores Hidalgo

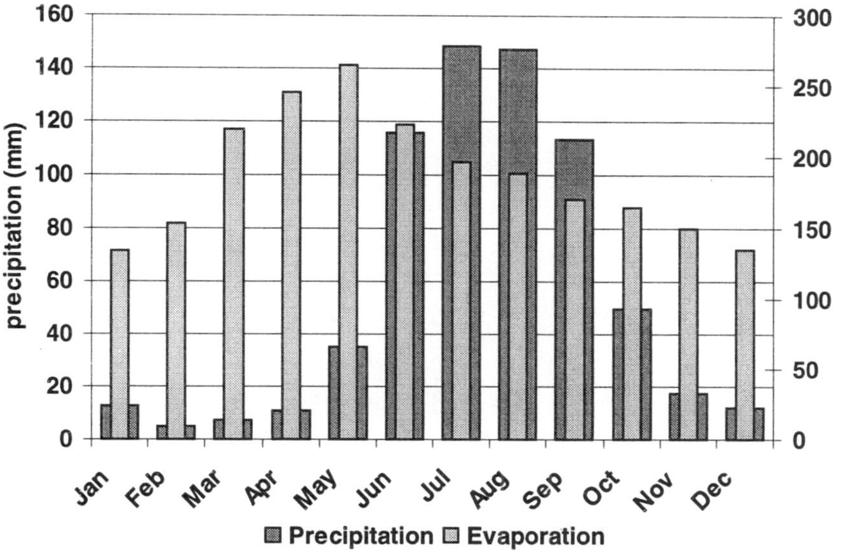

Figure 6b Mean monthly precipitation and evaporation in Moroleón

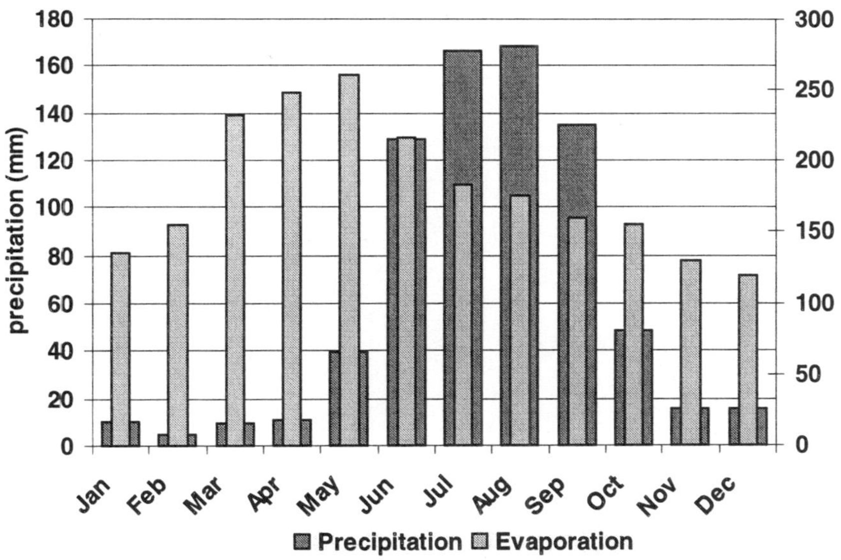

Figure 6c Mean monthly precipitation and evaporation in Valle de Santiago

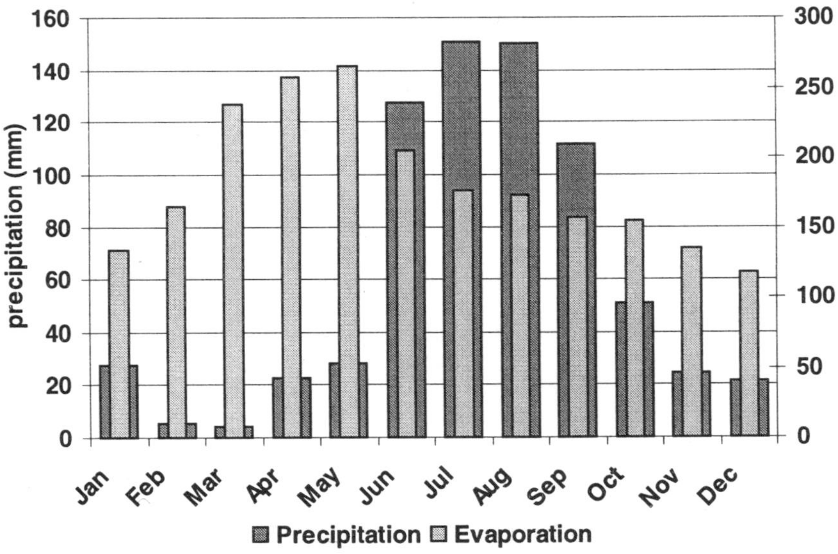

Figure 6d Mean monthly precipitation and evaporation in León

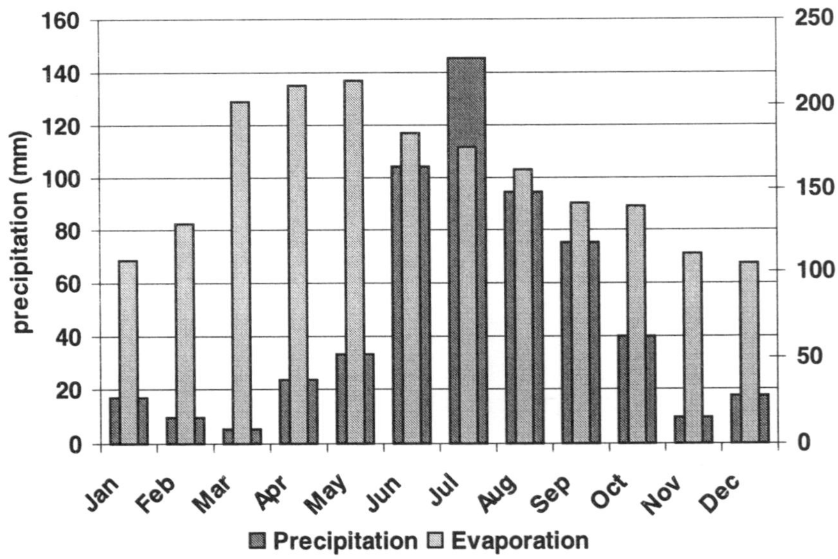

Figure 6e Mean monthly precipitation and evaporation in San Miguel de Allende

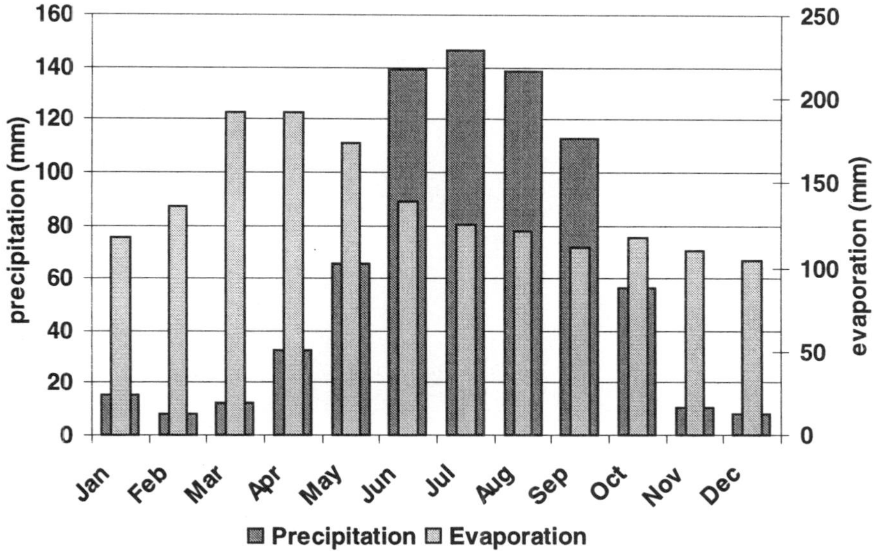

Figure 6f Mean monthly precipitation and evaporation in Toluca

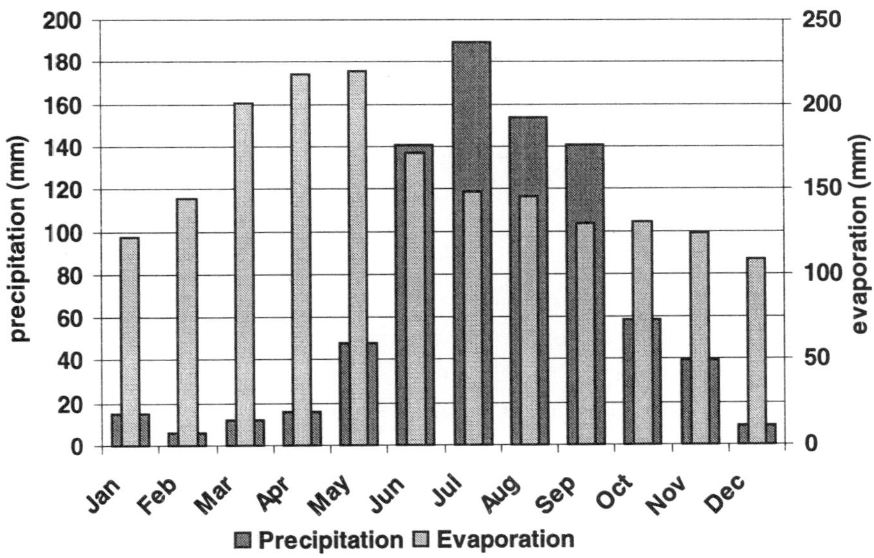

Figure 6g Mean monthly precipitation and evaporation in Morelia

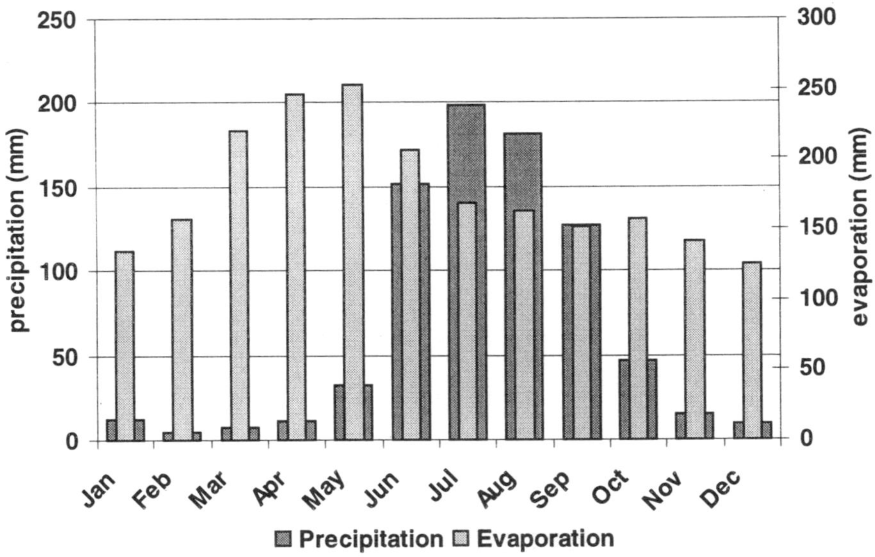

Figure 6h Mean monthly precipitation and evaporation in Yurécuaro

3.4 RUNOFF

The Lerma-Chapala basin is constantly surveyed by 110 hydrometric stations. According to the mean annual runoff, some of the most important are Yurécuaro, La Estanzuela, Pericos, La Begoña, and Solís. Annual runoff volumes recorded in these stations as well as the corresponding runoff frequency distribution functions are illustrated in Figures 7a to 7h.

Total mean annual runoff in the watershed to Lake Chapala is approximately 1,160 million m^3 (Mm^3), which is barely 4% of the total precipitation volume. This is due to a significant use of water in this region, which is the most highly developed in Mexico (Consejo Consultivo de Evaluación y Seguimiento del Programa de Ordenación y Saneamiento de la Cuenca Lerma-Chapala, 1991).

Hurricanes originated in the Atlantic and Pacific oceans have effects on the climate in the basin. About 90% of the cyclonic rainfall comes from the Pacific Ocean. The watershed has always suffered from flooding problems. Severe floods took place in 1926 and 1958. A brief description of the major recent flood events since 1967 is made in the following sections.

1967 flood (Secretaría de Recursos Hidráulicos, 1974). It was originated in the higher Lerma basin and produced discharges both in Tepuxtepec and Solís dams spillways. The Laja river did not have any significant discharges.

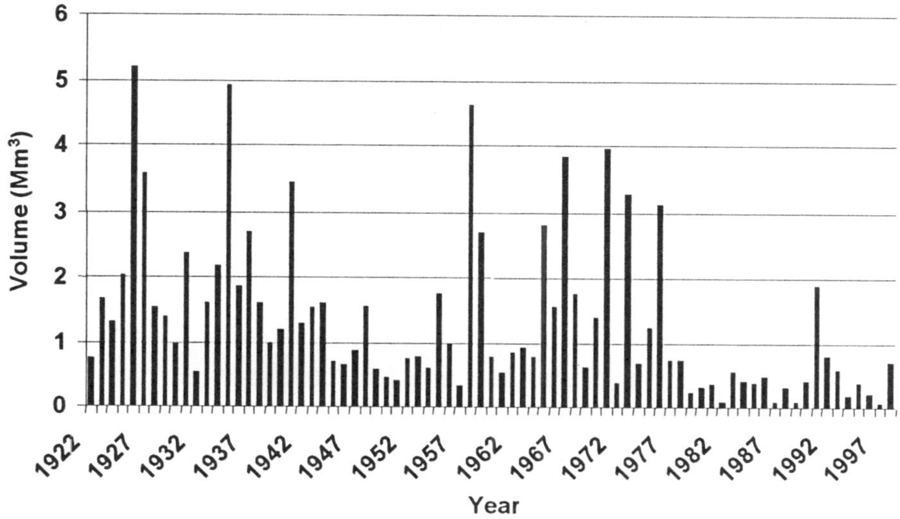

Figure 7a Annual runoff volumes in Yurécuaro

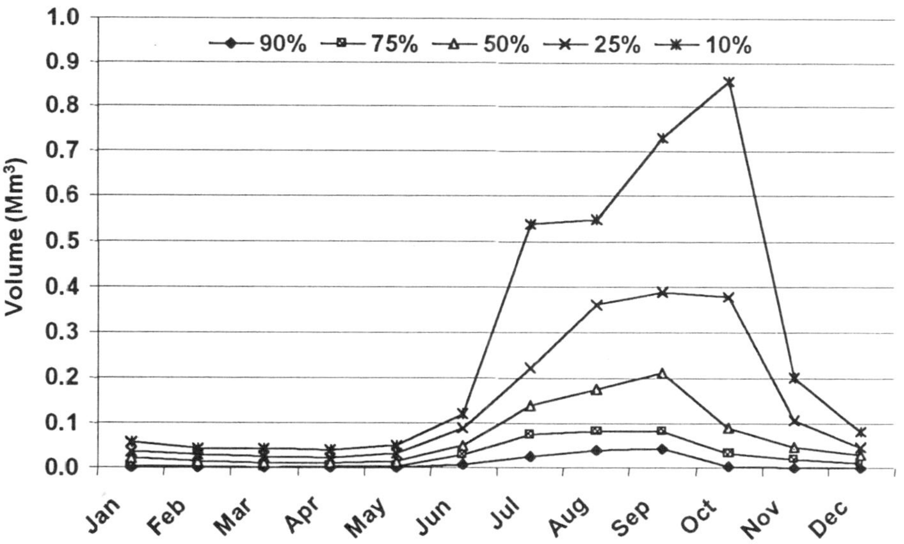

Figure 7b Runoff frequency distribution functions in Yurécuaro

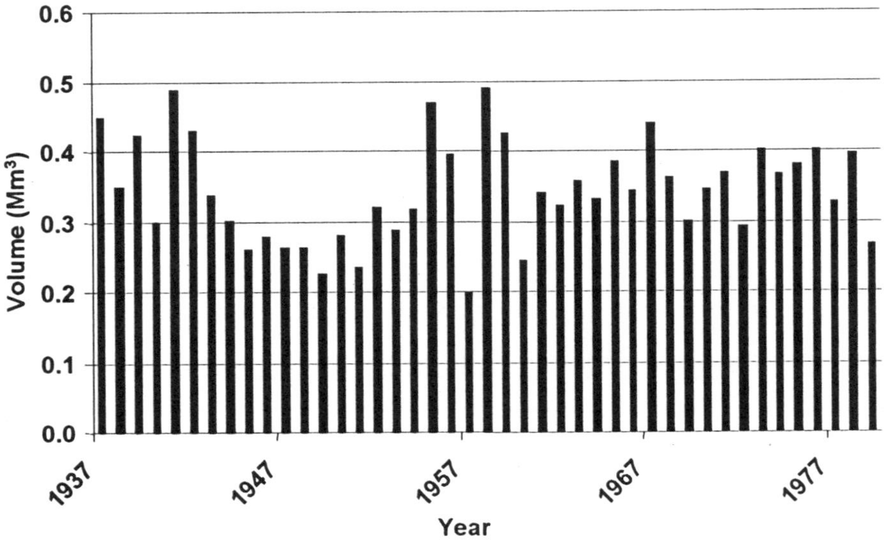

Figure 7c Annual runoff volumes in La Estanzuela

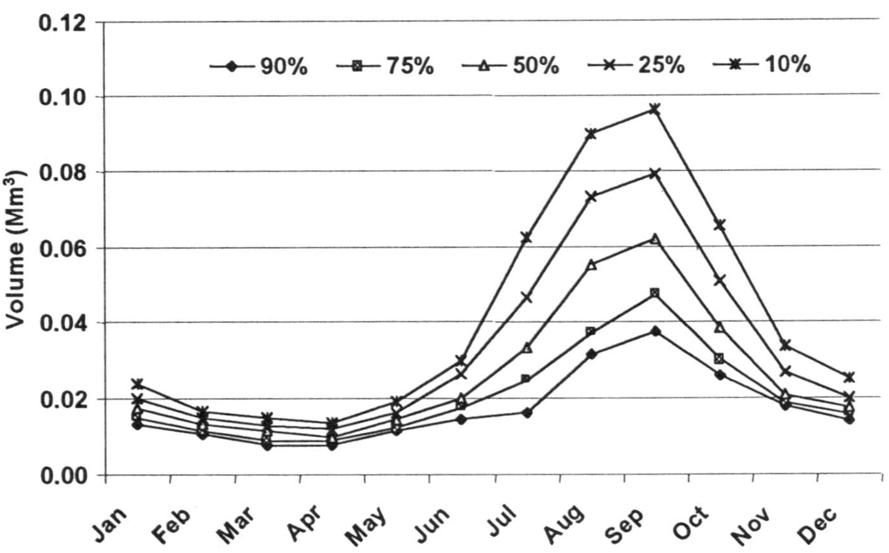

Figure 7d Runoff frequency distribution functions in La Estanzuela

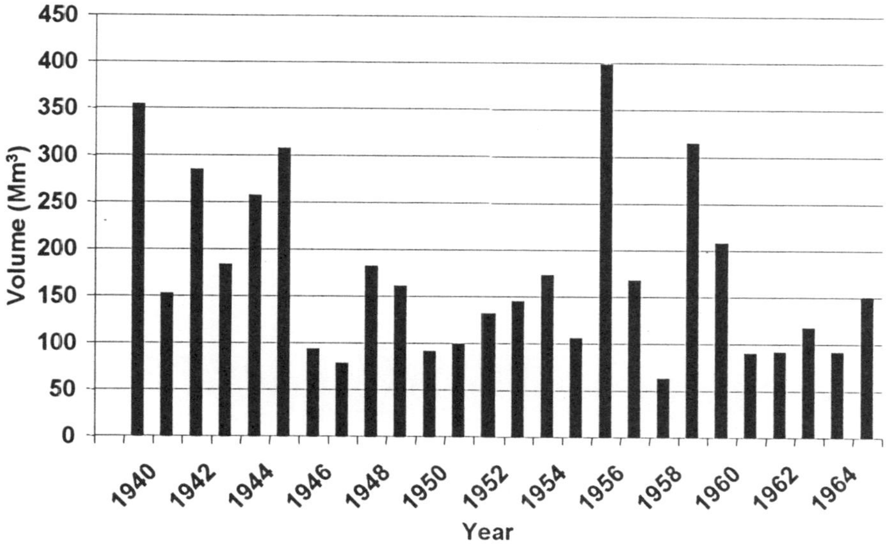

Figure 7e Annual runoff volumes in La Begoña

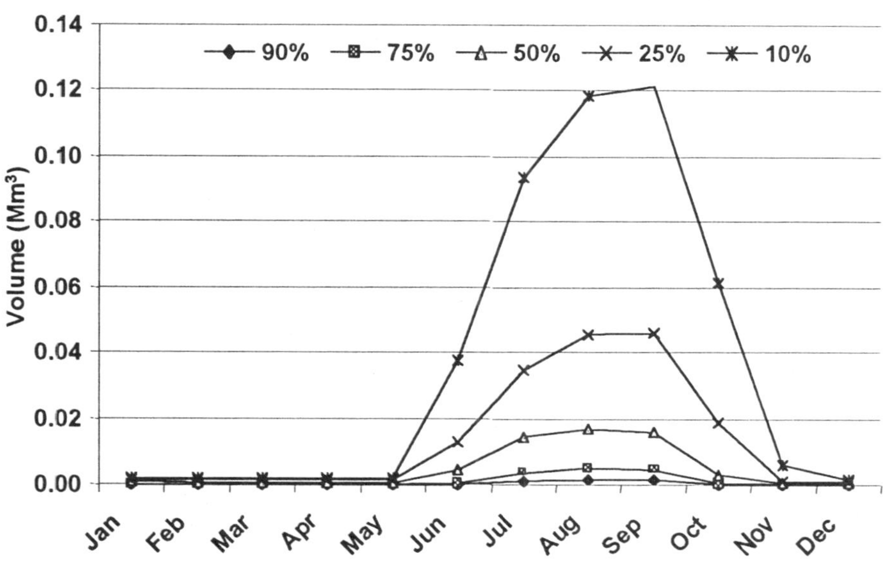

Figure 7f Runoff frequency distribution functions in La Begoña

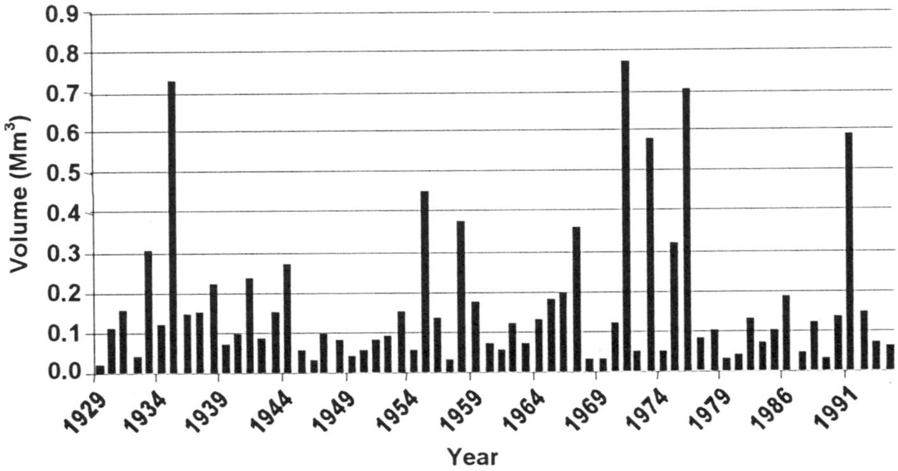

Figure 7g Annual runoff volumes in Pericos

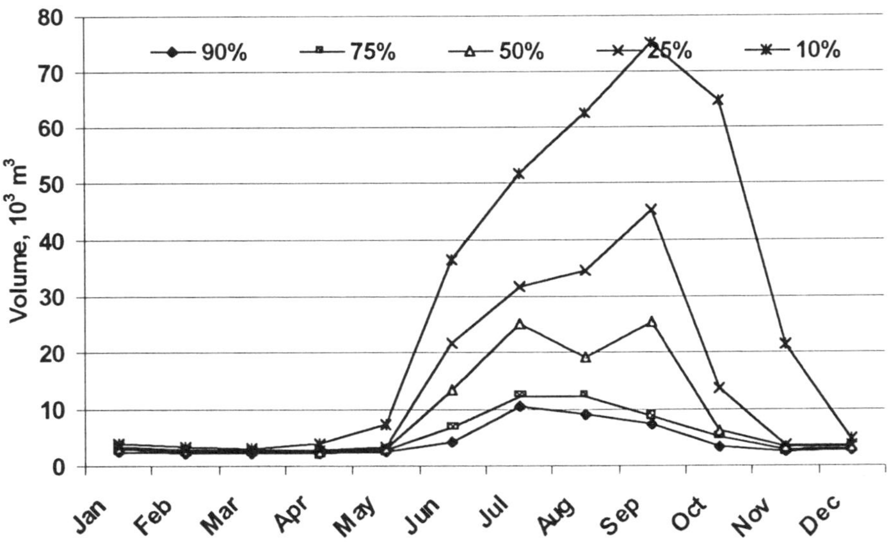

Figure 7h Runoff frequency distribution functions in Pericos

1971 flood (Secretaría de Recursos Hidráulicos, 1974). In this flood, severe damages were produced in irrigation district No. 11, downstream from Solís dam. This flood consisted of discharges from the Tepuxtepec, Solís and Ignacio Allende dams.

1973 flood. Ignacio Allende Dam spilled 400 $m^3 s^{-1}$, causing floods in the city of Celaya, partially due to coincident discharges through Querétaro river. El Conejo Dam failed and caused severe damages in the city of Irapuato. Numerous irrigation areas were damaged as well, a discharge of 1,100 $m^3 s^{-1}$ was observed at Yurécuaro station and another of about 1,500 $m^3 s^{-1}$ at La Barca station. The return period of the discharges at Corrales and Yurécuaro stations are estimated at 40 and 30 years, respectively.

1976 flood. A tropical wave prevailing for approximately 15 days over many parts of the watershed produced cumulative precipitation from 4000 to 7000 mm, which in turn provoked floods in the Turbio, Querétaro, Lerma and Duero rivers, among others. Hurricanes Liza and Madeleine also had effects on the watershed. Several storage bodies became overfilled and registered spills, including Lake Chapala, which discharged 268.5 $m^3 s^{-1}$ over the Santiago river, causing additional flooding downstream.

1987 flood. Intense precipitation produced flooding in more than a few urban areas of the states of Mexico (in the city of Toluca) and Guanajuato (in León, Salamanca and Silao).

1998 flood. Flooding and avalanches affected 24 counties in the Toluca valley. Intense damage was also produced in extensive cultivated areas in the state of Guanajuato, in extensive cropping zones, and in the cities and surrounding areas of San Miguel Allende, León and Comonfort.

Figure 8 shows the numbers of fatalities and affected population caused by floods in the watershed during the last three decades. Separately from the severe flooding episodes occurred during 1976 and 1987, there has been a steady increase in affected population figures during the last decade. This is due to the accelerated population growth in the watershed.

4. HYDRAULIC INFRASTRUCTURE

From the perspective of hydraulic infrastructure, the Lerma-Chapala watershed is one of Mexico's most developed. It has 12 major dams, with storage capacity equal to or greater than 10 Mm^3 (see Figure 9) and thousands of small to very reservoirs. These dams and reservoirs are mainly used to irrigate about half a million hectares. The overall storage potential of the system is around 10,000 Mm^3. Considering the watershed area and mean annual precipitation, as well as a mean runoff coefficient of around 0.3, it is observed that the watershed has a storage potential almost equal to the total mean annual runoff. Generally speaking, this means that additional storage capacity within the watershed would not be cost-effective.

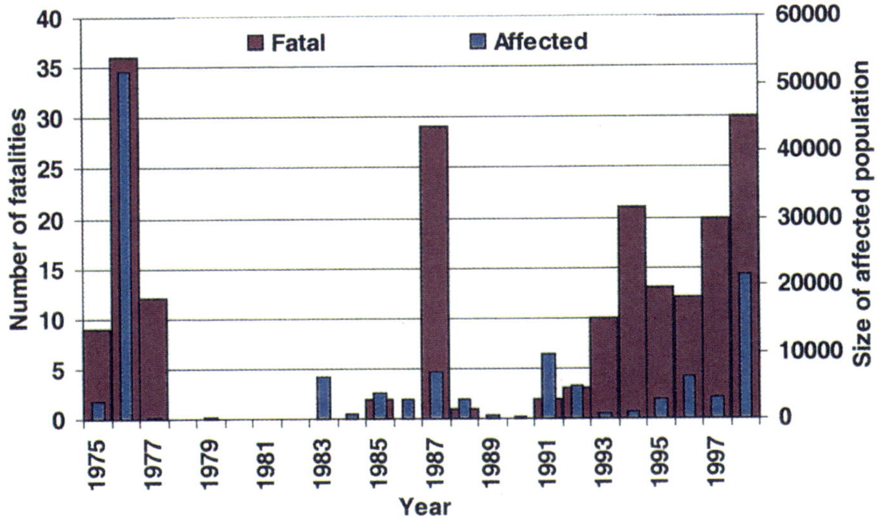

Figure 8 Fatalities and affected population due to floods in the Lerma-Chapala basin during three decades. Since the severe flooding episodes in 1976 and 1987, there has been an increase in these numbers. This is due to the accelerated population growth, which has taken place in the watershed

5. GROUNDWATER

Forty subsurface valleys that have been identified in the Lerma-Chapala watershed (CNA, 2000). Approximately 13% of Mexico's groundwater availability is found in this watershed (Consejo Consultivo de Evaluación y Seguimiento del Programa de Ordenación y Saneamiento de la Cuenca Lerma-Chapala, 1991).

Four geomorphologic provinces have been identified (Secretaría de Agricultura y Recursos Hidráulicos, 1977): plains, consisting of valleys formed by plateaus, volcanoes and sierras. The predominant rock formations are of igneous and continental sedimentary origin.

According to CNA (2000), there are nearly 7,300 wells in the watershed, about 22,600 dug-wells and almost 500 springs, which account for a groundwater exploitation of over 4,500 Mm^3 per year, whereas recharge is estimated in about 4,000 Mm^3 yr^{-1}.

However, exploitation is concentrated in some of the aquifers, where extraction surpasses recharge and therefore the resource its being overexploited. This is the case of the Toluca, Celaya, Laguna Seca, Ixtlahuaca, Querétaro, León, Pénjamo, Aguascalientes and Cuerámaro aquifers, among others (Secretaría de Agricultura y Recursos Hidráulicos, 1974, CNA, 2000).

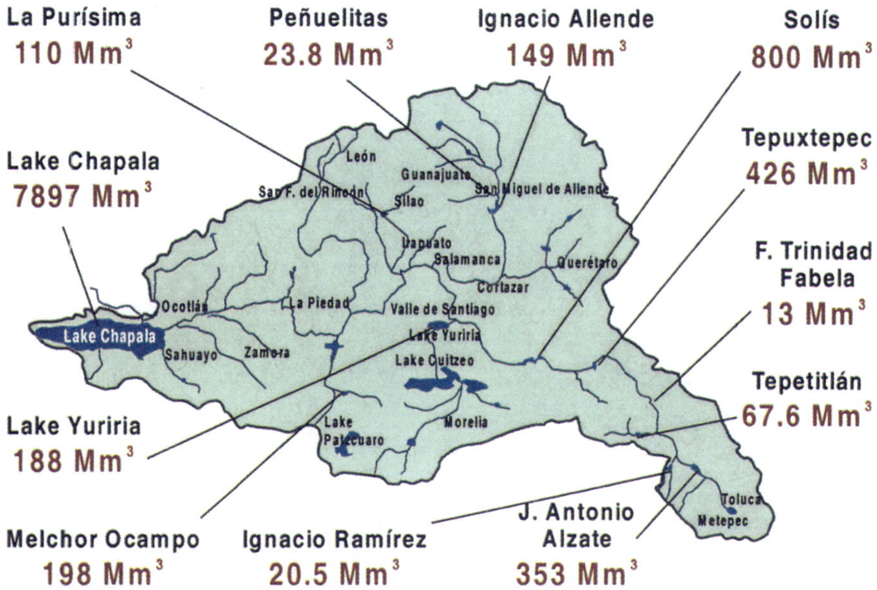

Figure 9 Hydraulic infrastructure and maximum storage capacities of the reservoirs in the Lerma-Chapala watershed.

6. HYDROLOGICAL BALANCE

6.1 Hydrological balance of the watershed

The hydrological balance in the watershed shows that the average precipitation of 730 mm per year (or about 36,000 Mm^3) generates a runoff of 4,740 Mm^3 yr^{-1}. (Consejo Consultivo de Evaluación y Seguimiento del Programa de Ordenación y Saneamiento de la Cuenca Lerma-Chapala, 1991).

From this runoff, 3,240 Mm^3 yr^{-1} is used for irrigation and the rest of it, *i.e.*, about 1,100 Mm^3 yr^{-1}, runs into Lake Chapala. Inputs to Lake Chapala come basically from the Lower Lerma river basin and from direct precipitation. Figure 10 illustrates the mean hydrological balance in the entire watershed.

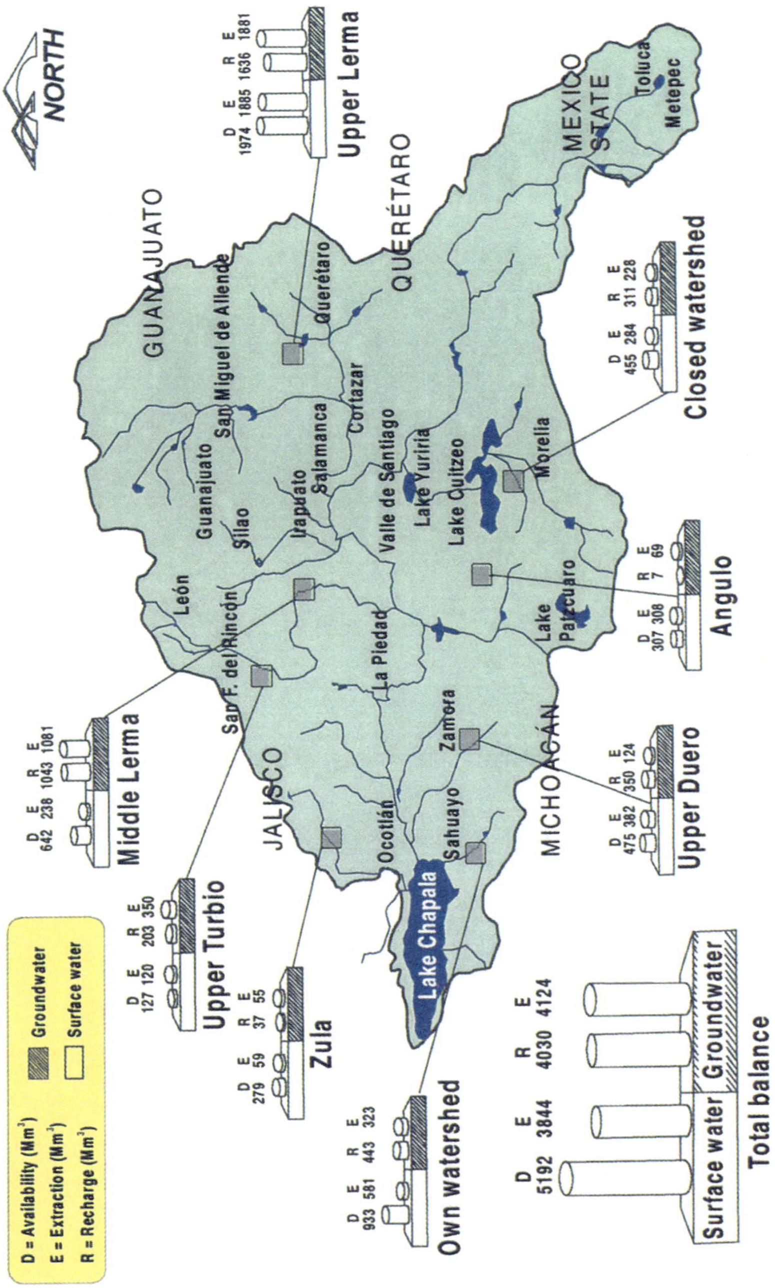

Figure 10 Annual hydrological balance of the Lerma-Chapala basin

6.2 Hydrological balance for Lake Chapala

Figure 11 shows the hydrological balance for Lake Chapala. It is observed that inflow from the Lerma river is relatively small compared to other sources, such as direct precipitation and contributions from the lake's own watershed. It is also worth noting the considerable loss of water as a consequence of evaporation in the lake, because of its vast surface area (approximately 700 km^2). From the total inflow to the lake, over 1,400 Mm^3 are lost by evaporation, 192 Mm^3 are used for water supply in the city of Guadalajara and about 70 Mm^3 goes to irrigation.

In the historical record of water levels in Lake Chapala (Figure 12), a long-term cyclical variation is observed, leading to critical low levels, such as observed in 1955, 1990-1992 and 2000. It is said that another critical low level took place in the nineteenth century. These periodic low water levels in the lake occur each time more frequently and severely because of the increasing use of water in the Lerma river basin. In Figure 13, the annual average values for the Lerma river inflow and annual mean precipitations in the region are shown. A considerably low inflow to Lake Chapala during the last 20 years is observed, which has lead to low water levels and storage volumes in the lake during the two most recent decades. Table 3 shows the geohydrological conditions of the aquifers in the basin. It is clearly observed that the Middle Lerma is the most over-exploited sub-region and that it accounts for the overall groundwater deficit in the watershed.

7. FINAL CONSIDERATIONS

A general description of the hydrology and hydraulics of the Lerma-Chapala watershed has been made. This watershed is one of the most developed in Mexico from the perspective of hydraulic infrastructure and has a complex water management status. The hydrological balances in the watershed and in Lake Chapala reveal, that careful management of the hydraulic resources of the watershed is required.

The relatively limited water resources in the watershed and its large population make water-related conflicts likely. Also, intensive groundwater exploitation jeopardizes future water availability. Agreements among different water users in the watershed should be continuously reviewed, in order to achieve rational and sustainable water use.

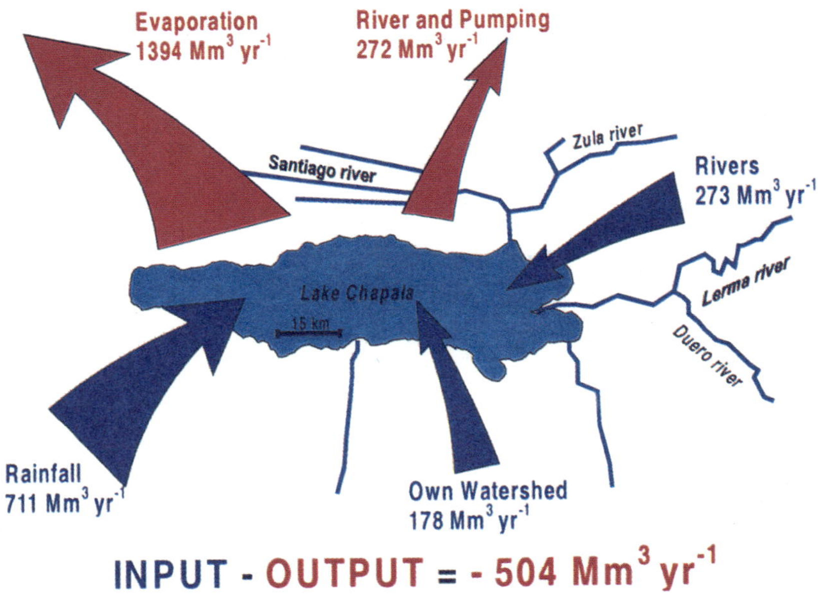

Figure 11 Annual hydrological balance of Lake Chapala

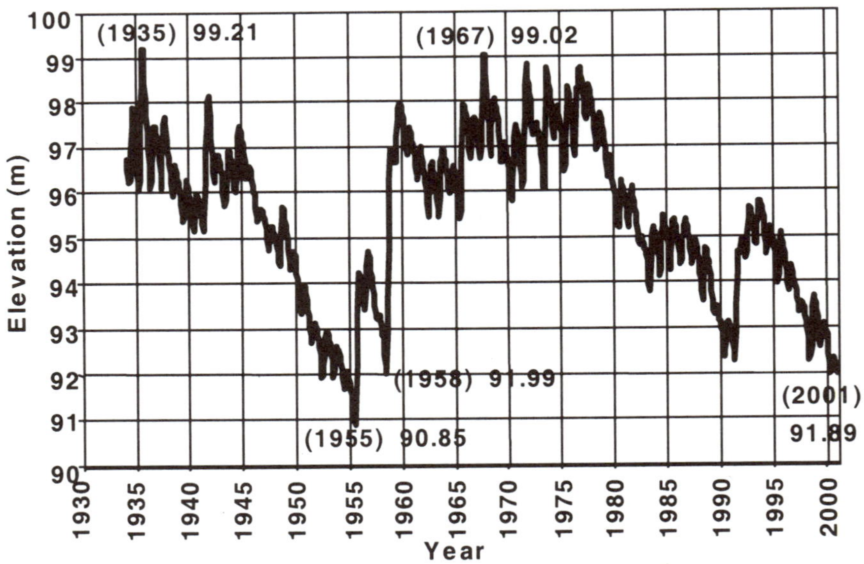

Figure 12 Historical water levels in Lake Chapala

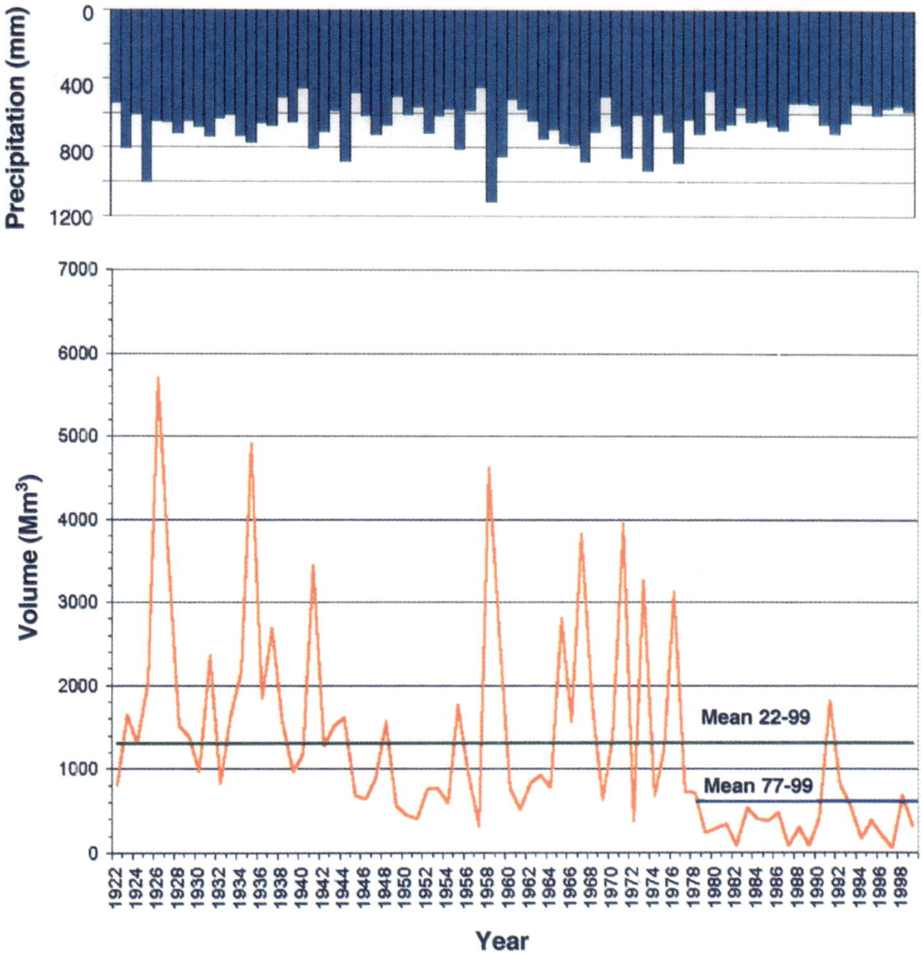

Figure 13 Annual inflow from the Lerma river and annual mean precipitation in the watershed

Table 3 Geohydrological conditions of aquifers in the watershed (Source: CNA, 2000)

Region	No. of aquifers	No. of wells	Extraction ($Mm^3\ yr^{-1}$)	Recharge ($Mm^3\ yr^{-1}$)	Availability ($Mm^3\ yr^{-1}$)
Higher Lerma	8	4,013	1,088	1,103	15
Middle Lerma	20	16,613	2,8861	2,173	-687
Lower Lerma	12	3,809	602	734	132
TOTAL	40	24,435	4,551	4,010	-540

ACKNOWLEDGEMENTS

Much of the information in this chapter was provided by the *Centro de Información de la Cuenca Lerma-Chapala (Information Center of the Lerma-Chapala watershed)*. The kind attention of the Information Center, particularly of the Center's Director-in-charge, Alejandro Calixto Herrera, is gratefully acknowledged.

The following persons collaborated in the elaboration of this chapter as well: Ernesto Aguilar Garduño, Rafael Equihua Medina, Juan Francisco Gómez Martínez, Jaqueline Lafragua Contreras, Roberto Mejía Zermeño.

REFERENCES

Consejo Consultivo de Evaluación y Seguimiento del Programa de Ordenación y Saneamiento de la Cuenca Lerma-Chapala (1991) Acuerdo de Coordinación de Aguas Superficiales, Colección Lerma-Chapala, Vol 1, No 5, March 1991.

CNA (Comisión Nacional del Agua, 2000) Programa Hidráulico de Gran Visión 2001-2020, Región VII Lerma-Santiago-Pacífico, Mexico.

García E (1964) Modificaciones al Sistema de Clasificación Climática de Köppen para Adaptarlo a las Condiciones de la República Mexicana. Offset Larios, México.

Instituto Mexicano de Tecnología del Agua (1996) Banco Nacional de Aguas Superficiales, Mexico.

Medina GG, Ruiz CJA, Martínez PRA (1998) Los Climas de México. INIFAP, Libro técnico No 1.

SARH (Secretaría de Agricultura y Recursos Hidráulicos) (1979) Subsecretaría de Planeación, Dirección General de Protección y Ordenación Ecológica. Clasificación del río Lerma hasta el Lago de Chapala, Mexico.

Subsecretaría de Construcción, Dirección General de Irrigación y Control de Ríos (1974) Resumen de las condiciones actuales en la cuenca del río Lerma, Mexico.

Secretaría de Recursos Hidráulicos (1973) Subsecretaría de Planeación, Dirección General de Estudios. Boletín hidrológico No 51, Región hidrológica No 12 (parcial), Mexico.

Secretaría de Agricultura y Recursos Hidráulicos (1977) Comisión del Plan Nacional Hidráulico. Inventario Regional de Aguas Subterráneas, Documentación del Plan Nacional Hidráulico, No 19.

Chapter 2

Geology, Sediments and Soils

PEDRO F. ZÁRATE-DEL VALLE[1*], FRANÇOIS MICHAUD[2], CLAUDE PARRÓN[3], GABRIEL SOLANA-ESPINOZA[4], ISABEL ISRADE-ALCÁNTARA[5], HERMES U. RAMÍREZ-SÁNCHEZ[1,3,6], AND FRANÇOIS FERNEX[6]

[1*]*Guadalajara University-CUCEI-Chemistry Dept. P.O. Box 4-021. Guadalajara, 44410 Jal., Mexico. email: pzarate@ccip.udg.mx, corresponding author*

[2]*Universite Pierre et Marie Curie La Darse, B-P 48, 06235, Villefranche sur Mer, cedex, France. email: micho@obs-vlfr.fr*

[3]*Centre européen de recherche et d'enseignement de géosciences de l'environnement (C.E.R.E.G.E) UMR CNRS 6635 Europôle méditerranéen de l'Arbois BP 80, 13545 Aix-en-Provence cedex 04, France. email: cparron@cerege.fr*

[4]*Guanajuato University, CIQI (Fac. Química), Noria Alta, Guanajuato, 36050 Gto., Mexico. email: gsolana@quijote.ugto.mx*

[5]*Universidad Michoacana de San Nicolás de Hidalgo, P.O. Box 52-J, Morelia, 58060 Mich., Mexico. email: aisrade@zeus.ccu.umich.mx*

[6]*Faculté des sciences, Parc Valrose, F-06108 Nice cedex 2, France.email: fernex@unice.fr*

Key words: Sedimentation rates, geochemistry, active tectonic control, geology

Abstract: Lake Chapala, the major Mexican fresh water reservoir, is located in western Mexico, is a good example of a neotectonic lake with a history linked eastwards to the development of the active Transmexican Volcanic Belt geological province. Lake Chapala crosses the E-W trending Citala rift, which forms with the Tepic-Zacoalco and Colima rifts the so-called Jalisco triple junction. The Lake Chapala basin is tectonically controlled and belongs to a regional intra-arc system of half-grabens. It is assumed that the current Lake Chapala basin has a depth of about 600 m. Shallow (1.40 m) sediments in Lake Chapala are dark and silty with granulometric median ranging from 10 to 50 μm. Sediment samples were studied by isotopic, geophysical, geochemical and micro-paleontological methods. Major oxides in sediments are SiO_2 (54.05%), Al_2O_3 (18.94%), Fe_2O_3 (6.75) and CaO (2.57%). The major mineralogical composition of sediments is as follows: plagioclases (25.86%), quartz (20.92%), clays (38.5%), carbonates (2.99%) and TiO_2 (1.05%). Sedimentation rates (**SR**) were evaluated from ^{210}Pb profiles assuming constant initial concentrations; and we can conclude that **SR** is at least two or three times higher on the East side of the lake than in the western part.

As water depths on both sides of the lake are similar, we assume there must be a tectonic compensation on the eastern part. This explains why Lake Chapala has a uniform depth in spite of the differences detected in **SR**. We have estimated an approximate 1 mm yr^{-1} subsidence at the eastern part of the lake that compensates the higher **SR**. This tectonic compensation also explains the fact that the lake's fluvial input and output are so abnormally close to each other. Scarcity of diatoms in shallow sediments of Lake Chapala is due to pH variations in both sediments and water, hence inhibiting their preservation. The major soil types around Lake Chapala are: Vertisol (50%), Pheozem (30%), Lithosol (15%) and Luvisol (5%).

1. INTRODUCTION

Lake Chapala, the largest Mexican fresh water reservoir, is a large, neotectonic, eutrophic and tropical lake at 1510 m above sea level (Figure 1). The lake belongs to a volcanic and tectonic region in western Mexico (Figures 1, 2 and 3). Despite its large surface, Lake Chapala is actually less than 4 m deep. The Lerma river used to be one of the major inputs to the lake and flows through more than 700 km (Figure 3). Lake Chapala receives water from a catchment area of about 54,400 km^2 whose present land uses include cropland, industrial and urban development and forest exploitation; the moorland use disappeared after 1908 (Sandoval, 1994). There is field evidence to postulate that the geometry of the basin of Lake Chapala has changed in time and space (Leyden, 1994). Since the Late Miocene-Early Pliocene to Early Pleistocene, there was a large lacustrine basin named Jalisco paleo-lake located a little North of current Lake Chapala. This paleo-lake was influenced by tectonic and volcanic activities in the West and South and by progressive isolation of different paleo-depressions, due to volcanic and tectonic activities, which fragmented and blocked them (Rosas and Fucugauchi, 1998). This resulted in the actual tectonic lakes located in rifts of the so-called Jalisco continental triple junction (**A**, **B** and **C**, Figure 2). Due to natural and anthropogenic factors, Lake Chapala has been reduced since 1559 in its morphometric features: bathymetry, insulation and shoreline development. As we will see, the tectonic origin of Lake Chapala basin partially explains the fact that the input and output of water (Lerma and Santiago rivers, respectively, Figure 5) are so close to each other.

In Figure 2, (**1**) represents Jalisco Batholithic geological province (pre-Cretaceous basement): **a**, meso-Cenozoic granite granodiorite, diorite; **b**, pre-Jurassic schistes and quartzite. (**2**): Guerrero-Colima - Orogenic Complex geological province (belonging to Guerrero suspect terrene) composed of Late Jurassic-Early Cretaceous volcano-sedimentary arc sequences: **a**, calc-alkaline basalts, "ocoite" facies included: basic lava with big phenocrysts of plagioclase affected by burial metamorphism (prehnite-pumpellyte facies; Levi, 1969), andesite and rhyolite (tuff included), interceded with Aptian-Albian reefal limestones; **b**, Cretaceous silicic ash flows (Lange and Carmichael, 1991). (**3**): Ignimbrite Mexican Belt geological province: Calc-alkaline tertiary volcanic. (**4**): Transmexican Volcanic Belt geological province (undifferentiated Upper Miocene to Pleistocene andesite and minor basalt; lacustrine deposits included). (**5**): recent alluvium. (**6**): Pliocene (triangles) and Quaternary volcanoes of the **TVBgp** (1, Colima volcanic complex; 2, Primavera cauldron; 3, Tequila; 4, Ceboruco; 5, Tepiltic; 6, Sanganguey; 7, San Juan). (**7**): Quaternary rhyolite. (**8**): Laramide folds. (**9**): Mio-Pliocene normal faults. (**10**): Quaternary normal faults belonging to the Jalisco continental triple junction rift system. (**11**): Jalisco continental triple junction: (**A**) Colima rift; (**B**) Citala rift; (**C**) Tepic-Zacoalco rift. Inset left lower corner: velocity triangle of the Jalisco continental triple junction rift system: (Am) North America plate; (J) Jalisco Block; (Mch) Michoacán Block (after Barrier *et al.*, 1990). Horizontal lines show Lake Chapala and other tectonic lakes that belong to the Jalisco continental triple junction.

2. GEOLOGY

2.1 General geology of western Mexico

Western Mexico (Figure 2) has been divided into four geological provinces (**gp**) (Ortega-Gutiérrez *et al.*, 1991): a) Jalisco Batholithic (**JBTgp**; **a**ge: Meso-Cenozoic; **o**rigin: plutonic; **g**eotectonic or **g**eodynamic **f**ramework: arc base); b) Guerrero-Colima Orogenic Complex (**GCOgp**; **a**: Mesozoic; **o**: volcano-sedimentary; **gf**: submarine arc); c) Ignimbrite Mexican Belt (**IMBgp**; **a**: Cenozoic; **o**: volcanic; **gf**: continental arc) and d) Transmexican Volcanic Belt (**TVBgp**; **a**: Cenozoic; **o**: volcanic; **gf**: continental arc). One of the most spectacular "elements" of the geology of western Mexico is **TVBgp**, in which magmatic activity is related to north-eastward subduction of the Cocos and Rivera plates beneath the North America Plate along the Middle America Trench (Figure 1).

The general E-W orientation of the volcanic belt defined by the Colima, Popocatépetl and Citlaltépetl volcanoes (Figure 3), diverges from the trend of the Middle America Trench by about 15°S (Figures 1 and 3). This divergence correlates to an eastward increase in convergence rate and an eastward shallowing of the Benioff zone (Luhr and Carmichael, 1985; Pardo and Suárez, 1995).

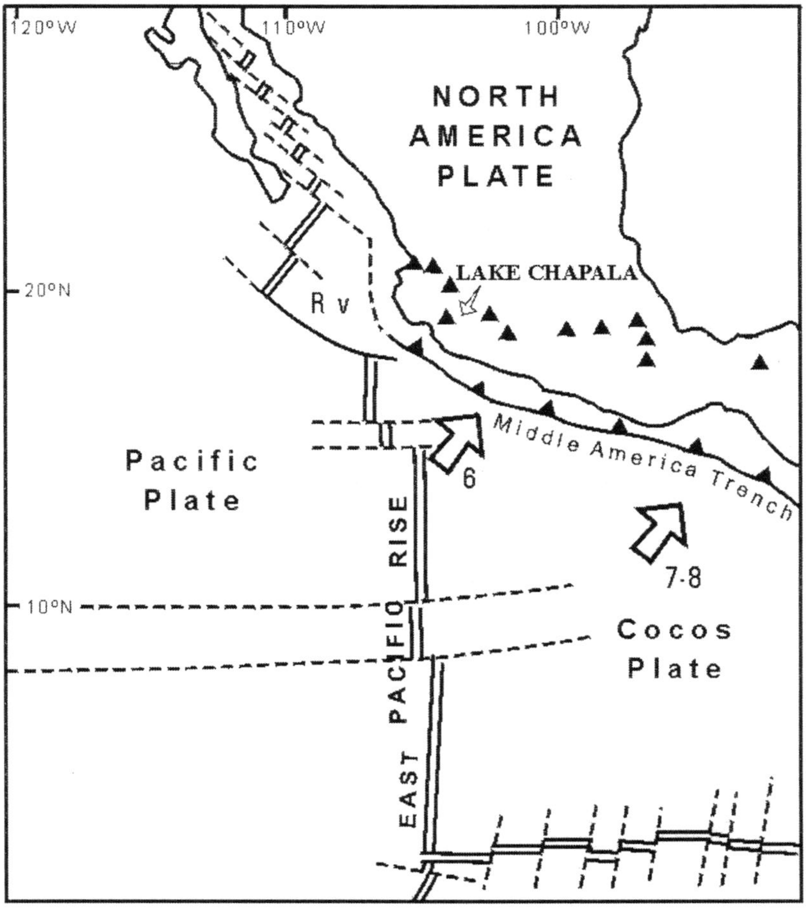

Figure 1 Geodynamic framework of western Mexico (modified after Barrier *et al.,* 1990). Arrows indicate motion of Cocos Plate relative to North America Plate (cm yr^{-1}); **Rv** = Rivera Plate; triangles indicate Quaternary volcanoes of **TVBgp**

2.2 Geology and neotectonics of the Lerma-Chapala basin

TVBgp is a complex continental arc extending from the western Pleistocene Volcano San Juan (Figure 2) to the eastern, recently (1982) active, Volcano El Chichón (Luhr and Carmichael, 1990). Intense, widespread Late Cenozoic tectonic and volcanic activity in central Mexico has produced numerous volcanic centers, cones, grabens and crater lakes (Suter *et al.*, 1995; Rosas and Fucugauchi, 1998), so it's numerous lakes and volcanoes characterize **TVBgp**.

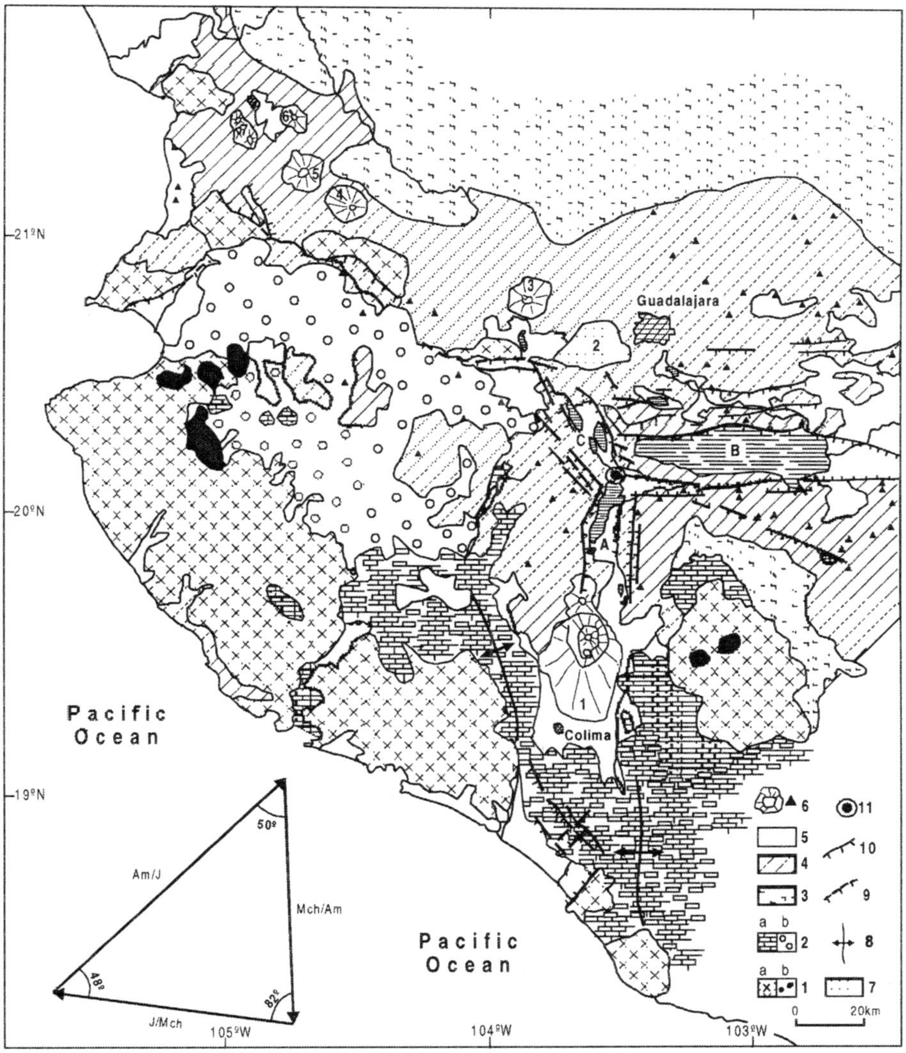

Figure 2 Geology of western Mexico (modified after Zárate-del Valle and Maldonado-Reyes 1990; Zárate-del Valle and Aguirre-Barragán 1993; Tardy *et al.,* 1994; Zárate-del Valle *et al.,* 1999; Calmus *et al.,* 1996; Zimmerman *et al.,* 1988)

Central Mexico contains a wide diversity of eruptive centers. Three large, historically active andesitic composite volcanoes occur along the volcanic front: the western Volcano Colima (one of the most active volcanoes in the

American continent), Popocatépetl (currently active) and Citlaltépetl (**1**, **4**, **7**, Figure 3). Monogenetic cinder and lava cones dominate other portions of the Michoacán-Guanajuato volcanic field, which includes Jorullo (1759-1774; **11**, Figure 3) and Paricutín (1943-1952; **10**, Figure 3) and may be the most extensive monogenetic cone field on the Earth. Four large Pleistocene cauldrons of rhyolitic domes and pyroclatics deposits occur far behind the volcanic front (Luhr and Carmichael, 1990): La Primavera (state of Jalisco; **12**, Figure 3), Amealco (Mexico State; **13**, Figure 3), Huichapan (state of Hidalgo; **14**, Figure 3) and Los Humeros (state of Veracruz; **15**, Figure 3). Many of these tectonic and volcanic depressions located in **TVBgp** have been or are still occupied by both graben and volcanic lakes (Israde-Alcántara, 1999b). Cuitzeo and Yuriria in the state of Michoacán and near the triple junction (Figure 2): Sayula, San Marcos and Chapala in the state of Jalisco (**A**, **B** and **C**, Figure 2) are examples of tectonic lakes. Some volcanic lakes are Valle de Santiago in the state of Guanajuato and the Sol and Luna lakes at Volcano Nevado de Toluca (**2**, **16**, Figure 3). Almost all of these tectonic lakes have field evidence (i. e. outcrops of lacustrine deposits made of a volcano sedimentary sequence) of the migration of their basins in time and space such as Lakes Chapala and Cuitzeo; some of them are nowadays only dry depressions. Tectonic lakes located in the eastern Jalisco Triple Junction, generally reflect the E-W tectonic orientation of **TVBgp** (Figures 1 and 3).

2.3 Structural geology and neotectonics of the Lake Chapala region

The western extreme of the Transmexican Volcanic Belt presents neotectonic features (Figures 2 and 3), which defines a continental block called the Jalisco Bloc (**JB**, Mosser, 1972; Ferrari *et al.*, 1997). The **JB** basement consists of the following: **JBTgp** and **GCOgp**, being the first one the basement of the second (Zárate-del Valle and Maldonado-Reyes, 1993). Both **gps** are partially covered by rocks belonging to **TVBgp**. The Tepic-Zacoalco rift (C, Figure 2) and Colima rift (A, Figure 2) located to the North and East of the **JB**, respectively. The Lake Chapala basin is located in the E-W trending Citala rift (B, Figure 2 and CG, Figure 4), which forms the so-called Jalisco continental triple junction with the Tepic-Zacoalco and Colima rifts (O, Figure 2; Allan, 1986; Michaud *et al.*, 1994). Surrounded by large tilted blocks, the depressions related to these active tectonic features have favored the development of Lake Chapala (Figure 4; Barrier *et al.*, 1990; Quintero-Legorreta, 1995; Rosas-Elguera *et al.*, 1997; Rosas-Elguera and Urrutia-Fucugauchi, 1998; Michaud *et al.*, 1999).

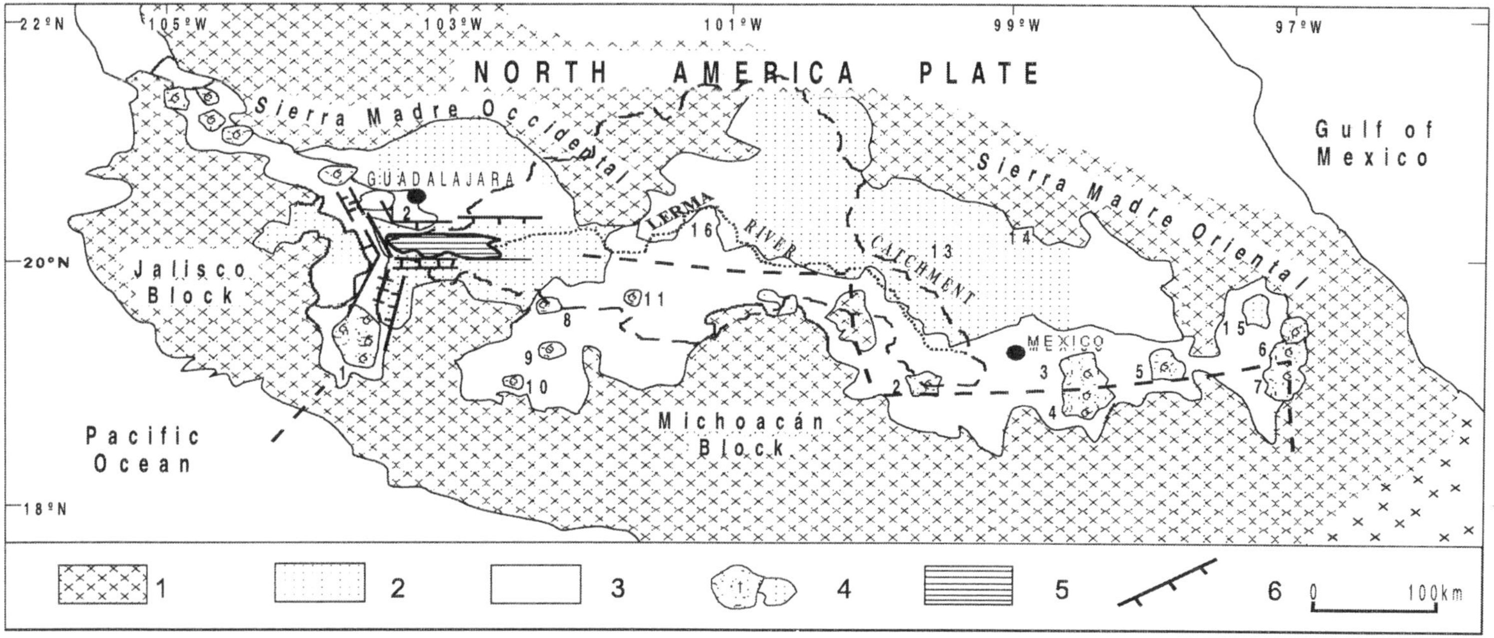

Figure 3 Schematic geological map of the Transmexican Volcanic belt geological province (Modified after Demant, 1981 and Ferrari et al., 1997). (**1**) Pre **TVBgp** rocks. (**2**) Upper Miocene to Plio-Quaternary volcanics and continental sediments of **TVBgp**. (**3**) Recent formations of **TVBgp**. (**4**) Main Quaternary volcanoes. (**5**) Chapala tectonic lake. (**6**) Normal faults. Lerma-Chapala hydrological basin is indicated with dashed line and dotted line shows the bed of the Lerma river. The inferred boundary among tectonic blocks: Jalisco Block, Michoacán Block and North America Plate is indicated by bold discontinuous line. Figures1-11 show main Quaternary volcanoes: **1** Colima, **2** Nevado de Toluca, **3** Iztaccíhuatl, **4** Popocatépetl, **5** Malinche, **6** Cofre de Perote, **7** Citlaltépetl, **8** Patambán, **9** Paricutín, **10** Jorullo and **11** Capaxtiro. Figures 12-16 show following calderas: **12** La Primavera, **13** Amealco, **14** Huichapan, **15** Los Humeros and **16** Valle de Santiago

As a result of a stratigraphic and tectonic study Rosas-Elguera and Urrutia-Fucugauchi (1998) concluded that the volcano sedimentary lacustrine deposits at the Lake Chapala basin as well as at the Zacoalco and Colima tectonic depressions are of about 900-1000 m thick and are made of a sequence of lacustrine sediments and pyroclastic units of ash and pumice. Conglomerates, sandstones, siltstones, claystones and diatomite horizons are observed in the sediments. However, the lake does not directly lie on the Citala rift axis (CG, Figure 4), except for its eastern part, where the axis is hidden under the delta and prodelta area sediments. The large flat depression near the Lerma river inflow (Figure 5) is called "La Ciénega" and used to be a marsh until 1908 (Sandoval, 1994). The Lake Chapala basin (Figure 4) is limited to the North by a more or less continuous 300-900 m high scarp, which corresponds to southward dipping normal faults. In the Southwest it is bordered by northward dipping gentle slopes corresponding to the backs of tilted blocks, and in the Southeast by 300 m high scarps related to northward dipping normal faults. The Lake Chapala basin has recorded both the tectonic events related to the rifts and the volcanic activity of the western **TVBgp**.

Following the October 9, 1995, M = 8 earthquake along the Rivera subduction zone (**Rv**, Figure 1) Global Positioning System (**GPS**) measurements indicate that sites in Jalisco have moved up to 200 mm in response to post-seismic slips along the Rivera plate subduction fault. Sites close to Lake Chapala have moved 40-50 mm Southwest (Zárate-del Valle *et al.*, 2000). Displacements due to the subduction fault slip are far larger than the expected horizontal rifting rates across the Citala rift: 0-1 mm yr^{-1}, thereby illustrating the difficulty in extracting local fault slip from a velocity field that is strongly contaminated by strains from the nearby Rivera subduction fault (Figure 1). **GPS** measurements from 1999 suggest that steady-state inter-seismic strain accumulation is resuming along the coast, but the transient post-seismic slip continues near Lake Chapala. We assume that this transient phase will end in the next few years. Determining vertical rates will be extremely challenging and may require decades given the present level of accuracy and expected slow vertical rates.

The Los Alacranes and Mezcala islands in the lake are made of vesicular basalt; in Los Alacranes island hydrothermal activity is pervasive in basalt and linked to the N20°/70°SW fracture system filled with calcite and quartz. Two main volcanic centers are located near the Lake Chapala basin (Figure 2). The most recent activity of the alkaline Caldera La Primavera, located less than 50 km NW of the Lake Chapala basin (Caldera 12, Figure 3) took place 30000 years BP (Mahood 1980). Volcano Colima, situated 90 km South of Lake Chapala (volcano 1, Figure 3) is currently active and had an important eruption about 500 years ago (Demant, 1981; Medina-Martínez, 1983; Stoopes and Sheridan, 1992).

2.4 Calcareous sinter

On the SE shore and five km NW from Regules (Figures 4 and 5) outcrops a non-marine carbonate deposit named “La Calera” that has been mistaken with marine Cretaceous limestone (Ortega-Gutiérrez *et al.*, 1991) and with lacustrine limestone (Rosas-Elguera and Urrutia-Fucugauchi, 1998). It is a calcareous sinter that extends 2 km towards the E-W and 600 m towards the N-S. This deposit overlays andesitic rocks and belongs to the **TVBgp** (Figure 4). Spring deposit (Gierlowsky-Kordesch, 1998) “La Calera” is very limited in area extent and is the only calcareous sinter known around Lake Chapala. Being approximately 5 m thick, the calcareous sinter material is characterized by both massive and banded structures. When massive, it is brownish yellow and gray colored and elsewhere a pseudo-brecciate structure can be appreciated. On the other hand, when banded, alternated yellow and dark millimetric bands can be distinguished, the calcareous sinter material is characterized by vuggy porosity and silica (quartz and chalcedony) veinlets. Under a microscope, a pseudo-micritic texture is observed; vugs coated by iron oxides, are filled with calcite, and/or quartz, chalcedony and clay minerals. An incipient process of pervasive silicification is observed as well as calcite recrystallisation.

The accumulation of carbonates in continental environments is highly dependent on the input source (Gierlowsky-Kordesch, 1998). According to López-Ramos (1979) the input source of carbonate to this calcareous sinter is identified as deep (>2,348 m) Aptian-Albian reefal limestone, which outcrops 50 km South and is covered by undifferentiated Upper Miocene to Pleistocene andesite and minor basalt from **TVBgp** (Figure 2).

2.5 Geothermal activity

In the tectonically active zones of the earth’s crust, geothermal systems are very abundant and the Citala rift, where Lake Chapala is located, is not the exception. The Lake Chapala basin is characterized by its geothermal activity, which includes: thermal springs, hydrothermal alteration outcrops (San Juan Cosalá) and sinter deposits (*cf.* 2.4). The spring water around Lake Chapala is carbonated (Medina-Heredia A, 1986). To the NE area is San Luis Agua Caliente (69°C; $[HCO_3^-]$ ~ 240 mg L^{-1}) in the NW at Jocotepec (36°C; $[HCO_3^-]$ ~ 263 mg L^{-1}); in the South we find Tuxcueca and Tizapán El Alto (30°C; $[HCO_3^-]$ ~ 193 mg L^{-1}). However, there is an exception, the spring water at the San Juan Cosalá sector (North), which is sulphated (64-83 °C; $[SO_4^{2-}]$ ~ 479 mg L^{-1}).

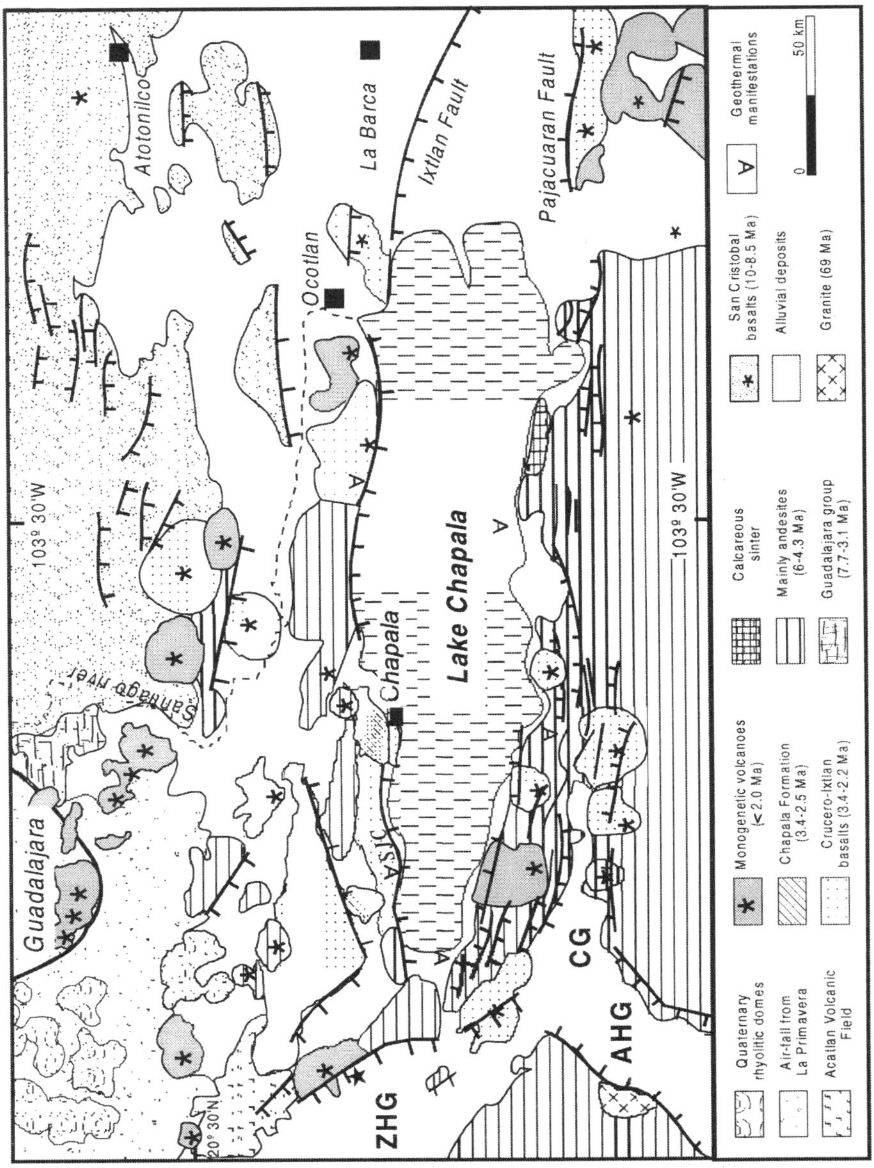

Figure 4 Geology of the Lake Chapala area (after Rosas-Helguera and Urrutia-Fucugauchi, 1998)

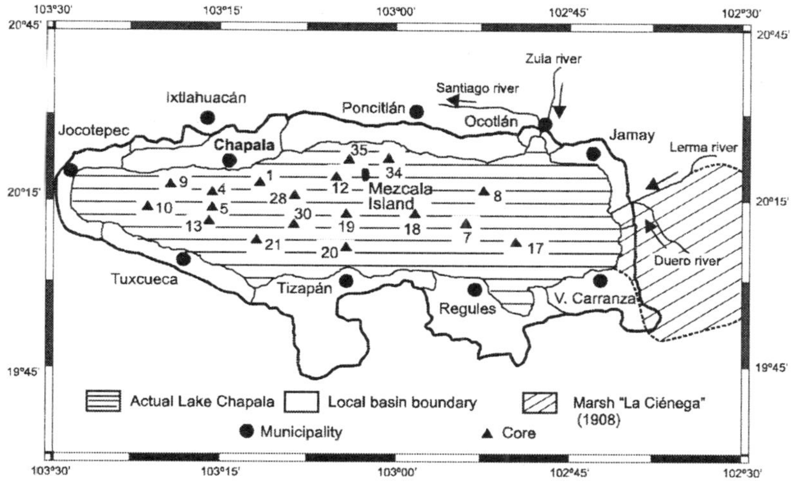

Figure 5 Lake Chapala basin, boundaries and core sample stations

Thermal springs occur both inside and outside the lake Figure 4): Examples of out-shore thermal springs are: San Juan Cosalá (N); San Luis Agua Caliente (NE); Jocotepec (W); Tuxcueca and Tizapán El Alto (S) (Figure 4). An example of a thermal spring in the lake is "Los Gorgos" with temperatures approximately 2°C higher than the lake's average). The input source of carbonate for this calcareous sinter is also identified as deep (>2,348 m after López-Ramos, 1979) Aptian-Albian reefal limestone.

2.6 Hydrocarbon manifestations

Near the sub-lacustrine spring "Los Gorgos" (Figure 5) there are some small land spots which look like islands (<3-4 m^2) made of asphalt, black viscous bitumen and which are linked to the bottom of the lake. Because of these hydrocarbon manifestations the company Petroleos Mexicanos drilled, without success, an oil exploration well 2,348 m deep (López-Ramos, 1979). We identify the source of asphalt as deep (>2,348 m) Aptian-Albian reefal limestone. It was concluded that deep geothermal fluids have traversed rich organic matter limestone formation and a remobilization of asphalt, which reached the bottom of the lake through E-W oriented fractures, occurred.

2.7 Physiography

Natural and anthropogenic morphometric variations in Lake Chapala since the sixteenth century, including bathymetry, isolation, and development of shoreline, have been documented by Zárate-del Valle (2000)

and Zárate-del Valle and de Anda (2000). In 1579, Abraham Ortelius published a map (Figure 6A). According to this map, the lake was approximately 116 km long and 32 km wide. The mean depth was unknown. At the eastern part of the lake an important offshore island, named *Paxacoran*, was located (Figure 6A). Comparing the cartography of the *Chapalicum Mare* (1579) with another published by Grasshopper (1999) and Jose María Narvaez (1816) the integration of *Paxacoran* island (Pajacuaran Fault, Figure 4) to the inland and the reduction of the lake to the East can be observed, (Ixtlan and Pajacuaran faults, Figure 4) maybe due to natural causes. In Narvaez' map, the lake is 20.5 leagues (114 km) long, 5.5 leagues (31 km) wide, and a mean depth of about 6.5 braces (10.8 m). In 1889, the eastern marsh of Lake Chapala was known as follows: the northern section near the town named La Barca (Figures 4 and 6b) was called the "Qumuato" marsh and the southern section was named the "Duaracha" marsh, near the towns of Jiquilpan and Sahuayo. Part of the Qumuato marsh is actually called La Ciénega (Pajacuaran and Ixtlán Faults, Figures 4 and 5).

In 1897 a historical drought temporarily reduced the lake´s surface to local bench mark 90.50 (Figure 6C; Sandoval, 1994). At this level, the length of the lake is reduced to 60 km, a width of 14 km and a mean depth of nearly 6 m. For comparison, the mean lake level is 96.11 (local benchmark 100 correspond to 1,526 meters above mean sea level, mamsl). The Maltaraña earth embankment was built in 1908 to aid agriculture at the eastern part of the lake, and resulted in the loss of 50,000 hectares of the Lerma river prodelta area, La Ciénega (Figure 5), which at that time was considered one of the most important areas for migratory birds coming from North America. Additionally, it acted as a wetland to purify the Lerma river waters before entering Lake Chapala. The bathymetries presented on maps from 1816 (Narvaez' map; UNED, 1990, Figure 6B), 1929 (Ballesteros' map; SARH, 1981), and 2000 (CNA, Figure 6C) have also revealed that the morphology of the lake bottom has changed The maximum depth is reported near the Los Alacranes island (103° 5.71' W; 20° 15' N; Figure 5).

2.8 Geological hazards

Geological hazards in the Lake Chapala basin are: a) E-W fissures in soils by shallow aquifer withdrawal (in ex. Ajijic); b) mass movements, like subsidence (in ex. Cerro Grande Hill) and landslides (Cerro San Miguel); c) debris- and mud-flows preceded by waterspouts (San Luis Soyatlán and San Juan Cosalá at Las Vigas-El Tecuán range, 1997-1998); d) floodings (1956, 1973, 1976), droughts (1897, 1950, 1998-2001); e) differential movement of clay soils (town of Chapala); f) volcanic eruptions (Volcán de Colima) and g) earthquakes.

Figure 6 Morphometric changes in Lake Chapala **A. XVI century**: *Chapalicum Mare* (detailed view of Ortelius' map, 1579;). **B. XIX century**: Grasshopper map 1998 (modified after Narvaez map, 1816). C. **XX century**: modified after bathymetry (CNA, 2000) (isolines represent 1 m depth variations; local datum level: 100 = 1525 mamsl)

3. SEDIMENTS

On the basis of gravimetric modeling, Allan (1986) estimated a thickness of ~900 m for the infilling of the northern basin of the Colima graben (**A**, in Figure 1). From drilling data Venegas *et al.* (1985) and Casarrubias (1995) concluded that the thickness of sediments is of 700 m at Lake San Marcos (**C**, in Figure 1) and of 275 m at Ixtlán de los Hervores geothermal field, respectively. On the basis of this data, we suppose that the silt infilling in some places of Lake Chapala is more than 500 m thick. Terrigenous minerals dominate the sediments corresponding to Lake Chapala. Particularly, shallow sediments in the lake are loose, dark, and silty with granulometric median ranging from 10 to 50 μm. The water content is generally higher than 80% in weight, at a depth of 50 cm. In the surface (approximately 0-2 cm) sediments are rich in water (> 85%) and are therefore difficult to sample. In the upper 10 cm the total organic carbon contents varies from 4 to 6% (Fernex *et al.*, 2001).

3.1 Geochemistry

Mineralogy of Lake Chapala sediments was studied by induced coupled plasma mass spectrometry (ICP-MS), X-ray diffraction (XRD) and also by scanning electron microscope (SEM) techniques. The major oxides detected in the sediments were SiO_2, Al_2O_3, Fe_2O_3 and CaO. The composition of sediments by SEM and XRD is the following: plagioclase, quartz, clays, carbonates and a minor fraction of TiO_2. The presence of halloysite was confirmed by differential thermal analysis (DTA). Additionally, minor fractions composed of glassy volcanic ash (pumice) were identified by SEM and XRD analyses (Sheridan and Marshall, 1987). Chemical contents are shown in Table 1 and mineralogical contents, in Table 2. The presence of post-depositional authigenic minerals, like framboidal pyrite, was observed in core 13 (Figures 5 and 7-H) and fluorapatite as a nodule of 1 cm in diameter at a depth of 66-68 cm in core 28 (Figure 5). The presence of fluorinated minerals is probably due to geothermal activity in Lake Chapala.

To understand the importance of the sediment-water interactions in the bottom of Lake Chapala, interstitial water samples were obtained in 13 stations distributed in the lake and analyzed by Ríos-Donato (2001). The results suggested that interstitial water quality is highly influenced by lixiviation of pollutants from contaminated sediments. The resuspension of sediments and associated contaminated pore-water may result in the liberation of sediment-associated pollutants, which may affect water quality and ecosystem health, especially at low water levels in the lake.

Average P concentrations in cores 30 and 35 were of 858 mg kg^{-1} (dry weight; Solana-Espinosa *et al.*, 1998). The bottom sediments may therefore represent an important phosphorus source for the water column above the water-sediment interface.

Table 1 Geochemical composition of sediments in Lake Chapala

ZONE	EASTERN	CENTRAL	WESTERN
OXIDES	%	%	%
SiO_2	48.6	56.3	57.3
Al_2O_3	23.1	16.8	16.9
Fe_2O_3	7.6	6.5	6.2
CaO	3.3	2.0	2.4
MgO	2.0	2.0	1.9
K_2O	1.3	1.6	1.5
Na_2O	0.7	0.9	1.8
TiO_2	1.0	1.1	1.2
P.P.C.	12.5	12.3	10.6
TOTAL	99.8%	99.7%	99.8%
Average trace metal concentrations (mg kg^{-1})	Ba 403, Co 17, Cr 58, Cu 16, Nb 19, Ni 32, Rb 66, Sr 194, V 103, Y 37, Zr 229, Zn 96		

Table 2 Mineralogical compositions of sediments in Lake Chapala

ZONE	EASTERN	CENTRAL	WESTERN
MINERALOGY	%	%	%
Andesine: $[Si_2(Si,Al)O_8]$ (Na,Ca)	14.2	13.9	23.2
Orthoclase: $[Si_3AlO_8]K$	7.5	9.7	9.1
Clays: Halloysite, hectorite	46.2	36.9	32.3
Quartz: (SiO_2)	14.2	25.6	23.0
Carbonates: ($CaCO_3$ + $MgCO_3$)	5.3	1.7	1.9
Fe_2O_3	7.1	6.4	6.2
TiO_2	0.9	1.1	1.2
Organic Matter	4.1	3.5	2.2
TOTAL	99.5%	98.9%	99.1%

3.2 Micropaleontology

The scarcity of diatom frustules at a depth of 1.40 m in Lake Chapala sediments (cores 22, 30, 31 and 35; Figure 5) is possibly due to variations in e.g. lowering in pH in sediments and water, hence inhibiting their preservation. Degradation is enhanced by the reduction of pH in the lake water (see previous section). Nevertheless, at core 13 (Figure 5) the following Pleistocene micro-fauna was determined (Figure 7), at a depth of 1.5 m: *Stephanodiscus aff. nemanensis* (Kursevich) or *Stephanodiscus minutus* (the most common species), *Synedra Ulna, Cyclotella meneghiniana, Synedra ulna, Cyclotella menenghiniana, Surirella turgida, Surirella robusta, S. ovalis* and *Amphora ovalis, Amphora lybica, Navicula gastrum* and *N. criptocephala* and *Campilodiscus clypeou, Stephanodiscus minutus* and *Nitzschia subacicularis*, they represent the most common population in the lake today. The presence of *Synedra ulna* testifies the eutrophication of the lake and the presence of *Surirella* and *Campilodiscus clypeou* may be related to water level fluctuations and/or variations in salinity (Israde-Alcantara, 1999a) in the lake.

Pollen was found and species were identified by Solana-Espinosa *et al.* (1998): Pinos, Quercus, Chenopodiaceas, Amarantaceas; Pediastrum, Coeleastrum, Botriococcus and Curvularia, Fusarium and Alternaria. The provenance of Chenopodiaceas and Amarantaceas (evidence of arid conditions) is located 15 km to the Northwest of the lake (Chazaro-Bazañez, personal communication) in the San Marcos seasonal endorheic saline basin (**C**, Figure 1).

3.3 Magnetic susceptibility

Magnetic susceptibility is the capacity of an object to be magnetized in the presence of a magnetic field. It was measured in cores 5, 8, 9, 12, 19, 20 and 34 (Figure 5) by means of a Kappabridge KLY magnetic susceptibility meter (Ramírez-Sánchez, 1998). Vertical profiles from different sites can be compared, when similar stratigraphic correlations are made. Particularly, a layer characterized by a high susceptibility peak can be used as a stratigraphic marker (Figure 8). With SEM and energy disperse spectral analysis (EDS) study of core 40 particles of titano-magnetite and authigenic framboidal pyrite (Figure 7H at depth 116-118 cm) were identified. Through magnetic susceptibility measurements and stratigraphy, irregularities were identified in the sedimentation processes occurring in different parts of the lake. These may partly be linked to bioturbation and fishing activities.

4. SEDIMENTATION RATES

Knowledge of the sedimentation rate is an important tool to understand processes in lakes; variations have been related to processes in cases of basins controlled by tectonics (Gasse, 1990). Sedimentation rates in Lake Chapala were evaluated from the isotopic study of fifteen cores (Figure 5).

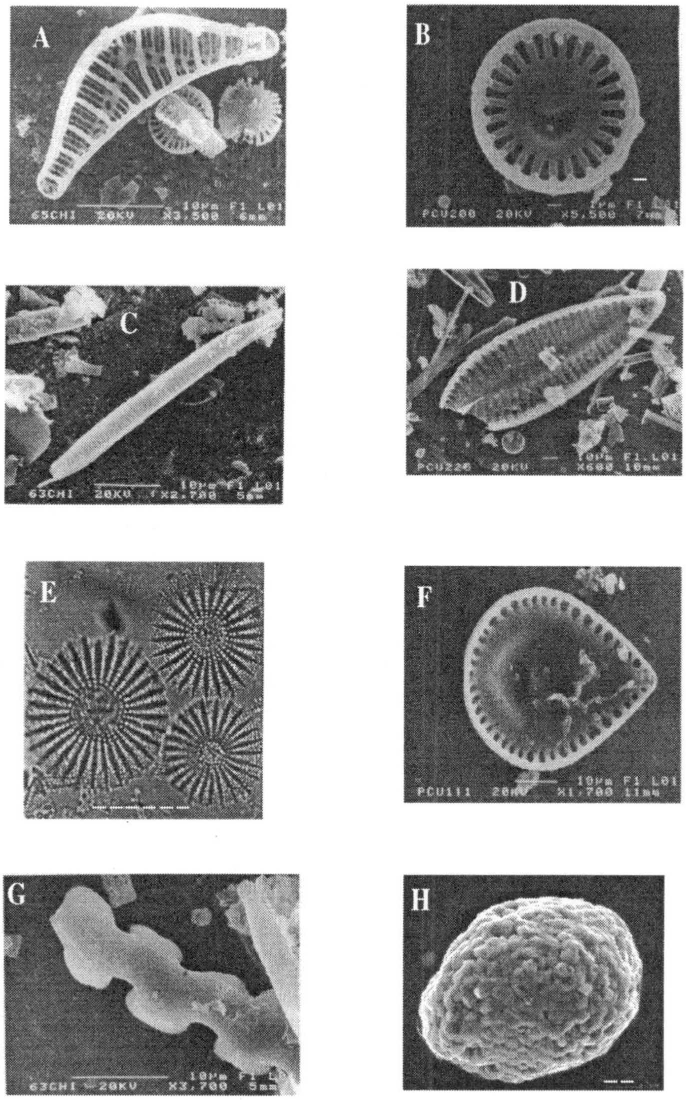

Figure 7 Scanning electron micrographs of organic and inorganic materials in core 13 (depth 1.40 m). A-F: common species of diatoms (A = Ephitemia sorex; B = Cyclotella menighiniana; C = Synedra ulna; D = Surirella splendida; E = Stephanodiscuss sp.; F = Surirella hoefleri; G = phytolite rest (depth 1.16 m); H = framboidal pyrite. The bars represents 10 microns in all the graphs except B, where it represents 1 micron and H, where it represents 2 microns

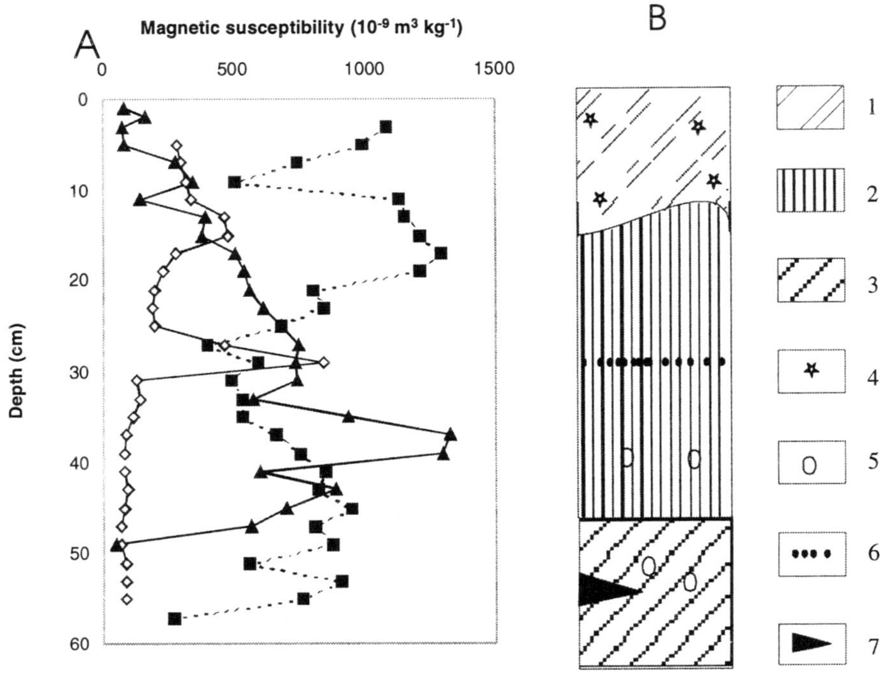

Figure 8 Profiles of shallow sediments in Lake Chapala. (For location of sites, please refer to Figure 5). **8-A** Magnetic susceptibility profiles: Diamonds = core 5. Squares = core 9; Triangles = core 19. **8-B**. Lithological log of core 5: 1 = dark brown fluid silt; 2 = dark brown fine silt; 3 = compact silt; 4 = small dark spots; 5 = bubble; 6 = thin sandy layer; 7 = silt. Notice the correlation between the magnetic susceptibility peak (diamond in **8-A**) and the thin sandy layer (6 in **8-B**) with magnetic particles of iron oxides in core 5

4.1 Analytical

^{210}Pb activities were analyzed by alpha counting of ^{210}Po deposited on silver plate and the standard used was ^{208}Po. ^{137}Cs activities were measured using gamma-spectrometry and those of $^{239-240}$Pu by alpha counting following the method described by Ballestra (1980), IAEA (1989), and Michel *et al.* (1999).

4.1.1 ^{210}Pb method

Seven profiles were analyzed for their ^{210}Pb activities (cores 1, 4, 9, 17, 18, 19 and 21; Figure 5). Radioactive equilibrium between ^{210}Pb and ^{210}Po is often not achieved within the first few centimeters of sediments, since these do not react in the same way in the water body before they are deposited.

Inaccuracy in the measurements of activities of fallout radionuclides in surface sediments (0-2 cm) may also be caused by sediment mixing due to bioturbation or to bottom water currents. Therefore, sediment accumulation rates have generally been evaluated by considering mainly deeper levels (from 2 to 15-20 cm).

The ^{210}Pb profiles show exponential forms typical of radioactive decay, with a period of 22.3 yr (Fernex *et al.*, 2001), but declination is irregular with respect to depth. It can be assumed that these are related to variations in sedimentation rates. The later may originate from climatic variations with successively humid and dry epochs, and perhaps also from the construction of dams in the watershed.

Five profiles (cores 7, 10, 17, 18, 19 and 21, Figure 5) were analyzed for ^{137}Cs activity (Figure 9), in order to locate the depths of maximum concentration. This layer corresponds to sediments deposited during the years 1962-64, when fallout due to atmospheric nuclear tests was the highest (Walling and He, 1992, He *et al.*, 1996). $^{239-240}$Pu was analyzed in sediments of core 7 (in the eastern part of the lake) and core 13 (to the West). Similar to ^{137}Cs, layers with highest Pu concentrations are assumed to be composed by sediments deposited during the 1962-1964 period. As a matter of fact, maximum ^{137}Cs fallout occurred in 1963, mainly because of atmospheric atomic tests carried out by the former USSR.

4.1.2 ^{137}Cs and $^{239, 240}$Pu methods

The highest ^{239,240}Pu values were found at 19 cm depths near the Lerma river inlet (core 7) and at 3 cm in the western part of the lake (core 13); activities were 0.9 and 0.7 Bq kg^{-1} (= 0.054 and 0.042 dpm g^{-1}), respectively. The slight discrepancies between the rate values calculated from the ^{137}Cs and ^{210}Pb profiles in same points may result from bioturbation effects or storm deposition (Chumra and Kosters, 1994).

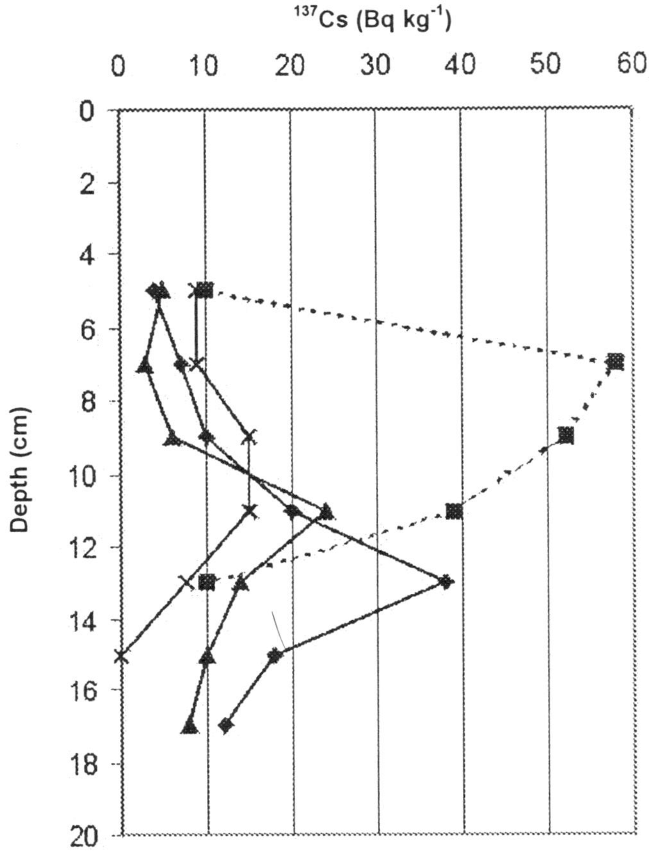

Figure 9 Caesium-137 activity profiles in Lake Chapala. Diamond = core 17; square = core 18; triangle = core 19; cross = core 21 (for location of sites, see Figure 5)

4.2 Results

Sedimentation rates (SR) throughout Lake Chapala are not uniform. In the eastern part of the lake (Figure 5), near the Lerma river inlet (core 7), the SR exceeds 5 mm yr^{-1} corresponding this site to the bottom of the prodelta slope (Figure 6C). On the slope and closer to the Lerma river inlet (Figures 5 and 6C), the SR was slightly lower, 4.0 mm yr^{-1} when measured by the ^{137}Cs method and 3.5 to 3.7 by the ^{210}Pb method. On the other hand, in the central part of Lake Chapala the SR ranged from 2.2 to 3.0 mm yr^{-1} in cores 18 and 21, respectively, and from 3 to 3.3 in core 19, where the La Pasión river seems to influence SR. Finally, on the western part (cores 4 and 9), the SR has been evaluated in less than 2 mm yr^{-1}.

5. SOILS

According to the FAO/UNESCO Soil Classification System (1970) the major soil types in Lake Chapala (Figure 10) are the following (CETENAL, 1974; INEGI, 1990; SEDUE, 1991): *Vertisol* (50%), *Pheozem* (30%), Lithosol (15%) and Luvisol (5%). They are *in situ* soils, colluvial soils (located on slopes greater than 8%) or transported soils (located on slopes smaller than 8%). The main use of soils is for agricultural purposes.

5.1 Soils

Vertisol

The predominant soil-type is pellic (black colored) associated with both chromic *Vertisol* (gray color) and eutric planosol of fine texture; it shows sporadic saline or sodic facies. Also the *Vertisol* presents the following characteristics: 1) deep duric (low levels); 2) stony (high layers) with medium to large boulders; 3) lithic (bedrock between 10-50 cm depth).

It is dominant in the regions of the eastern half of the lake's shoreline (Figure 10). This kind of soil is thin, its fertility varies from medium to high, and it has a clayey texture and is the most evolutionary soil in the Lake Chapala basin. Disadvantages of Vertisol are noticeable in drainage, aeration, and collapse and, because of this, it tends to become saline. It is difficult to use this soil for agriculture and grazing because, when it is wet it becomes doughy and when dry it cracks.

Pheozem

The predominant type is the haplic Pheozem associated with minor types such as: chromic Vertisol, eutric Regosol, chromic Luvisol and Lithosol. It is dominant in the mountains and as stony phases in foothills, whereas in the top of the mountains it becomes lithic with bedrock formations at depths of between 10-50 cm. This type of soil occurs to the North and Southwest of Lake Chapala (Figure 10). Its fertility ranges from medium to high and is rich in organic matter; it has medium texture and good drainage and aeration. This type of soil occurs in thin layers, which are stony most of the times. This soil is unstable and usually not used for agriculture but rather for pasture.

Lithosol

This type of soil is shallow, with a thickness of less than 10 cm over bedrock. Its agricultural application is limited by the presence of water and it is easily erosionable. It occurs in the Northwest and Southeast extremes of the Lake Chapala basin (Figure 10).

Luvisol

The predominant variety of this type of soil is chromic Luvisol, which shows red and yellowish red colors and is associated with chromic Vertisol. This type of soil occurs only at the southern borders of the Lake Chapala basin (Figure 10).

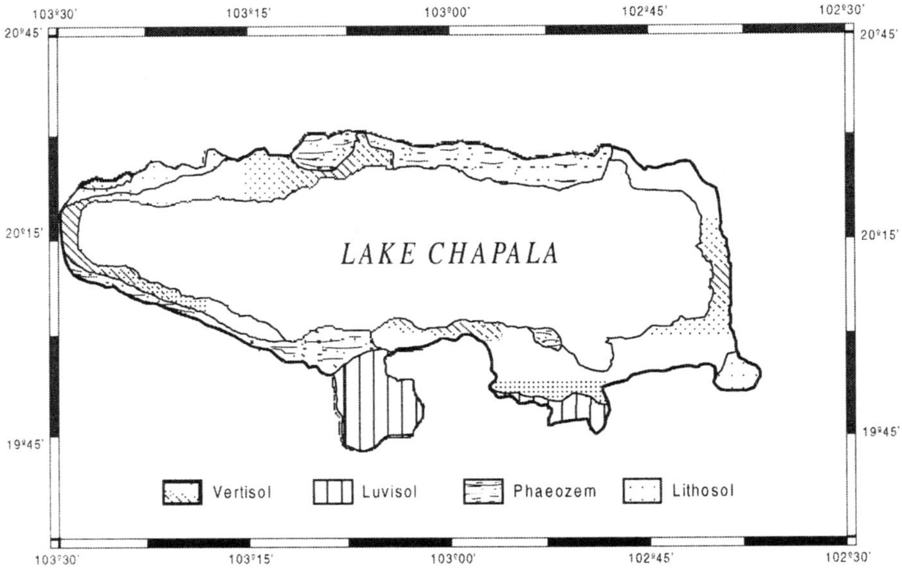

Figure 10 Edafology of the Lake Chapala basin (modified after CETENAL, 1977, INEGI, 1990 and SEDUE, 1991)

5.2 Soil erosion and infilling of Lake Chapala

Erosion involves physical removal of previously broken down rock and mineral material. Soil erosion is caused by the action of water and wind; slope constitutes the most important factor in erosive processes. We can classify slope of the land surrounding Lake Chapala in four categories: plain, slightly steep, steep and very steep (Table 3). The Lake Chapala basin is susceptible to a yearly loss of about 121,000 tons of soil due to erosive processes (SEDUE, 1991). The potential erosive rate and erosion risk are shown in Table 3.

Each year (SEDUE, 1991) Lake Chapala receives about 1'926, 500 m^3 (2'891, 243 tons) of silt transported by the major rivers (Figure 5) that flow into the lake (Sandoval, 1994): Lerma (81.4%); Duero (6%) and Zula (4.6%) as well as by streams within the Lake Chapala basin (8%).

Table 3. Slope classification of land in the Lake Chapala basin (SEDUE, 1991).

SLOPE (%)	CLASSIFICATION	% OF TOTAL BASIN AREA	EXAMPLES	LAND USE
< 8	Plain	38	Plains and ripian zones	Cropland and urban development
8-15	Slightly steep	25	Piedmont	Cattle
15-30	Steep	23	Foot-hill	Silviculture
> 30	Very steep	14	Highlands	Reserves

6. CONCLUSIONS

(1) The shallow sediments in Lake Chapala are dark and silty with granulometric median ranging from 10 to 50 μm.

(2) ^{137}Cs and $^{239,240}Pu$ activities were analyzed in six cores from different parts of the lake and the profiles show maximum activity peaks whose layers correspond to sediments deposited during the maximum activity of atmospheric nuclear tests in 1963. Sedimentation rates of up to 5 mm yr^{-1} were estimated. The SR in the eastern part of the lake is two to three times higher than in the western part. It was also hypothesized that the depth of deposited sediments in the eastern part of the lake was shallower; this has not yet been confirmed.

(3) We assume that a tectonic compensation (a relative subsidence) is responsible for the quasi-uniform depth in spite of the different SR. We have inferred an approximate one mm yr^{-1} relative tectonic rising at the West that compensates the lower SR rate.

(4) Based on stratigraphic facts and with the help of magnetic susceptibility values, we have identified irregularities in sedimentation throughout the lake, which partly seem linked to slumping, bioturbation and fishing activities.

(5) Micro-paleontological data confirmed that Lake Chapala is affected by eutrophication processes and is characterized by water level fluctuations. Scarcity of diatoms in shallow sediments from Lake Chapala (<1.40 m) is probably due to pH variations of sediments and water that inhibits the preservation of these microorganisms.

(6) The development of fluorapatite nodules in shallow sediments was recognized. This may be due geothermal activity in and outside the lake.

(7) The nonmarine carbonate deposit at Regules has a geothermal origin and the input source of carbonate for this calcareous sinter is identified as deep (> 2,348 m) Aptian-Albian reefal limestones.

ACKNOWLEDGEMENTS

Acknowledgements are made to both Mexican-French *ad-hoc* ECOS council (M95-U01 project) and SIMORELOS-CONACyT Trust (970306003 project) for the support of this research. We are grateful to Guadalajara University (CUCEI, MEXICO), the Urban Development Department (SEDEUR, Jalisco State Government, Mexico), Water National Commission (CNA, Mexican federal government), Aix-en-Provence University (CEREGE, FRANCE), CIATEJ (Guadalajara, Mexico) and the Geology and Mineralogy Department-UMNH (Michoacán University, Mexico) for their technical and logistic support. The authors wish to thank F. Gasse (CEREGE) for diatoms determination and J. Dunbar (Baylor University, USA) and M. Martínez (IMTA, Mexico) for their helpful suggestions to improve the organization and English in the text.

REFERENCES

Allan JF (1986) Geology of the Northern Colima and Zacoalco Grabens, Southwest Mexico: Late Cenozoic rifting in the Mexican volcanic belt. Geological Soc. America Bulletin, 97: 473-85.

Ballesteros (1929) Estudio batimétrico del Lago de Chapala (map). Secretaría de Agricultura y Fomento. Unpublished.

Ballestra S (1980) Radioactivité artificielle et environnement marin; étude relative aux transuraniens ^{238}Pu, ^{239,240}Pu, ^{241}Pu et ^{241}Am en Mediteranée. Thèse, Université de Nice (IAEA, Monaco), 250 p.Unpublished,

Barrier E, Bourgois J, and Michaud F (1990) Le point triple de Jalisco: vers un proto-golfe de Jalisco.C.R. Acad. Sci., Paris, 310-IIa: 1513-20.

Calmus T, Zárate-del Valle PF, and Michaud F (1996) Le Cretacé de San Sebastian del Oeste (Jalisco, Mexique): un jalon de l'arc Alisitos-Teloloapan, CR Acad. Sc, Paris, 321-IIa: 795-800.

Casarrubias Z (1995) Resultados y evaluación de la perforación en la zona geotérmica de Los Negritos, Michoacán, México. Geos, 15:92-99.

CETENAL (National Territory Study Commission) (1977) Manual para la Aplicación de las Cartas Edafológicas de CETENAL para fines de Ingeniería Civil. CETENAL.

CETENAL (National Territory Study Commission) (1974) Edafological charts: F-13-D-75 (Jocotepec), F-13-D-76 (Chapala), F-13-D-77 (Ocotlan), F-13-D-85 (Zacolaco de Torres), F-13-D-86 (Tizapán El Alto), F-13-D-87 (Sahuayo de Morelos). CETENAL.
Chumra G L, Kosters EC (1994) Storm deposition and ^{137}Cs accumulation in fine-grained marsh sediments of the Mississippi Delta plain. Estuar. Coastal Shelf Sci., 39: 33-34.
CNA (National Water Commission) (2000) Bathymetric map 2000.
Demant A (1981) L'axe néo-volcanique Transmexican. Etude volcánologique et pétrographique. Signification géodynamique. Thesis, Univ. D'Aix-Marseille III, France, 259p. Unpublished.
Fernex F, Zárate-del Valle P, Ramírez-Sánchez H, Michaud F, Parron C, Dalmasso J, Barci-Funel G, and Guzmán-Arroyo M (2001) Sedimentation Rates in Lake Chapala (Western Mexico). Possible Active Tectonic Control. Chem. Geology (accepted for publication).
Ferrari L, Nelson SA, Rosas-Elguera J, Aguirre-Díaz GJ, and Venegas-Salgado S (1997) Tectonic and volcanism of the western Mexican Volcanic Belt in Aguirre-Díaz GJ, Aranda-Gómez JJ, Carrasco-Núñez G and Ferrari L (1997) eds: Magmatism and tectonics in the central and northwestern Mexico – a selection of the 1997 IAVCEI Assembly excursions: Mexico, DF, UNAM-Geological Inst, Excursion 12: p 85-129.
Gasse F (1990) Tectonic and Climatic Controls on Lake Distribution and Environments in Afar from Miocene to Present. In: Lacustrine Basin Exploration, AAPG Mem. BJ Katz ed, 50: 19-42.
Gierlowsky-Kordesch EH (1998) Carbonate deposition in an ephemeral siliclastic alluvial system: Jurassic Shuttle Meadow Formation, Newark Supergroup, Hartford Basin, USA: Palaeogeogr, Palaeoclimatol, Palaeoecol, 140: 161-84.
Grasshopper Map (1999) http://www.grasshoppernet.com/usmexna/graphics/mapldc.jpg
He Q, Walling DE and Owens PN (1996) Interpreting particle size effects in the adsorption of ^{137}Cs profiles in several lakes and reservoirs in southern England. Chem Geol, 129: 115-31.
IAEA (1989) Measurements of Radionuclides in Food and the Environment. Internat. Atomic Energy Agency, Vienna. Technical report no. 295, 28 p.
INEGI (National Institute of Statistics, Geography and Informatics) (1990) Guías para la interpretación de Cartografía: Edafología. INEGI.
Israde-Alcántara I (1999a) Informe ECOS-CONACyT sobre registro de diatomeas en el núcleo 46 colectado en el Lago de Chapala, Mexico. Technical report. Unpublished.
Israde-Alcántara I (1999b) Los lagos volcánicos y tectónicos de Michoacán, *in* Corona-Chavez P, and Israde-Alcántara I eds: Carta Geológica de Michoacán, escala 1:250,000, pp 45-73, Univ. Michoacana de SNH.
Lange RA and Carmichael ISE (1991) A potassic volcanic front in western Mexico: The lamprophyric and related lavas of San Sebastian. Geol Soc of America bull, 103: 928-40.
Levi B (1969) Burial Metamorphism of a Cretaceous Volcanic Sequence West from Santiago, Chile Contr Mineral and Petrol, 24: 30-49.
Leyden J (1994), as presented in Mexico connect: *http://www.mexconnect.com/mex_/geology.html.*
López-Ramos E (1979) Geología de México. Ed Tesis Reséndiz, Mexico, 3: 28
Luhr JF and Carmichael ISE (1985) Contemporaneous eruptions of calc-alkaline magmas along the volcanic front of the Mexican Volcanic Belt. Geofis Int, 24: 203-16.
Luhr JF and Carmichael ISE (1990) Geology of Volcan de Colima. Instituto de Geología-UNAM, bull, 107.
Mahood, G (1980) Geological evolution of a Pleistocene rhyolitic center - Sierra La Primavera, Jal, Mexico. Journal Volcanology Geothermal Research, 8: 199-30.

Medina-Heredia A (1986) Reconocimiento y evaluación geoquímica de zonas termales del Estado de Jalisco. Technical report 04-86, Geothermal Research Dep, Comisión Federal de Electricidad. Unpublished.

Medina-Martínez F (1983) Analysis of the eruptive history of the Volcan de Colima, Mexico (1560-1980). Geofis Int, 157-78.

Michaud F, Bourgois J, Parrot JF, Taud H and Kasser M (1994) Le point triple de Jalisco (Mexique): évolution tectonique à partir du traitement de données MNT-SPOT. CR Acad Sciences, Paris, 318-IIa: 1117-24.

Michaud F, Gasse F, Bourgois J and Quintero O (1999) Tectonic controls on lake distribution in the Jalisco block area (western Mexico) from Pliocene to Present. Geological Soc. America, sp. paper 334.

Michel H, Barci-Funel G, Dalmasso J and Ardisson GA (1999) one step ion exchange process for the radiochemical separation of americium, plutonium and neptunium in sediments J of Radioanal Nucl Chem, 240-2.

Mosser F (1972) The Mexican Volcanic Belt: Structure and Tectonics. Geofísica Int. 12:55-70.

Ortega-Gutiérrez F, Mitre-Salazar LM, Alaniz-Alvarez S, Roldán-Quintana J, Aranda-Gómez JJ, Nieto-Samaniego A and Morán-Zenteno D (1991) Geologic Provinces of Mexico; a new proposal and bases for their modern definition in Rubinovich-Kogan R and Carreño AL (eds.) mem Convención sobre la Evolución Geológica de Mexico, Inst de Geología, UNAM.

Pardo M, Suarez G (1995) Shape of the subducted Rivera and Cocos plates in southern Mexico: Seismic and tectonic implications, Journal of Geophysical Research, 100-B7: 12357-73.

Quintero-Legoretta O (1995) Les frontières à terre du Bloc Jalisco, (Mexique). Etude néotectonique et structurale de la fracturation et reconstruction des paléocontraintes. Thèse Université, Paris VI : 255 p.

Ramírez-Sánchez HU (1998) Determination du taux de sedimentation dans le Lac Chapala (Mexique) par les methodes radiometriques. DEA report Univ D'Aix-Marseille III, France, 32p.

Ríos-Donato N (2001) Bioquímica del agua intersicial en la interfase sedimento-agua en el Lago de Chapala, Jalisco, México. MSc Thesis. CUCEI-Universidad de Guadalajara. 71 p.

Rosas-Elguera J, Ferrari L, López-Martínez M and Urrutia-Fucugauchi J (1997) Stratigraphy and Tectonics of the Guadalajara Region and Triple-Junction Area, Western Mexico. Inst Geology Rev, 39: 125-40.

Rosas-Elguera J and Urrutia-Fucugauchi J (1998): Tectonic Control of the Volcano-Sedimentary Sequence of the Chapala Graben, Western Mexico. Inst Geology Rev, 40: 350-62.

Sandoval FP (1994) Pasado y futuro del Lago Chapala. Unidad Editorial Guadalajara, Jal, Mexico.

SARH (Secretaría de Agricultura y Recursos Hidraulicos) (1981) Levantamiento Hidrográfico Lago de Chapala, Jal. y Mich. (map).

SEDUE (1991) Plan integral de desarrollo urbano de la Ribera del Lago de Chapala, Jal, (1ra. Etapa medio físico natural). Gobierno del estado de Jalisco, Mexico.

Solana-Espinosa G, Israde I, De Anda J and Zárate-del Valle PF (1998) Geochemistry data at Chapala lake, Jalisco western Mexico (ECOS M95-U01 and SIMORELOS 970306003 Projets, in progress). Proceedings of the Primera Reunión Nacional de Ciencias de la Tierra, Mexico City.

Sheridan MF and Marshall JR (1987) Comparative Charts for Quantitative Analysis of Grain-Textural Elements on Pyroclasts in Marshall J.R., (ed). Clastic Particles, Van Nostrand R Co, pp 98-121.

Stoopes RG and Sheridan MF (1992) Giant debris avalanches from the Colima Volcanic Complex, Mexico: Implications for long-runout landslides (>100 km) and hazard assessment. Geology, 20: 299-302.

Suter M, Quintero O, López M, Martínez, Aguirre G and Farrar E (1995) The Acambay graben, active intra-arc extension in the trans-Mexican Volcanic Belt, Mexico. Tectonics, 14: 1245-62.

Tardy M, Lapierre H, Freydier C, Coulon C, Gill JB, Mercier de Lepinay B, Beck C, Martínez RJ, Talavera MO, Ortiz HE, Stein G, Bourdier JL and Yta M (1994) The Guerrero suspect terrain (western Mexico) and coeval arc terrains (the Greater Antilles and western cordillera of Colombia): a late Mesozoic intra-oceanic arc accreted to cratonal America during the Cretaceous. Tectonophysics, 230: 49-73.

Venegas S, Herrera J and Maciel R (1985) Algunas características de la faja volcánica transmexicana y sus recursos geotérmicos. Geofísica Int. 24:47-81.

UNED (UNidad EDitorial) La Laguna de Chapala, Jalisco Government. Venegas S, Herrera JJ and Maciel R (1990) Algunas características de la Faja Volcánica Transmexicana y sus recursos geotérmicos: Geofís Int, 24: 47-81.

Walling DE and He Q (1992) Interpretation of Caesium-137 profiles in lacustrine and other sediments: the role of the catchment-derived input. Hidrobiología, 235:219-30.

Zárate-del Valle PF and Maldonado-Reyes JA (1990) Comentarios metalogenicos sobre el Estado de Jalisco, *in* Delgado-Argote and Martín-Barajas A (eds.): Contribuciones a la Tectónica del Occidente de Mexico, monografía num. 1, Un. Geofísica. Mex, pp 193-210.

Zárate-del Valle PF and Aguirre-Barragán J(1993) Los limites del Terreno Guerrero en el Estado de Jalisco. Observaciones de campo. Geos Un Geofísica Mex, 13-5.

Zárate-del Valle PF, Rosas-Elguera J, and Urrutia-Fucugauchi J (1999) Proyecto CONACyT 1667-PT: Evolución tectónica terciaria de la porción norte del Bloque Jalisco: Depresiones tectónicas Talpa de Allende y Mascota (implicaciones en la exploración minera). Technical report. Unpublished.

Zárate-del Valle PF (2000) Información indirecta del clima: Lagos. Información histórica (Códices y archivos parroquiales). Proceedings of the Seminario Internacional Predicciones climáticas para Mexico durante las próximas décadas: Una base Para la Estimación de la Oferta de Agua. Jiutepec, Mor, Mexico.

Zárate-del Valle PF, DeMetz C, Hutton W, Sánchez O and Stock J (2000) Using High-Precision GPS to constrain Vertical and Horizontal Crustal Movement near Lake Chapala, Mexico. Proceedings of First International Symposium on Lerma-Chapala Basin, Ajijic, Jal, Mexico, p. 37.

Zárate-del Valle PF and De Anda J (2000) Cartogtraphic Variations Between Chapalicum Mare (1579) and Lake Chapala (1999). Proceedings of First International Symposium on Lerma-Chapala Basin, Ajijic, Mexico, p. 38.

Zimmermann JL, Stussi JM, González-Partida E and Arnold M K (1988) Ar evidence for age and compositional zoning in the Puerto Vallarta-Rio Santiago Batholith (Jal, Mexico). J South Amer Earth Science, 1-3: 267-74.

Chapter 3

Natural Resources Management in the Lerma-Chapala Basin

BENJAMÍN DE LEÓN-MOJARRO[1], RAÚL MEDINA-MENDOZA[2], AND ARTURO GONZÁLEZ-CASILLAS[3]
Instituto Mexicano de Tecnología del Agua, Paseo Cuauhnáhuac 8532, Jiutepec, 62550 Mor., Mexico. email:[1] bleon@tlaloc.imta.mx; [2]rmedina@tlaloc.imta.mx; [3]agonzalc@tlaloc.imta.mx

Key words: Irrigation districts, irrigation units, soil conservation, living barriers, sustainable development

Abstract: This chapter describes the natural resources in the Lerma-Chapala watershed. The soil erosion is a problem that occur moderately in 45% of the Lerma-Chapala basin and in an accelerated way in 35% of the area. This is a consequence of the change in use of forest soil for agriculture, which generates the erosion of 25,000 ha per year and a sedimentary volume of 6 Mm^3 , which are dragged off the reservoir. The most elevated area is located in the sub-valley of Mexico State, where seasonal agriculture is practiced with low technology on highly inclined slopes, with limited vegetation, and deficient outlining of the terraces. To reduce erosion, several conservation programs have been applied, including filter dams, reforestation, stone barriers, high-density cultivation's and surrounding terraces, among others. The environmental problem in the valley derives from strong water demand due to the accelerated socio-economical development of the region. Industrial and agricultural production overpasses the national average, and availabilities of water is exceeded by the demands in all the region. Extracting more surface or groundwater at any point of the valley, invariably affects the downstream water use as the continuous decrease of the Lake Chapala water level reflects as a result of the regulation of the water in the Lerma river. This situation endangers the social fairness, the environmental equilibrium, and the economic efficiency in the valley. Remedial actions are proposed to preserve water quality, organization and regulation of water use, and valley management and conservation that would contribute to the sustainability of the region.

1. NATURAL RESOURCES INVENTORY

In this chapter, a soil description of the Lerma-Chapala basin is presented, followed by an explanation of the vegetation in the area and, finally the distribution of livestock inventory and raising.

1.1 Soils

Almost all of the soil units defined by the FAO/UNESCO classification system are present in the region. Feozem is the dominant soil type in the Upper Lerma region. This soil type is highly fertile and produces abundant harvest. Productivity may be increased even further with phosphate applications. During intensive cropping, the addition of other fertilizers such as lime, is required. The Feozem soil type is found in the surroundings of Toluca and Atlacomulco in Mexico State, in the southeastern part of Michoacán and to the South of Querétaro. Second in importance in the Upper Lerma region is the Vertisol soil type, followed by Andosol and Planosol (Figure 1).

In the Middle Lerma region, the dominating soil type, present in 51% of the region, is the Vertisol. The Vertisol is dark and contains high percentages of organic matter and is composed by more than 30% clay. It's most important feature is the dominance of expandable clays in its composition, usually montmorillonite, which causes the contraction of the soil as it dries and consequent production of cracks. This soil type are commonly found in arid and semi-arid regions; grass or forests grow readily on this type of soil, in plain areas or at moderately elevated slopes. Of the two known types of Vertisol only the pellic Vertisol is present, either alone, or in association with Feozem and Litosol. This soil type is distributed all over the basin: in the Querétaro and San Juan del Río Valleys in the state of Querétaro, in the central and southeastern parts of the state of Guanajuato, and in the northeastern part of the state of Michoacán, from the Lerma river to Lake Cuitzeo. The Feozem, Planosol and Litosol soil types follow the Vertisol in order of importance.

In the Lower Lerma region the Vertisol soil type also dominates, it is mainly found in the state of Michoacán (area from Lake Chapala to Zamora), and in the state of Jalisco (from Poncitlán to La Barca). The Feozem, Luvisol and Regosol soil types follow in importance.

The accelerated population growth, a consequence of urban and industrial developments, has led to radical changes in land uses over the past 40 years. The expansion of agricultural frontiers and of urban-industrial areas has permanently altered surface areas, which previously were natural ecosystems.

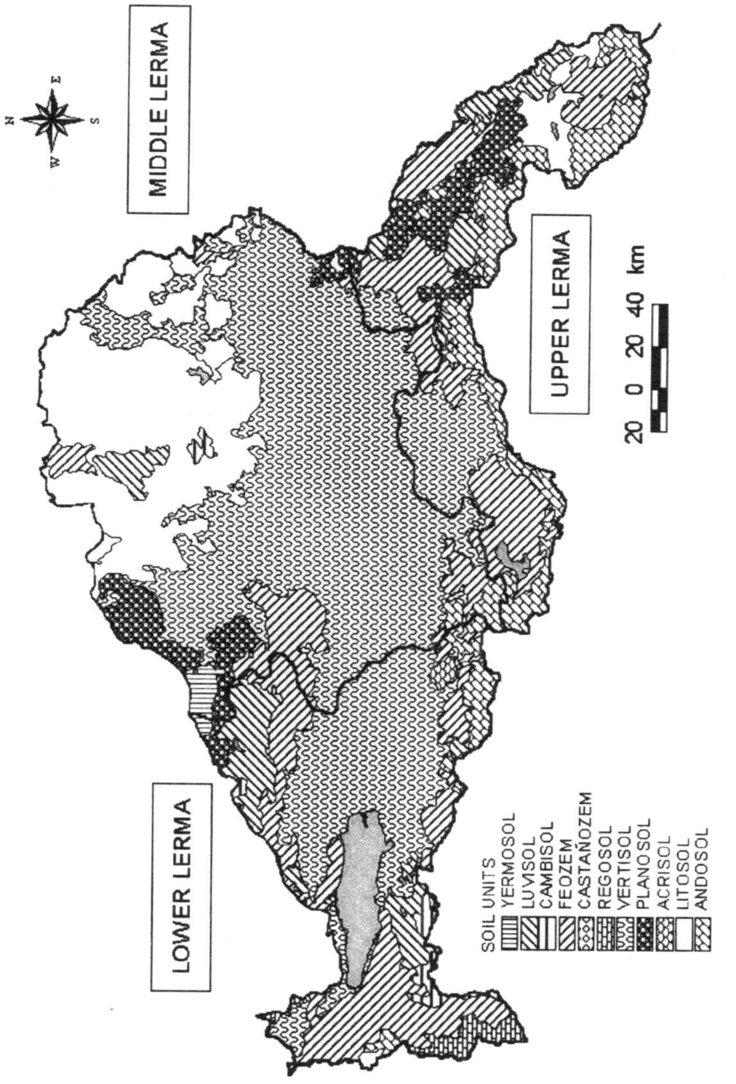

Figure 1 Soils units in the Lerma-Chapala basin

Without proper preservation and management, the risk of generating physical, chemical and biological effects is present; such an event would result in loss of soil fertility, that is, the loss of the substances that make it productive, such as nitrogen, phosphorous and potassium.

Soil degradation can lead to permanent desert conditions. Erosion is the main contributor to desertification, considered the most serious ecological problem in the basin, which affects natural renewable resources (Figure 2). Within the basin, the most affected regions are the states of Guanajuato, Jalisco, Mexico State, and Michoacán (Table 1).

In what refers to soils with salinity problems, there is minimum affectation in the Chapala area, in the states of Jalisco and Guanajuato. Another natural phenomenon affecting the basin is wind erosion; the most severely affected regions are localized in the state of Querétaro.

Figure 2 Erosion and the formation of gullies

Table 1 Water erosion in the Lerma-Chapala basin (CNA, 1994)

Erosion types	Upper Lerma (%)	Middle Lerma (%)	Lower Lerma (%)
Low <10 t ha^{-1} yr^{-1}	11.8	6.4	-
Moderate <10-50 t ha^{-1} yr^{-1}	8.2	-	11.5
Severe <20-200 t ha^{-1} yr^{-1}	2.0	30.2	10.3
Extreme <200 t ha^{-1} yr^{-1}	78.0	63.4	78.2
Total	100	100	100

1.2 Forest, shrub coverage and pasturage

The Lerma-Chapala basin has a complex topography, presenting large variations in latitude in altitude. This results in a macro and micro-climatic mosaic and establishment of the various vegetation communities.

In the Upper Lerma region pine-oak and oak-pine forests are dominant, they are located mainly towards the Northwest of the Mexico State and around Lake Pátzcuaro, in the state of Michoacán. *Quercus spp.* forests develop at altitudes of 500 to 2,000 meters above sea level (masl), with annual rainfall fluctuations between 600 and 1200 mm. This community is composed of bush- and tree-type oaks.

Pine forests are distributed as clumps and occur frequently on lands with deficiencies in certain mineral components. The heights reached by this type of vegetation species vary from 8 to 20 m. Sometimes only the upper branches persist, and they are susceptible to form large, roughly hemispheric, canopies. The underbrush accompanying the pines in these evergreen forests consists basically of ferns and herbaceous plants of the families *Leguminoseae, Labiattae, Srophulariacea, Rosaceae, Umbelliferacea, Gramineae and Compositae*, among others. These plants present changes in their flowering, fructification and defoliation periods, and they disappear during non-favorable periods. However, some of the underbrush persist, presenting color changes.

Vegetation classified as sub-tropical brush is also important in this region, it is found mainly around Lake Cuitzeo in the state of Michoacán, and in the northern section of Mexico State.

In the Middle Lerma region the pine-oak and oak-pine forests prevail, they are distributed to the South and West of the state of Querétaro and they are also located in the central and northern parts of the state of Guanajuato. Other important species found in this region are pasturelands, which are extended towards the western part of the state of Guanajuato and the Altos de Jalisco. Sub-tropical thicket-type vegetation is also found in the Irapuato and Pénjamo areas of the state of Guanajuato, from La Piedad to Zamora, in the state of Michoacán, and from Atotonilco to Tototlán in the state of Jalisco.

In the Lower Lerma region sub-tropical thickets are dominant, they are situated in the area from Jacona to the shoreline of Lake Chapala in the state of Michoacán, and they continue from Tepatitlán to Guzmán City in the state of Jalisco.

It should be noticed that the distribution of vegetation in the basin follows a defined pattern influenced by altitude and other ecological considerations. Pine and oak forests occur in areas with altitudes of more than 1000 masl and occupy similar areas, and developing one along the other. These forests consist of different oak species (*Quercus sp.)* and pines belonging to the families *Pinus Patula, P. Hedrerae, P. Oocarpa, P. Michoacana and P. Cembriodes.* Also cedars (*Juniperus sp., Rhus sp.), madrones (Arbustus sp.*), and crab apples (*Arctostaphylos sp.*) are found.

Another important type of vegetation in the basin is the so-called gallery vegetation, a classification in which clumps of trees located along river banks are recognized, they can reach heights ranging from a few to 50 m; among these, perennial, caduceus, and sub-caduceus individuals are found.

The gallery vegetation community includes, amongst its most notable species: *Ficus, sp., E. cicloocarpum*, P. dulce, *Astiantus vimanalis*, *Salix humboldtiana* (willow), and *Acacia macracautha.*

1.3 Basin and livestock-raising

Concerning livestock-raising and in accordance with information provided by INEGI (State Statistical Annuals of 1995), animal population increased considerably during 1991, to approximately 66 million (see Table 2), including 5,730,357 major and minor farm animals, of which 34.9% were bovine, 11.5% ovine, 13.5% caprine, 37.6% porcine and 2.3% equine.

The livestock-raising production system is conformed by two large divisions, separated by their technology, productivity and size: (1) specialized production, including stabling, semi-stabling and suburban backyard (intensive farming) with meat (fattening) and milk production; and (2) non-specialized production, with the purpose of producing meat and milk with pasturage (extensive farming).

Table 2 Animal inventory in the Lerma-Chapala region (1995)

Region	Number of animals					
	Bovine	Porcine	Ovine	Caprine	Equine	Birds
Upper Lerma	367,557	453,088	405,821	105,165	97,186	5,469,426
Middle Lerma	1,052,121	1,189,819	215,378	505,423	10,521	48,845,389
Lower Lerma	580,301	526,533	38,390	162,447	20,607	5,878,520
Total	1,999,979	2,169,440	659,589	773,035	128,314	60,193,335

The practice of extensive farming is typical of the marginal semi-arid regions of the plateau, with vegetation consisting of pasturage and thickets, acacias and coniferous as well as oak woods, with considerable orographic irregularities. In these areas, the limiting factors are rain and scarcity of fodder. This affects the livestock's survival capacities during periods when extreme meteorological conditions occur, such as droughts and frosts. In many cases, this type of farming is complementary to agriculture.

The feeding of livestock is conditioned by the over-use of wild vegetation, which consists of native grasses, such as: (*Bouteloua spp.*) razor grass, (*Bouteloua curtipendula*) banderita, (*Buchloe dactyloides*) buffalo grass, (*Aristida addscebsionis*) clover (*Lycuros phleoides*) fox tail, (*Leptochloa dubia*) giant and (*Sporobolus airoides*), alkaline pasture (*Muhlembergia sp.*), (*Stipa sp.*), (*Festuca sp.*) and (*Sporobulus sp.*), among others; there are also different species of tree and bush, comprising types of (*Prosopsis soo.*) mezquite, (*Acacia spp.*) huizaches, and palo dulce (*Esenhardtia polistachya*). The livestock's diet is complemented by backyard feeding with species of scorched prickly-pear (*Opuntia spp.*), chopped maguey (*Agave spp.*) and wastes of harvests such as corn stubble (*Zea mays*), and bean straw (*Phaseolus vulgaris*). Summer pasture coefficients range between 4 and 18 ha per animal unit per year.

The region of Los Altos de Jalisco outstands for the milk production in semi-intensive and intensive forms, a traditional livestock raising region, where Holstein cows are preferred. The state of Guanajuato outstands for its production of goat milk. The fodder is based on improved brands of ballico (*Lolium multiflorum* and *L. Perenne*), sorghum *(Sorghum vulgare x. S. Sudanense*), alfalfa achicalada (*Medicago sativa*), corn ensilage (*Zea mays*), barley (*Hordeum vulgare*), buffel (*Cenchurus ciliaris*), and rhodes, as well as improved and balanced agricultural remains.

El Bajío, a rich agricultural cereal area due to its closeness to the dairy regions of other states such as Hidalgo., Querétaro, Jalisco, San Luis Potosí, and Aguascalientes, assign a large part of its agricultural production to the production of lucerne (Medicago sativa). In the production of meat, pre-fattening takes place. In the more sophisticated cases, it is carried out in irrigated pastures with controlled pasturing and improved fodder. This is complemented with fattening in confinement where the main provisions are sorghum grains, corn ensilage, agricultural residues, minerals and agro-industrial by-products, among others. Livestock used for this purpose is mostly European, such as Holstein, Brown Swiss, and Angus, cross-breeded with Zebu and Creole cattle.

Production of pork and confined raise of poultry (eggs and fattened chicken) are important activities in the states of Jalisco and Guanajuato. Pigs are produced by: (1) technical raising (30% of the herd), (2) semi-technical raising, and (3) family or backyard raising. Meat, milk and egg production in the basin supplies mostly Mexico City and Guadalajara.

2. AGRICULTURAL AREAS

In this section, the organization of agriculture into irrigation districts (larger areas) and irrigation units (relatively smaller areas) has been described.

2.1 Irrigation Districts

The most important irrigation districts (ID) in the Lerma river basin are the ID 011 in the Upper Lerma region, in Guanajuato, with 125,898 hectares, and the ID 087, Rosario Mesquite, in Michoacán, with 63,143 hectares, followed by ID 020, in Morelia, Michoacán, with 18,021 hectares; ID 024, Ciénega de Chapala, Michoacán, with 18,800 hectares; ID 033, Mexico State, with 17,508 hectares; ID 061, Zamora, Michoacán; and ID 085, La Begoña, Guanajuato. The surface areas represent the average cropped surfaces during the last five years, and is lower than the surface areas recorded in the users' record of each irrigation district.

Table 3 Irrigation Districts in the Lerma Chapala basin

Irrigation District		Surface (ha)	Efficiency (%)
001	Pabellón, Aguascalientes	9,800	36
011	Alto Río Lerma, Guanajuato	125,898	36
013	Estado de Jalisco, Jalisco	27,839	33
020	Morelia, Michoacán	18,021	38
024	Ciénaga de Chapala, Michoacán	18,800	34
033	Estado de México	17,508	38
034	Estado de Zacatecas, Zacatecas	18,360	32
043	Estado de Nayarit, Nayarit	45,189	37
045	Tuxpan, Michoacán	19,643	38
053	Estado de Colima, Colima	22,000	32
061	Zamora, Michoacán	11,895	40
085	La Begoña, Guanajuato	6,390	41
087	Rosario Mezquite, Michoacán	63,143	48
093	Tomatlán, Jalisco	22,450	32
94	Jalisco Sur, Jalisco	11,797	30
Total (15 Irrigation Districts)		439,733	37*

In Figure 3 and Table 4 the land use in the watershed is given for forest, pasture, seasonal agriculture and irrigation.

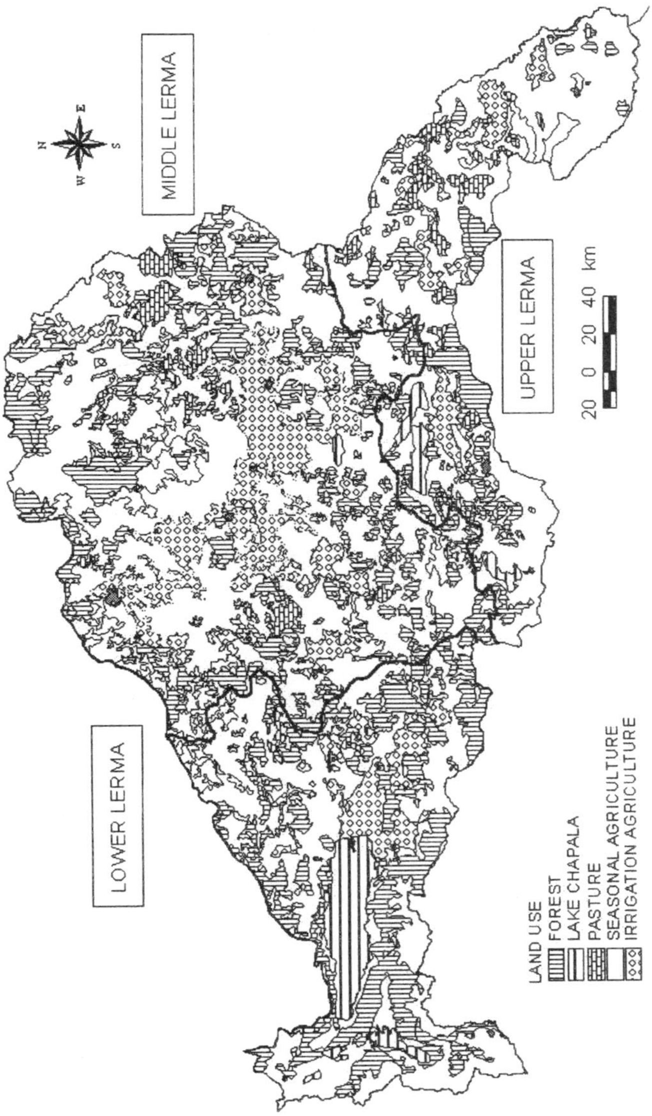

Figure 3 Land use in the Lerma-Chapala basin

Table 4 Soil uses in Rural Development Districts (DDR)

State	DDR	Surface (ha)					Fraction of each state
		Seasonal	Pasture	Forest	Others	Total	%
Mexico State	Toluca	132,370.8	20,796.0	81,072.0	47,273.0	281,511.8	
	Atlacomulco	148,428.0	28,065.0	76,672.0	23,771.0	276,936.0	42
Querétaro	Querétaro	53,582.0	124,981.0	7,653.0	8,817.0	195,033.0	48
Guanajuato	Dolores, Hgo.	136,808.0	373,238.0	94,000.0	55,786.0	659,832.0	
	S.L. de la Paz	30,401.0	329,749.0	101,076.0	55,211.0	516,437.0	
	León	81,067.0	84,250.0	80,175.0	116,052.0	361,544.0	95
	Celaya	96,270.0	70,013.0	26,750.0	41,361.0	234,394.0	
	Cortázar	162,140.0	191,823.0	34,500.0	141,020.0	529,483.0	
Michoacán	Uruapan	10,013.0	9,913.0	27,814.0	3,171.0	50,911.0	
	Zamora	99,422.0	131,386.0	137,145.0	18,595.0	386,548.0	34
	Sahuayo	23,229.0	64,280.0	5,655.0	33,230.0	126,394.0	
	La Piedad	90,028.0	43,752.0	52,011.0	10,242.0	196,033.0	
Jalisco	La Barca	132,903.0	164,045.0	37,830.0	125,571.0	460,349.0	
	Cd. Guzmán	46,732.0	100,670.0	102,888.0	34,481.0	284,771.0	24
	Lagos de Moreno	48,787.0	182,215.0	7,268.0	17,583.0	255,853.0	
	Total (ha):	1,292,180.8	1,919,176.0	872,509.0	732,164.0	4,816,029.8	

The main source of water used for irrigation is surface water, which accounts for 83%. The average efficiency in the 15 irrigation districts is of 37%, which implies considerable losses during conduction and lot-level application. The irrigation districts have the following hydraulic infrastructures: 43 storage reservoirs with a useful capacity of 2,685 Mm^3, 45 diversion dams, 2,685 deep wells and 213 pumping facilities (Figure 4). Water is distributed through a main-channel network with a total length of 5,000 km and 3,500 km of open-air drains.

2.2 Rural Development Districts

A Rural Development District includes Irrigation Units (see section 2.3) and seasonal agriculture. The Lerma-Chapala basin contributes with 44% of the seasonal crop surface area (2,891,805 hectares) in the five states within which the region is located. It comprises 10% of the surface area at national level, and is the chief supplier of grain corn. Within the basin, corn is grown in 73% of the surface area, followed in importance by beans, sorghum and corn varieties, oats, chickpea, wheat, barley and vegetable fodder.

Figure 4 Water distributions in the agricultural areas

In the Lerma-Chapala basin, an average of 2,058,480 metric tons of grain corn is produced every year, representing 19% of the national production, and 442,676 metric tons of grain sorghum, corresponding to 13% of the national production. The yield of corn of the Lerma-Chapala basin is higher than the national average of 1.78 metric t ha^{-1}. The rural development districts of Jalisco and Mexico State yield more than 3 metric t ha^{-1}, and the average in the basin is of 1.81 metric t ha^{-1}.

2.3 Irrigation units

The irrigation units located in Lerma-Chapala basin represent 28% of the national total, and include both organized and unorganized units, the states of Guanajuato and Michoacán being those with the larger part of their surface areas occupied by this form of agriculture. Spatially, irrigation units are distributed mostly in the Upper, Middle and Lower Lerma basins, with 74% of the total number of units located in this region, and 46% of the total corresponding to the Middle Lerma. The estimated extraction of these irrigation units is 4,112 Mm^3, 64% of which is obtained from groundwater, counting with a hydraulic infrastructure consisting of 13,000 works distributed throughout the different developments.

2.4 Water availability

The Lerma-Chapala basin consists of a main collector, the Lerma river, which is 750 km long, it runs into Lake Chapala and its formed by the Lerma, Jaltepec, Laja, Silao, Guanajuato, Turbio, Angulo, Duero, La Pasión and Zula rivers. The sub-basins are described in Table 5.

Table 5 Water availability in the Lerma-Chapala basin

Sub-basin	Surface (km)	Precipitation (mm yr^{-1})	Runoffs (Mm^3 yr^{-1})
Upper Lerma	23,132	675	1,974
Middle Lerma	9,203	720	642
Upper Turbio	2,913	760	127
Zula	2098	830	279
Lake Chapala watershed	6,644	820	933
Upper Duero	2,198	910	475
Angulo	2,027	920	307
Closed watersheds	6,085	810	455
Basin total	54,300	735*	5,191

* Average (CNA, 1993)

The mean average rainfall in the basin is 735 mm, mainly concentrated from June to October, with significant variability in time and space. Rainfall in the northern part of the basin is around 500 mm yr^{-1}, while in the Upper and Lower Lerma and Lower Santiago it increases to over 1000 mm yr^{-1}, a rainfall of between 500 and 700 mm is detected on the Middle Lerma and Upper Santiago areas. Long drought periods and exceptional rainy years occur, with a recorded historical cycle in the range of 20 years.

The mean annual evaporation in the basin is 1,977 mm, with the highest rate in the Upper Santiago and Middle Lerma regions and the lowest in the Upper Lerma region. The average temperature is 21 °C. Temperature is irregularly distributed throughout the territory, and the lowest temperatures occur at the Upper Lerma region.

Under the new regionalism established by the National Water Commission (CNA), the region of the Santiago-Pacífico was added to the Lerma-Chapala region, resulting in the Lerma-Santiago-Pacífico region. This region is characterized by a main river course, which is divided in two sections, the Lerma and Santiago rivers, and several of tributaries on the left bank, such as the La Gavia, Jaltepec, Angulo, and Duero rivers; the Laja, Guanajuato, Silao, Turbio, Zula, Calderón, Verde, Juchipila, Bolaños and Huaynamota rivers discharge into the right bank. The total river course is 1,194 km long, 20% flows through regions of over 2,200 masl, 50% between 2,000 and 1,500 masl, and one-third through areas lower than 500 masl.

Regarding the basin's surface water, the average annual runoff is of 5,191 million cubic meters (Mm^3), 44% of which is generated in the Lerma river basin and the rest of it in the Santiago river basin, 68% of the runoff occurs during the rainy season. However, the natural runoff has been modified, mainly in the Lerma river basin, primarily due to exploitation for irrigation purposes and urban water-supply, and water storage in large dams. Springs have disappeared due to the over-exploitation of groundwater, and tributaries have no runoff in some periods of the year. The water use in the basin is distributed as follows: 82% is used for agriculture, 14% for urban water supply, and 4% for the industry.

3. WATER DEMAND

3.1 Agricultural use

In agriculture, the surface area, which can be irrigated, is distributed mainly in the Upper, Middle and Lower Lerma river regions, and the water requirements are not fully met, it has been calculated that the available volume (7,832 Mm^3) only fulfills 70% of current needs. The principal crops are corn, which covers 58% of the useful area, beans with 15%, sorghum with 14%, wheat with 8%, and the rest of the land is used for other crops, such as potatoes, alfalfa, and rice.

3.2 Urban and industrial use

Yearly water extraction in the Upper Lerma region is 534 Mm^3, of which 88% is obtained from the aquifers, and the remaining, 64 Mm^3, from surface water. The major cities in this area are Toluca, Metepec, Morelia and Acámbaro. Industrial use accounts for 117 Mm^3, of which 82% is obtained from groundwater and 18% from surface water.

In the Middle Lerma region, water use is of 346 Mm^3 per year, 95% is extracted from aquifers and 5% from surface water, which provides the water supply in the main cities of La Piedad Cabadas, León, Irapuato, Salamanca and Silao, among others. Industrial use accounts for 127 Mm^3, 85% is extracted from groundwater.

The water use of the Lower Lerma region is similar to that of the Middle Lerma basin. However, the main supply in the former is surface water, 71%, while the remaining is obtained from aquifers. The principal municipal water supply in this area is for the cities of Zamora and Sahuayo. The industrial use of 32 $Mm^3\ yr^{-1}$ is mainly obtained from groundwater supplies.

The industrial establishments in the region represent 19% of the national total; the main industries in the region are manufacturing, food production, textile production, and paper production. Most of these industries are concentrated in the Middle Lerma region. Total water use in the industrial sector is estimated in 412 Mm^3, 67% of which is concentrated in the Upper Lerma basin and 31% in the Middle and Lower Lerma. Hansen and van Afferden (2001) elaborated an industrial inventory for the basin and estimated the sources of 12 industrial pollutants and indicators in 65 municipalities between the Solís Dam and Lake Chapala.

3.3 Hydrological balance

The hydrological balance of surface waters was presented by Aparicio (2001). Of the total runoffs in the region (14,019 Mm^3), 44% correspond to the Lerma river basin and the rest to the Santiago-Pacífico region. It should be mentioned that one-third of the runoffs are generated in the Santiago river before it flow into the Pacific Ocean. The hydraulic balance of surface waters shows, that there is equilibrium in the region; however, a large part of the available volume is located in the Santiago-Pacífico Region, while in the Lerma river basin, requirements exceed availability, so this region is classified as being water deficient in the Lerma river basin, and having abundant water supply (even surplus) in the Santiago river basin.

The groundwater balance shows that of the total estimated recharge of the 85 aquifers in the Lerma-Santiago-Pacífico region, 5,293 Mm^3, users in the agricultural, urban and industrial sectors extract 5,773 Mm^3. Therefore, the subsurface water balance is not in equilibrium, this is seen mostly in the Lerma river basin, where extraction levels is higher than recharge, particularly in the Middle Lerma region.

4. OVERALL BALANCE

4.1 Degradation of naturals resources

The dynamics of land use in the Lerma-Chapala basin has been subject to considerable fluctuations over the 1980-1997 period. (Table 6, Figure 5). Among the changes detected in land use, seasonal agriculture has decreased 10%, and the surface area used for irrigated agriculture has increased; this expansion corresponds to 20% more surface area for cultivation than that in 1980, along with the conversion of traditionally-seasonal areas to irrigated areas, part of the affected woodland and forest surface area of -10%, 20% and -5%, respectively.

Table 6 Overall land use in the Lerma-Chapala basin (SEMARNAP, 2000)

(1980-1997)	Change in use of the soil (ha)			Variation (ha)
Type	1980	1997	%	
Agriculture				
Seasonal	1,421,398.00	1,292,180.80	-10	-129,217
Irrigated	707,750.00	2,049,300.00	20	341,550
Subtotal	3,129,148.00	3,341,480.80	7	212,332
Woodland	474,642.51	420,825.00	-11	-53,817
Disturbed woodland	107,507.73	161,325.24	50	53,817
Subtotal	582,150.24	582,150.24	0	0
Bodies of water	176,466.75	155,973.94	-12	-20,492
Forest	923,029.85	872,509.00	-5	-50,520
Other uses	25,552.65	732,164.00	27	706,611
Total	5,308,209.55	5,895,992.94	11	587,783

The original 582,150 hectares of woodland, of which 107,507 hectares were disturbed already prior to 1980, were reduced in 1997 due to the increasing human intervention for the purpose of adding 53,817 hectares to seasonal agriculture, this signified the switch of the productive system from seasonal to irrigated agriculture, and the clearing and felling of woodlands and forest species. Water bodies, on the other hand, have decreased by 20,492 hectares principally due to over-exploitation of aquifers, recharge areas and to natural phenomena, producing leakage to groundwaters and the diminution of initial volumes recovery potential.

The surface area subject to natural or induced forestry has been noticeably reduced due to the advance of agriculture for subsistence on the hillsides, which are mainly covered by forest species, over a surface area of approximately 50,520 hectares; the change in use of plain areas from seasonal to irrigated agriculture, and because of excessive felling for commercial purposes, without a renewal of the resources.

Within the basin, livestock raising activities occupy considerable land areas (1,919,176 hectares). This results in changes in soil use, which are observed by an increase of 68% in pastureland from 1980 to 1997; intensive irrigation pasturage increased in approximately 418,698 hectares, whereas ranches and summer pasture account for approximately 706,611 hectares (including 25,552.65 hectares of areas either with aquatic or humid soil plants or without any vegetation).

Uses identified as the more dynamic and subject to change are agricultural and livestock raising, the latter evolving most rapidly.

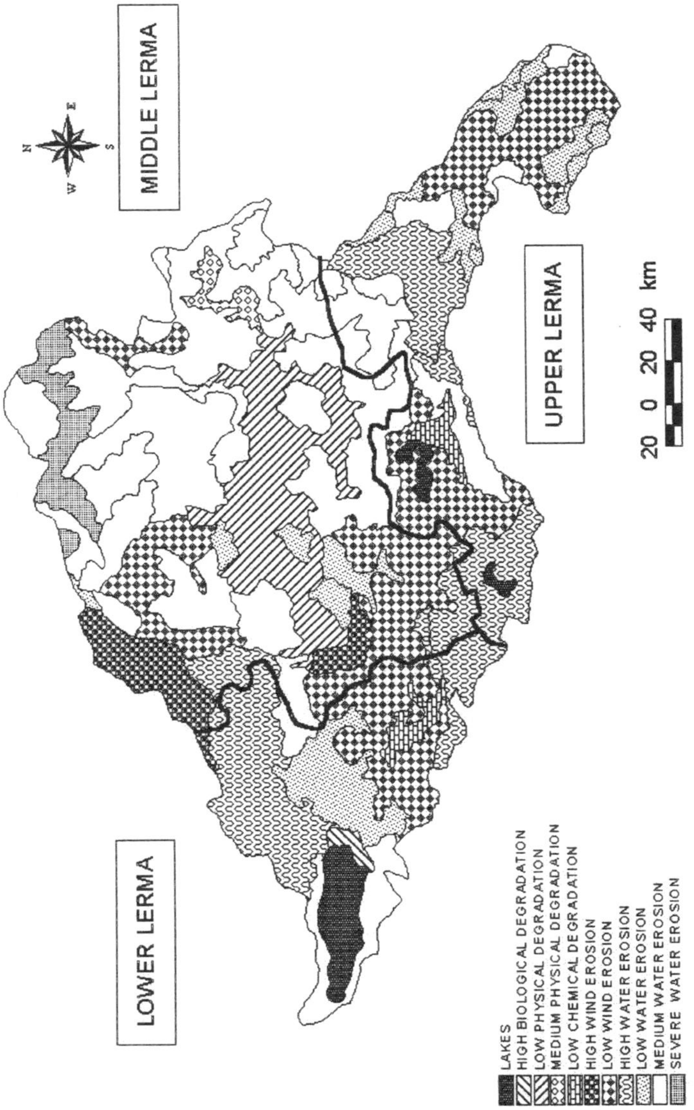

Figure 5 Degradation of natural resources in the Lerma-Chapala basin

4.2 Soil conservation practices

The soil conservation practices described in this section are the following: (1) Vegetative practices (Living barriers, Pasture, and Basin reforestation; (2) Mechanical Practices; and (3) Terrace leveling.

4.2.1 Vegetative practices

To improve the protection of soil against erosion, it is necessary to have permanent vegetation, fields and woods. To facilitate this, vegetative practices must be implemented to protect soil from the impact of rain and surface runoff. When covering with trees, grass or any other kind of vegetation, erosion control and ecosystems' recovery is secured. Recommended vegetative practices are the following: The formation of living barriers (with prickly-pear, maguey and ditch-reed), pasture (adjustment of talus), and reforestation.

4.2.1.1 Formation of living barriers

When the process of erosion begins in an incipient manner due to laminar flow, the formation of living barriers is recommended to prevent water from acquiring enough speed to dislodge and transport particles. The barriers can be made by prickly-pear, maguey, and ditch-reed (Figure 6).

Figure 6 Barriers against erosion, consisting in cultivation of maguey and green cover

The use of ditch-reeds is important to retain soil in filtering dams. Because of the type of reproduction, the development of roots is extensive and they interweave to form a mesh, and thereby retain fine soil particles deposited in the dams. Ditch-reed is planted along non-overflow dams, once the storage basin is full, by placing rows of rhizomes or tubers. The first rows should be located one meter away from the non-overflow dam, with separations of 45 cm. It is suggested that the following rows are placed with separations of 30 cm between the plants, and, in the last rows, the separation should be 15 cm. The ditch-reed requires little humidity and can grow in low-nutrients soils. The disadvantage of using this vegetation is that it is not soil-forming, and must be replaced as the soil gradually consolidates.

4.2.1.1.1 Prickly-pear

This type of vegetation can retain and fasten the soil, due to the development of roots, which can form an extensive network. Its type of root is not so different from that of the ditch-reed. Planting is carried out on land prepared in small furrows and with a slight inclination, using the common root system. Plants are introduced during the dry season; the fleshy leaves must be previously dehydrated to promote the growth of roots during the first rains. The previously prepared prickly-pears are planted in the furrows, dykes and pits prepared with a blow from a grub-hoe (Figure 7). This type of plantation is preferably used in the Middle Lerma river basin, in places with little runoff, since humidity would rotten the leaves.

Figure 7 Barriers prepared with prickly-pear and different shrub species

During growth, prickly-pear leaves fail to protect the soil from the physical impact of erosion; however, they can protect it from runoffs. They are not expected to be soil-forming, but will gradually be replaced by other species, such as pines, which allow soil formation to take place through the integration of nutrients.

4.2.1.1.2 Maguey

Another recommended form of retaining soil in erosionable areas is cultivation of maguey. These plants increase hydraulic conductivity and water infiltration. This is achieved by the formation of roots, which are almost superficial and with large extensions, thus promoting water infiltration.

Plants are introduced in places where the terrain's slope is greater than 30%. They are placed on small mounds made of soil extracted from the furrow, with the purpose of avoiding rottening of the roots. On gentle-slope terrains a pit can be made with a grub hoe. The limitations of maguey planting are that it does not protect the soil against the impact of the rain, it is not soil forming, and does not greatly increase the amount of nutrients, so it should be replaced as the soil becomes more stable (Figure 8).

Figure 8 Planting of maguey in areas with erosion problems

4.2.1.2 Pasture

In small areas suffering extreme erosion, and difficult to reach with heavy machinery, and where the slope is greater than 45%, pasture is planted to avoid further erosion. These areas are commonly known as "piloncillos". The grass planting takes place on land treated by adjustment of talus; a guide shoot is planted on each step. Once planted, it is covered with a small amount of soil to avoid the ventilation of the root so that the crop will prosper (Figure 9).

Figure 9 Sowing of stripp-grass in eroded areas with a slope higher than 15%

4.2.1.3 Basin reforestation

Reforestation is a practice used for erosion control; it protects the soil from the impact of rain and runoffs. It also allows the formation of soil within a short time period through the integration of nutrients. A thin layer is formed from decomposing leaves of the reforested trees and it protects the ground from runoff, the trees' foliage intercepts raindrops. For this reason it is important to have a clear concept on the purpose of reforestation to select the most suitable species, without losing sight of the site's ecological conservation (Figure 10).

4.2.2 Mechanical practices.

Mechanical practices are those carried out with special agricultural implements or heavy machinery and labor. To avoid the erosion of territories with inclinations exceeding 35%, soil is removed to reduce slopes and surface runoffs. The land can then be used for planting of shrub, bush and herbaceous species, since the soil porosity is altered and the water infiltration is increased, thus decreasing surface runoff. The mechanical practices employed in the Lerma-Chapala basin are: terrace leveling, talus refining, and terrace formation with stakes, sub-soil treatment, and a common cutoff wall.

Figure 10 Reforestation of basins with pine trees

The purpose of terrace construction is: to reduce the inclination of the land, to increase water infiltration, to decrease the runoffs' volume, to slow down runoff water and provide a suitable land for reforestation. The use of this type of construction is recommended for places where rainfall is low or moderate, not exceeding 750 mm annually.

An overall analysis of the waters' balance in the three regions of the Lerma river shows that the availability of surface and groundwater is not sufficient to meet the requirements of the agricultural areas under irrigation and receiving seasonal rainfall. Urban and industrial water demands have also increased during the past decades. Competition for water in the basin has been discussed by Scott *et al.* (2001) in a different chapter of this book.

The dams and reservoirs system in the region have an estimated storage capacity of 11,325 Mm^3 (the third largest in Mexico). However, due to erosion problems in the basin, the reservoirs has become clogged, representing this a total volume of 215 Mm^3 and leaving a useful storage capacity of only 11,110 Mm^3, for the distribution in irrigation districts and units.

The reservoirs intended for water supply to population centers (Figure 11) have a capacity of 8,495 Mm^3, 1.8 Mm^3 of which is clogged, leaving a useful capacity of 8,493 Mm^3.

Dávila-Poblete and Treviño-Carrillo (2001) and Wester *et al.* (2001) have discussed the institutional and social arrangements in the water management in the basin in different chapters of this book. Measures to solve the conflicts due to competition for water in the basin have been proposed by Scott *et al.* (2001).

Figure 11 El Batán reservoir in Querétaro

5. ADMINISTRATION OF WATER RESOURCES

5.1 The Basin Councils

Under the Mexican National Water Law, the establishment of Basin Councils is accounted for and to promote the definition of objectives, goals, strategies, policies, programs, projects and activities in basins. Basin Councils are constituted with the participation of Federal Water Authorities, duly accredited water users and social groups and organizations. The purpose of this type of organization is to facilitate coordination between the federal, state and municipal governments and to illustrate coordination and agreement between government institutions, water users and society, for the purpose of achieving better management of water in hydrological basin.

5.2 The Lerma-Chapala Basin Council

The basin councils have auxiliary organizations at the accomplishment level:

- ·Sub-basin > basin commissions
- ·Micro-basin > basin committees
- ·Aquifer > technical groundwater councils (COTAS)

These organizations are hierarchical subordinates to the Basin Councils; their activities consist on promoting the users' organization by region, basin and aquifer. Basin Councils form a link between government authorities and water users, and they play an important role in regional water management.

Basin Councils are the modern expression of overall water management and promote the participation of society in defining, directing and carrying out water management policies.

Outstanding among established Basin Councils in Mexico is the Lerma-Chapala Basin Council, which operates since 1989. This Council is strengthened by a Coordination Agreement between the Federal Government and the Guanajuato, Jalisco, México, Michoacán and Querétaro state governments, for the purpose of implementing a program for the regulation of hydraulic exploitation and wastewater treatment in the basin.

5.3 State Water and Treatment Commissions

State Water Commissions have been established to support the basin councils and to implement activities under the National Hydraulic Plan. These Commissions pursue the following objectives:

- Coordinate actions taken by the federation, the state and municipalities to resolve the supply of drinking water and sewage system services.
- Establish technical standards in hydraulic subjects.
- Prepare and keep State Water and Water Treatment Plans updated, in coordination with Federal Government guidelines.
- Prepare State Hydraulic Plans.
- Administrate and operate drinking water and sewage services.
- Promote administrative, technical and economical independence of drinking water and water treatment companies.
- Promote programs for the efficient use of water and sewage.
- Provide support to water companies in their updating and establishment of water tariffs, in accordance with the costs involved in providing the service.
- Keep the information system and inventories updated.
- Allocate an annual percentage of finantial resources to foster social participation in water culture.
- Plan, study, design, rehabilitate and build drinking water, and sewage water treatment systems.

The drinking water and water treatment public services are the responsibility of municipalities, which can either provide the service directly or through public semi-official entities (operating organizations, drinking water committees, drinking water councils) through concession and coordination agreements.

The urban-public sector, which includes city dwellers, is represented by the operating organizations.

5.4 Water Operation Organizations

Water Operation Organizations are institutions connected to water resources at a municipal level, as public semi-official entities. Their main purpose is to implement public hydraulic works for the construction, management, operation, maintenance and development of drinking water and sewage systems, including suitable water treatment.

5.5 Technical Groundwater Committees (COTAS)

In the case of groundwater, agricultural users who were previously organized in irrigation units, the COTAS were established as social reconciliation mechanisms, they are designated to prepare and implement activities oriented towards the recovery and conservation of aquifers.

COTAS are organizations that support the Basin Councils, which basically consist of the groundwater users of each aquifer. Represented on the COTAS are members of the civil society and governmental institutions, these last ones only for the purpose of providing technical assistance and advisory services.

In the Lerma-Chapala basin, the state of Guanajuato is a pioneer in this type of organization. During 2001 the establishment of 14 COTAS is expected to take place as well as State Hydraulic Councils, which directly depend on the State Water and Water Treatment Commission.

The specific goals of water users organizations are to:

- Participate in overall water management.
- Reverse over-exploitation and recover the groundwater level.
- Promote efficient use of water.
- Reverse the problems of water pollution.
- Promote water education.
- Foster sustainable development.

Among strategies to achieve these goals, the following are included:

- Involve users in a better water management and achieve sustainable development in the state.
- Propose regulation of the aquifer to save water for cases of shortage or over-exploitation. This includes efficient water use, and an equitable balance between recharge and extraction.
- Establish and implement mechanisms for the optimization of the use of water resources.
- Foster water treatment and conservation of water quality programs.
- Create the necessary tools for achieving sensitizing, awareness and change in attitude of members of the society, by considering water as an element of life.
- Engage in activities, which contribute to the preservation and improvement of our surrounding environment.

5.6 State Hydraulic Council (CEH) in the state of Guanajuato

An interesting example is represented by the water-management system in Guanajuato, which is based on the responsibility of the public in adopting decisions and acting on water matters through administrative entities made up of users, counting with official support on technical and budgetary matters.

The plan is to organize groundwater users, and later add surface water users from in irrigation districts units. The State Hydraulic Council is the maximum organization of water user representatives in the state. Its general objective is to represent the groundwater and surface Water Technical Councils for an overall water management in the state.

6. FUTURE OF THE LERMA-CHAPALA BASIN

6.1 Present situation

The Lerma-Chapala basin, from Mexico State to Lake Chapala, reveals a shortage in water, competition for this resource, conflicts between users and organizations and, most of all, water pollution.

The agricultural, urban and industrial sector's growth in the basin has practically exhausted the water resources, both surface and groundwater while, at the same time, shortage and pollution problems have worsened and augmented.

Most point-source pollution comes from urban areas and from industries, while non-point pollution occurs in irrigated and seasonal agricultural areas.

The following actions have been proposed to ease the problems of shortage, pollution and soil erosion:

- Preservation of water quality and basin water treatment within the states.
- Ordering and regulation of water use.
- Efficient water use by means of utilization programs at urban, livestock raising and industrial levels, together with the promotion of sewage water recycling and reuse.
- Soil management, preservation and recovery.

6.2 Water quality and treatment

The institutions in charge of the water sector have initiated a process of transformation in their programs, aimed towards more efficient water management with a higher level of participation, contributing to the achievement of sustainable development in the basin. The restructuring currently in process takes into account the decentralization and expansion of water management. An example of federalization is the progress achieved in the state of Guanajuato, the dominant situation and the measures adopted to ease pollution problems are shown below.

In the state of Guanajuato, a large part of sewage waters are used for irrigating high stalk crops. The construction of treatment plants is limited not only by the high cost of investment but also by the elevated cost of maintenance, as well as by a lack of technical-administrative capacity to assume water treatment programs in some of the operating organizations, and of administrative methods for costs recovery.

Table 7 Volume of sewage waters generated in the state of Guanajuato

Year	Urban sector	Rural sector	Total
	($Mm^3 yr^{-1}$)		
1995	190.0		190.0
1997	206.7	0.44	207.1

(CEASG, 1999)

6.3 Arrangement and regulation of water use

The concurrence in water use in the basin makes it necessary to ensure rational, equitable and fair distribution, for the purpose of maintaining and supporting economic development of the different regions of the basin. Due to a disordered growth in the user system, it became necessary to introduce a process for ordering and regulating the hydraulic resources of the basin to determine the real volumes required by the water users.

Overall demand in the basin was calculated by considering the principal user, irrigation, and the water volumes needed for urban and industrial use. Under irrigation, urban development districts and units (URDERALES), used by the National Water Commission for 1998 statistics, were considered. Organization for agricultural use considers, first of all, surface water requirements. The process is carried out using basic irrigation district and unit information.

The policy for volume distribution between the states of the basin is defined in the beginning of November each year. At this time, volumes assigned to each entity for the next November-October cycle are determined. The degree of exploitation is determined by the ratio between demands and draining present during an average year in a specific area. Considering the fact that Lake Chapala forms part of the hydrological system under study, the contribution of water from the Lerma river must be considered to establish the minimum volumes entering the lake and to ensure its preservation. The determination of distribution policies endeavors to minimize restrictions on use in each state, and to maintain a minimum contribution to Lake Chapala to ensure adequate functioning for its preservation. In this way, the hydrological balance of the basin and an equitable distribution of available water are guaranteed.

Studies carried out by the CNA show that the quality of water in Lake Chapala suffers serious deterioration when this is below 2,000 Mm^3. The lake has a maximum capacity of 8,125 Mm^3. When this volume is exceeded, water spillage will occur. Three operating policies are established to maintain lake volume. These correspond to the following situations: critical, average and abundant water availability in the basin. The operating policies give rise to three curves for each group of users, which relate the actual surface water availabilities to the overall water management and promote the participation of Society in defining, directing and carrying out water management policies for extraction volumes during the following year.

6.4 Efficient water use

The rehabilitation and modernization of irrigation districts and units is the strategy to be implemented at the short- and medium-terms for water volume recovery, particularly in the Upper and Middle Lerma regions where the water deficit is most severe. With this in mind, the following programs should be intensified: Rehabilitation of the principal irrigation districts in the Lerma-Chapala region, the introduction of irrigation technology, water delivery by volume and use of treated waters from urban centers and industries.

Within this plan, the rehabilitation of irrigation districts will allow, on one hand, for a more efficient use of the hydro-agricultural infrastructure currently operating under deficient conditions (Figure 12). On the other hand, the introduction of technology in irrigation districts and units stabilize aquifers or support urban and industrial use. This is an economically feasible alternative to increase food production and rescue water resources, which may be applied to increase secondary crop surface areas. The incorporation of technology is proposed in combinations in order to have greater impact on water saving and crop productivity, such as: soil leveling, lot drainage, low and high pressure irrigation systems, pumping plants, well rehabilitation, low-pressure piping, canal surfacing, structure installation in main channels, together with the measuring and delivery of water by volume.

Figure 12 Rehabilitation of hydraulic infrastructure

Priority activities are proposed for each irrigation district and unit, agreed upon with the producers in accordance to their specific problems, such as availability and water demand in their irrigation districts and units, or in the corresponding federal entity.

6.4.1 Lot irrigation projects

In every basin rehabilitation or modernization project, water volumes supplied by the concession should be analyzed and an adequate crop-pattern plan should be put into effect in accordance to the amount of available water, or alternative crops could be proposed to cover specific surface areas.

According to each case, technical projects should be proposed for lot irrigation by gravity and gate piping as basic options, together with pressurized irrigation methods, so that the user can select the most suitable option depending on his specific interests within production systems capacity (Figure 13).

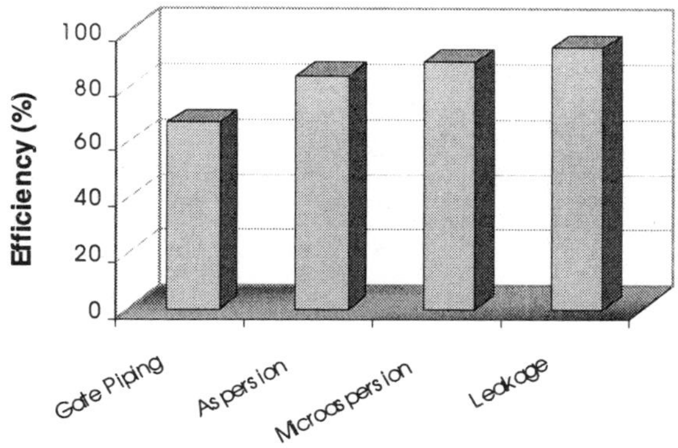

Figure 13 Overall efficiency in water uses for different irrigation systems (CNA, 1997)

Water consumption for each crop should be estimated for every irrigation method, as well as the costs of investment, operation and maintenance. Figure 14 shows the costs of investment for gate piping and aspersion irrigation. In this Figure a ratio of 1 to 2.25 in investment costs between technically aided gravity and aspersion can be appreciated, while efficiency in water use is similar.

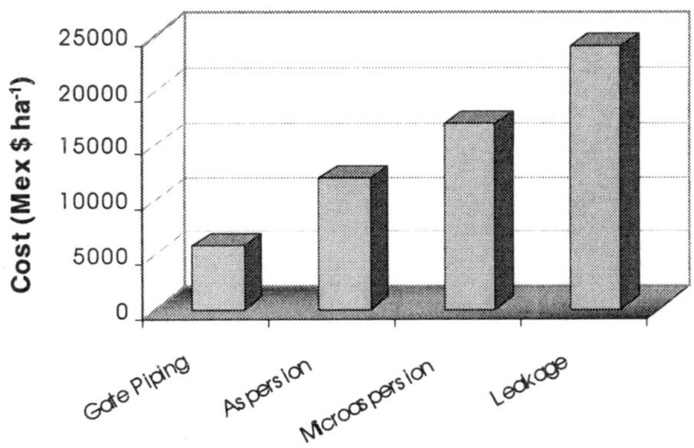

Figure 14 Investment costs for irrigation systems (CNA, 1997)

The annual cost of operation is higher when employing aspersion techniques, due to higher their energy consumption levels (Figure 15).

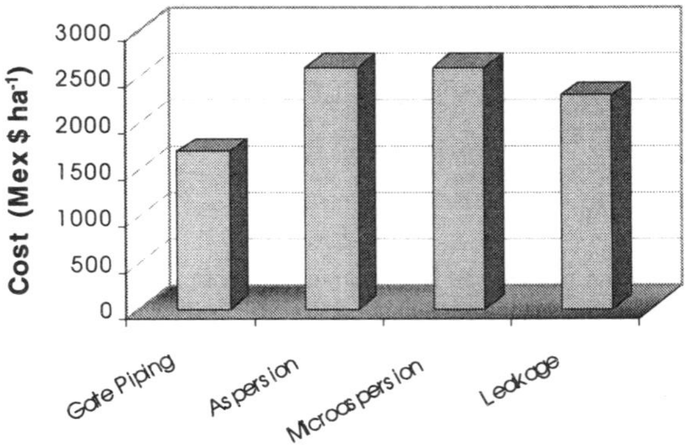

Figure 15 Annual operating costs for irrigation systems (CNA, 1997)

6.4.2 Inter-lot irrigation projects

In inter-lot networks the use of new materials should be considered, such as PVC, polyethylene and geo-membranes, to replace soil channels or concrete facings, since piped networks for capacities of less than 500 lps and slopes exceeding 0.3% are cheaper than concrete or masonry facing, they also provide 100% conveyance efficiency and facilitate operation. Depending on the topography, each piped networks may be considered for both low and high pressure. Low pressure is obtained from the channel network itself, or by means of pumping equipments distributed along the main channels. This design has become widespread in the states of Guanajuato and Michoacán.

6.4.3 Sewage water use in agriculture

Actually, due to the scarcity of water and the need to increase food production to satisfy the needs of an ever-increasing population, the possibility of using raw and treated sewage water for agricultural irrigation has been considered.

There are risks associated with the use of raw sewage water for agricultural irrigation, mainly pollution of agricultural soils (increased salinity, presence of pathogens, heavy metals), and groundwater pollution (pollutant lixiviation, heavy metals, nitrates, organo-chlorine pesticides, etc.), and from the perspective of public health (exposure of agricultural workers and the consumers of irrigated harvests to pathogens). However, treating sewage waters in primary, secondary or advanced systems, according to different water quality requirements, can decrease these risks.

Options for sewage water reuse are:

- Recharge of aquifers
- Agricultural irrigation
- Growth of aquatic plants
- Growth of salt-water plants
- Fish cultivation
- Lakes for recreational use
- Generation of humid areas as wildlife habitats
- Industrial cooling systems

Numerous investigations have been carried out in recent years to evaluate the possibility of safely using treated sewage water for agricultural irrigation. Among other aspects considered in these investigations are water quality, area setting and climate, nature of the soil, amount of water to be used, crops and irrigation management and post-harvest agricultural practices.

Although no specific technology has been developed on the use of sewage water for agriculture, much progress has been made in technologies for sewage water treatment, which have directly benefited the agricultural sector, since their results allow defining suitable conditions for sewage water use in irrigation.

6.4.3.1 Pollutants

Some of the studies carried out in connection with sewage water irrigated crops indicate that, if the sewage water contains pollutants affecting the microbial activity of the soil, this could increase respiratory activity and lower the removal of nitrates. A decrease in the removal of nitrates would result in the lixiviation of nitrogenous compounds to the water-bearing strata. It is well known that a reduction in irrigation time and an increase in number of irrigations fail to decrease the lixiviation of nitrates. Important amounts of nitrogen is absorbed by the crops or released to the atmosphere, thereby reducing the risk contaminating groundwater.

The use of sewage water to irrigate woodland has produced satisfactory results when irrigation is regulated (frequent and low-volume irrigations), allowing a large number of nitrites and nitrates to be absorbed by the trees, and averting lixiviation of these compounds to the sub-surface water-bearing strata. When using sewage water to irrigate woodland, one should remember that, at the beginning, the soil absorbs a large amount of N from the water, and that saturation of the soil with N must be avoided, otherwise lixiviation of pollutants to the water-bearing strata will occur.

It is known from experiments that the utilization of nitrogen for sundry crops (corn, soy) is greater when irrigation is limited during the growth stage and abundantly applied during the crops' reproductive period.

6.4.3.2 Heavy metals

Concerning heavy metals, it has been reported that the vertical displacement of Cd and Zn in sewage water irrigated soil is highly variable, and is influenced by the type of irrigation, by the organic carbon content of the soil and by its pH. In studies of over 50-year period, it is considered that sewage waters have contributed with 31% of the metals found on the soil's surface. However, alfalfa grown on these soils does not have high metal contents (Cd, Pb, Ni, Zn, Cr and Co), and therefore an estimated limited risk to animal or human health exposure.

6.4.3.3 Microorganisms

Some studies show that, on irrigating different crops by leakage, bacteriological pollution was not found in the crops. Helminthes eggs and fecal coliform were not found below 25 cm soil depth.

Experiments were carried out where the soil was considered as a biofilter, and it was found that a polio-virus added to the irrigation water (leakage irrigation) had very limited penetration in tomato plants, via the roots. No virus was detected on the leaves or in the fruit.

6.4.3.4 Biomass

Even though risks of pollution of the water-bearing strata (mainly by nitrogen) exist, results obtained from using sewage water for irrigation show a halophyte biomass production of up to 30 t ha^{-1} has been obtained by using sewage water for irrigation. In the case vegetables (eggplant) up to twice the production obtained with fresh water has been achieved, and in wood-producing forests (willow), an increase in the arboreal biomass has been reported. In another aspect, lentils' crops (which contain up to 30% protein in dry weight) and seaweed in tanks containing sewage water, efficiently remove the nutrients (N, P) from the water, allowing this water to be used for recreational purposes or irrigation. In addition, the biomass produced can be used as a food complement for other species.

6.4.3.5 Regulations

The use of sewage waters is governed by the Official Mexican Standard NOM.ECOL.001 (NOM, 1997), which indicates the maximum limits of pollutants permitted for sewage water discharges in Mexican waters and natural resources. The Federal Rights Law establishes the conditions for granting concessions and permits to use this type of water.

6.5 Management and conservation

The technical guidelines recommended in transferring basin conservation technology in the Lerma-Chapala basin are given below. Generally speaking, this transfer consists of a physical and socio-economical process, using specific technologies and undertaking multidisciplinary actions applied to the different elements such as soil, vegetation, water bodies, drainage networks and infrastructure, for the purpose of neutralizing the main degradation processes.

The specific technologies and actions recommended are the following:

- Diagnosis and planning using Geographic Information Systems (GIS) to describe the physical environment, identifying, locating and quantifying problems and establishing cause-effect relations. The "GIS" is then applied to provide objective priorities in different areas based on economic, social and environmental factors and to quantify liquid and solid (silt) flows in the basins.
- Once an overall diagnosis of the Lerma-Chapala region is obtained, actions will be proposed for intervening and counteracting problems. It should be noted that, in many cases, solutions will be very specific, although conservation technology can be identified to solve the main problems of the basin:
- Reforestation in compact areas, both in regions suitable for forests and in degraded sectors.
- Establishment of areas, which exclude pasturage and permit a natural recovery of the vegetation covering, inducing re-vegetation in degraded areas with native species.
- Lot design suitable for permanent soil conservation, such as the establishment of wind-breaking curtains on lot boundaries and intermediate vegetation lines in agricultural areas.
- Agricultural practices for soil conservation and maintenance of moisture, such as contour line fencing and use of plant material at soil surface (mulch).
- Rehabilitation of permanent and intermittent riverbeds and ditch control through reforestation, covering of taluses and riverbanks with retaining paste and building of semi-permanent (accommodated stone) or permanent (gavión) filter dams.

Activities in the social sector should include:

- Diagnosis and shared planning.
- Geographic conditions for application: Restoration and conservation technology applies to the Lerma-Chapala basin, mainly in the areas inhabited by rural population.
- Highly rewarding experiences were recently obtained at a micro-basin level in the state of Guanajuato with El Gusano, municipalities

of Dolores Hidalgo (Laja river basin) and Mesa de Escalante, Municipality of San Luis de la Paz (Laguna Seca basin). At the present time, work is being successfully done in some basins of the Cutzamala System, such as the Valle de Bravo and Villa Victoria reservoirs, both in Mexico State.

- As a result of these experiences (see 3), the estimated water volume that can be saved in the Lerma-Santiago region is in the order of 1,500 m^3 per hectare for each rainfall event.
- Restoration activities also promote soil conservation and improvement of this non-renewable resource, providing stability and high production level in the livestock-raising and forestry production.

REFERENCES

Aparicio J (2001) Hydrology of the Lerma-Chapala Watershed *In: The Lerma-Chapala Watershed: Evaluation and Management* (Hansen AM and van Afferden M eds) Kluwer Academic/Plenum Publishers, London.

CNA (1998) Diagnóstico Preliminar sobre Superficies Regables y Volúmenes Requeridos en Las Unidades de Riego Organizadas y Sin Organizar. 39 pp

CNA (1999) Rehabilitación y Modernización de Los Distritos de Riego en México. Informe del proyecto conjunto CNA-IMTA, 190 pp

CNA (1994) Memoria del Consejo de la Cuenca Lerma-Chapala, Vol II, No 3,4,5

CEASG (1999) Plan Estatal Hidráulico de Guanajuato 2000-2025, Fase I: Diagnóstico de la Situación Hidráulica del Estado de Guanajuato y Fase II: Estrategia Estatal en Materia de Agua.191 pp.

Dávila-Poblete S and Treviño-Carrillo AH (2001) Social Characterization of the Lerma-Chapala River basin *In: The Lerma-Chapala Watershed: Evaluation and Management* (Hansen AM and van Afferden M eds) Kluwer Academic/Plenum Publishers, London.

Hansen AM and van Afferden M (2001) Toxic Substances, Sources, Accumulation, and Dynamics *In: The Lerma-Chapala Watershed: Evaluation and Management* (Hansen AM and van Afferden M eds) Kluwer Academic/Plenum Publishers, London.

NOM-001-ECOL-1996 (1997) Norma Oficial Mexicana que establece los límites máximos permisibles de contaminantes en las descargas de aguas residuales en aguas y bienes nacionales *In* Diario Oficial de la Federación, January 6, 1997.

Scott C, Silva-Ochoa P, Florencio-Cruz V and Wester P (2001) Competition for water in the basin *In: The Lerma-Chapala Watershed: Evaluation and Management* (Hansen AM and van Afferden M eds) Kluwer Academic/Plenum Publishers, London.

Wester P, Melville R and Ramos-Osorio S (2001) Institutional Arrangements for Water Management in the Lerma-Chapala Basin *In: The Lerma-Chapala Watershed: Evaluation and Management* (Hansen AM and van Afferden M eds) Kluwer Academic/Plenum Publishers, London.

SECTION II

POLLUTION AND HEALTH

Chapter 4

Toxic Substances

Sources, Accumulation, and Dynamics

ANNE M. HANSEN[1] AND MANFRED VAN AFFERDEN[2]

Instituto Mexicano de Tecnología del Agua, Paseo Cuauhnáhuac 8532, Jiutepec, 62550 Mor., Mexico. email: [1]ahansen@tlaloc.imta.mx; [2]afferden@tlaloc.imta.mx

Key words: Industrial inventory, contaminant sources, accumulation, limits, risks

Abstract: The objectives of the present work were to evaluate the environmental health in the Lerma-Chapala watershed through the characterization of organic and inorganic contaminants in environmental samples from different areas in the basin, which are affected by both point and non-point pollutants; and, based on the dynamics of contaminants in the environment, establish the risk of accumulation in Lake Chapala. Information on industrial discharges was compiled, including over 3,500 enterprises. Mass balances for 12 parameters and indicators were estimated and the main sources of industrial pollution were identified. Water samples, suspended and deposited sediments were collected in 36 sampling stations along 600 km of the basin. Samples were analyzed for metals and organic contaminants, and the results evaluated according to different norms and guidelines. The results indicated that mean concentrations of almost all metals (Cd, Cr, Cu, Hg, Mn, Ni, Pb, and Zn) in sediments of 15 river and 21 lake stations were below severe effect levels (SEL) for heavily polluted sediments. Nevertheless, metal concentrations in all samples were higher in at least one parameter than the low effect levels (LEL) for slightly contaminated sediments. Polycyclic aromatic hydrocarbon (PAH) contents in sediments were high, especially in the Turbio river downstream from the city of León (47 mg kg^{-1}) and in the Lerma river station downstream from La Piedad (24 mg kg^{-1}). These concentrations exceeded LEL by one order of magnitude and by a factor of six, respectively. Stations with highest amounts of total organochlorine pesticides were Salamanca and La Piedad with concentrations higher than LEL by factors of three. Water in the river stations was affected by industrial discharges, whereas no significant levels of agricultural pesticides were found. Finally, applying hydrological and chemical equilibrium modeling, the risk of contaminant accumulation in Lake Chapala was assessed. It is shown that decreasing water levels in the lake will probably lead to accumulations of dissolved heavy metals, which in the future may affect water quality and the ecosystem's integrity.

The Lerma-Chapala Watershed, Edited by Hansen and van Afferden,
Kluwer Academic/Plenum Publishers, New York, 2001

1. INTRODUCTION

1.1 General overview

The Lerma river is important in several contexts. It serves as a water supply for irrigation of agricultural land, as a receptor of wastewater discharges from municipalities, industries, and return flow from agricultural land. It is a significant source of water, sediment, and nutrients to Lake Chapala, which serves as drinking-water supply to the city of Guadalajara and represents a unique ecosystem. A sustainable water management should therefore be based on knowledge of not only the water balance but also of the contamination and pollutant migration.

Considering the varied sources of pollution, it is important to determine which contaminants are contributing most to the overall toxic load to the river and then ascertain the origin of these chemicals. Regulators who develop and adapt water management plans can then use this information.

In many countries, the conventional approach to develop water management plans is obtaining chemical specific information from the analysis of industrial effluents and river water samples for priority pollutants. These data are then reviewed with respect to water quality standards and excesses are noted and underscored with respect to current discharge regulations.

Other countries have incorporated requirements to perform environmental effects monitoring (Canada) or ecological risk assessments (USA). In most cases, toxic responses are observed in the absence of priority pollutants. Furthermore, the frequent toxic response of suspended sediments suggests that water alone is not an adequate medium for sampling for toxic chemicals. The development of effective national and regional strategies for water quality management therefore requires knowledge about the sources, movement, and storage of contaminants and their reactions, such as degradation, accumulation and lixiviation.

1.2 Behavior of contaminants in aquatic systems

Metals in rivers come from natural as well as from artificial sources. Those that are naturally introduced into rivers may come from rock weathering, soil erosion or dissolution of water-soluble salts. Naturally occurring metals move through aquatic environments independently from human activities, usually without any detrimental effects. However, as river basins are settled and industrialized, the metals added by human activities may affect water quality. Humans consume metallic elements through both water and food. Some metals such as sodium, potassium, magnesium, calcium, iron, molybdenum, manganese, cobalt, copper, and zinc are found in living tissue and are essential to human life. Biological anomalies arise when these metals are depleted or removed but nevertheless they are toxic at high concentrations. Several heavy metals, like cadmium, lead, and mercury, are non-essential for human health but are highly toxic at relatively low concentrations and can accumulate in body tissues over long time periods.

The free dissolved ions are the most common toxic forms for metals and the most chemically reactive species as well. Metal speciation depend on a diverse factors, such as pH, redox-potential, sorption matrixes, organic and inorganic ligands, cationic competition and salinity, among others.

While metal-sediment interactions can be described by surface complexation equilibria (Hansen and Maya, 1997), biologically induced changes in metal speciation and metal binding by organic or inorganic ligands are still poorly understood (Rozan and Benoit, 1999).

Recent findings indicate that even in fully oxygenated surface waters trace levels of sulfides that form metal sulfide complexes persist. These sulfides can amount to half or more of the metal concentrations, both in colloidal and dissolved fractions (Benoit and Rozan, 1998). Therefore, in aquatic systems the characterization of the behavior of metals (bioavailability, toxicity, reactivity, transport and fate) contributes to a better estimation of the risk associated with water and sediment contamination by metals.

Pesticides include a wide variety of synthetic organic and inorganic compounds used to control weeds, insects, nematodes, and other pests. The physiochemical properties of the pesticides themselves, as well as other factors such as agricultural use, rainfall, and farming practices, are important in controlling the amounts and binding forms that actually occur in streams and lakes (Goolsby and Pereira, 1996). Among the main physiochemical properties that influence the distribution of pesticides and other organic contaminants in the environment are water solubility, soil half-life, and organic carbon sorption coefficient (K_{oc}) (Becker *et al.,* 1989).

Water solubility determines how easily pesticides wash off the soil and crop residues and how easily they leach through the soil. Pesticides with solubility's greater than 30 milligrams per liter (mg L^{-1}) are considered to be soluble and more likely to wash off the soil during rainfalls.

The half-life is the time length required for pesticides to degrade in the soil to one-half of their previous concentration. The longer the soil half-life the more persistent the pesticide, and the longer it will be available to wash off in storm runoff. Many pesticides are highly persistent in stream water and reservoirs, but very little information is available on pesticide half-lives in natural water bodies.

The sorption coefficient is a measure of the pesticide tendency to attach to the organic matter present in soil particles. The larger the K_{oc} value the more strongly the pesticide will be adsorbed to the organic soil fraction. Those with K_{oc} values less than 500 tend to be transported primarily in a dissolved form in the aqueous phase. Those with K_{oc} values of more than 1000 are transported primarily on suspended sediment particles.

High water solubility and low K_{oc} values characterize the majority of pesticides in use today. Most of the older generation chlorinated insecticides, such as DDT, chlordane, and dieldrin are less soluble in water, have long half-lives, and large K_{oc} values and consequently, they are transported primarily on sediment particles. Also, they partition into the organic coatings of sediments or accumulate in the fatty tissues of fish and other stream biota (Moore and Ramamoorthy, 1984).

Polycyclic aromatic hydrocarbons (PAH) represent another group of environmental contaminants. Most PAH are thermodynamically stable; they are characterized by low aqueous partitioning and high K_{oc} values and therefore tend to be associated with particle surfaces in the environment. Generally they are persistent to many natural processes, such as volatilization, photolysis, and biodegradation. This persistence, coupled with their potential carcinogenicity and toxicity, make PAH problematic environmental contaminants (Cerniglia and Heitkamp, 1989). PAH have been released into the environment from three main sources: biogenic, geochemical, and anthropogenic. While aromatic chemicals are commonly found in the environment as a result of biosynthesis by plants, "natural" PAH are mainly present in petroleum and coal deposits and are mainly liberated by natural pyrolytic processes such as bush and forest fires and volcanic activities. The anthropogenic sources of PAH, which may lead to contaminations of soil, sediments and water, originate primarily from the following products and wastes: creosote, coal tar, petroleum, and industrial effluents and gases. In sites affected by these sources, often-other chemicals such as aliphatic hydrocarbons, benzene, toluene, methylbenzene, xylene (BTEX), and phenols commonly co-contaminate the water and sediment environments (Mueller *et al.,* 1996).

No official guidelines or quality criteria for heavy metals and organic pollutants associated with suspended and deposited sediments, have been established in Mexico. In other countries, several efforts have been made to assess ecological risks by elaborating guidelines and effect levels for metals and organic contaminants, that define intactnesses of sediment ecosystems and, eventually, to recommend remediation actions (The Netherlands – List, 1994; Hellyer and Balog, 1999; CCME, 1999; Persaud *et al.*, 1993).

The different guidelines and sediment quality criteria have been developed using a variety of methods. Some are derived through analysis of data from a combination of spiked sediment bioassays, chemical analysis, and bulk sediment toxicity testing. Others are derived using a combination of sediment chemistry and benthic species presence/absence data. In general, these measures involve a review of biological effects associated with various concentrations of single substances or groups of chemicals.

Other guidelines are derived using models that predict the exposure of benthic organisms based on sediment chemistry data. These models predict the interstitial (pore) water concentration of sediment based on the bulk sediment concentration and the total organic carbon (TOC) of the sediment. Especially for organic contaminants, which are usually of hydrophobic character (high K_{oc}, see above), a TOC-adjustment is often recommended (Persaud *et al.*, 1993), since the dissolved contaminants mainly contribute to the toxicity of the *in situ* sediments. In other words, high-TOC sediments show less ecotoxicological effects than low-TOC sediments having both the same concentration of organic contaminants.

Since free dissolved metal ions are the most toxic forms, risk assessment of metal contaminated sediments should consider chemical forms and the reactivities of metal species. However, there is not an official guideline that includes a specification such as that described for the TOC-adjustment of hydrophobic organic contaminants. A risk assessment based on total metal concentrations can therefore be considered as a worst-case consideration that may overestimate the actual risks.

The objectives of the present work are to evaluate the state of pollution of the Lerma-Chapala basin (Figure 1) through the elaboration of the industrial inventory in the watershed and characterization of organic and inorganic contaminants in water and sediment samples from different sites in the study area. The sampling sites were selected to reflect both point and non-point source pollution. Additionally the probability of future contaminant accumulation in Lake Chapala is assessed, in order to deduce recommendations for appropriate monitoring and assessment programs.

During the time period from 1992 to 1994, the watershed was sampled on three different occasions between the Solís dam in the state of Querétaro, and Lake Chapala. Parallel laboratory experiments and numerical modeling were conducted to understand the dynamic behavior of some of the most important metals and pesticides found in sediment phases throughout the river basin (Hansen and Maya, 1997; Hansen *et al.,* 1998).

Given that samples were collected sometimes only once at different sampling sites, it was not possible to establish any time trend (whether contaminant levels were increasing or decreasing with time). This chapter represents therefore a snapshot rather than a chronicle. Also, the evolution of contaminants levels in Lake Chapala is prognosticated, by focusing on the accumulation of salts and contaminants (cadmium) at different water levels without specifying when this could happen.

2. INDUSTRIAL INVENTORY OF THE LERMA-CHAPALA BASIN

Information on 3,509 industrial enterprises (category, address, and water volumes) was obtained through the former Ministry of Urban and Ecological Development (SEDUE), the CNA state offices and the former Lerma-Balsas regional office (now Lerma-Santiago-Pacífico regional office), and from the Water Quality and Environmental Impact office of the CNA. The companies were grouped according to the following industrial categories and linked to the geographical coordinates of the municipalities of their location according to 1:250,000 topographical maps published by the National Institute of Geographical Statistics and Informatics (INEGI).

1. Animal-breeding (Cattle, pork, poultry, and fish)
2. Chemical and pharmaceutical (Dye, drug, pigment, and other chemical industries)
3. Distilling (Breweries and distilleries)
4. Electromechanical (Electrical, domestic and industrial electronical)
5. Food-processing (Provisions and bottling)
6. Furrier (Tanning, hardening, and furring)
7. Lumber and wood (Paper, cellulose, timber, and wood)
8. Metallic (Transformation of metals)
9. Metallurgical (Smelting of metals)
10. Mining and stone pitting (Porcelain, ceramics, bricks, cement, and mineral-extraction)
11. Pesticides (Herbicides, fungicides, insecticides, and other pesticides)

12. Petrochemical (Plastic, solvents, and other oil-derived productions)

13. Services (Hotels, restaurants, and office buildings)

14. Textile (Thread, yarn, fiber, cloth, fabric, and leather stitching)

Within four states of the watershed (Jalisco, Michoacán, Guanajuato, and Querétaro) 65 municipalities were selected and classified according to the number of enterprises present in each branch. According to this classification, the industries are distributed in the following manner: furrier (1095), animal-breeding (645), textile (554), services (454), food-processing (329), metal-mechanical (110), petrochemical (99), chemical and pharmaceutical (74), mining and stone-pitting (41), lumber and wood (25), electromechanical (17), distilling (16), metallurgical (13), pesticides (4), and others (33).

Hansen *et al.* (1995) compiled reported concentrations of twelve pollutants and indicators that are found in the effluents of the types of enterprises identified in the Lerma-Chapala basin. By multiplying the volumes of water discharged by the different industries in each municipality with average concentrations of these contaminants for the various branches, the industrial mass balances for each pollutant were estimated for the 65 municipalities in the watershed.

Estimated industrial contributions of metals in the basin were 0.5 kg day^{-1} for cadmium (Cd); 11 kg day^{-1} for copper (Cu); 464 kg day^{-1} for chromium (Cr), 5.9 kg day^{-1} for nickel (Ni), 13 kg day^{-1} for lead (Pb), and 46 kg day^{-1} for zinc (Zn). Total industrial discharges of nutrients, solids, and organic loads were estimated in 8 t day^{-1} of total nitrogen (totN), 275 t day^{-1} of total dissolved solids (TDS), 70 t day^{-1} of total suspended solids (TSS), 75 t day^{-1} of biochemical oxygen demand (BOD), 141 t day^{-1} of chemical oxygen demand (COD), and 33 t day^{-1} of fats and oil (F&O).

Among 65 municipalities the five main sources for each pollutant have been identified (Figure 2). This evaluation revealed industrial pollution sources, which in the case of the municipality of León in the state of Guanajuato, accounted for 26% of TSS, 24% of totN, 29% of COD, 20% of BOD, 15% of Ni, 48% of Pb, 22% of Cu, 92% of Cr, and 10% of industrially discharged Cd (Figure 2). According to the same estimation, the city of Querétaro is a main contributor of industrially discharged Zn (31%), F&O (28%) and TDS (22%).

3. CONTAMINANT CONCENTRATIONS IN ENVIRONMENTAL SAMPLES

To confirm the impact of point and non-point discharges in the watershed, a sampling program and evaluation of water and sediment qualities in the river basin and Lake Chapala were carried out as described in the following section.

3.1 Sampling and parameters

Water and sediment samples were obtained in different sampling stations in the river and lake in the time period from 1992 to 1994. The sampling program and the evaluated parameters are presented in Table 1. Figure 1 shows the Lerma-Chapala watershed with sampling stations.

Dewatered suspended sediment samples were collected using a portable continuous-flow centrifuge (Alfa-Laval model WSB 103B). Water in the rivers and in the lake was pumped to the centrifuge by a 5C-MD March submersible pump oriented in the direction of flow (depth and location varied with site). The centrifuge and pump were powered by a Honda 3500 generator, which was always placed as far away as possible and down-wind from the centrifuge.

Table 1 Numbers, types, and analysis performed on environmental samples from the Lerma-Chapala watershed (1992-1994)

Year	Sampling area (Number of samples)	Water	Suspended sediments	Deposited sediments
1992	Lerma-Chapala I (16)	-	-	Metals, PAH´s, pesticides
1993	Lake Chapala (20)	-	-	Metals
1994	Lerma-Chapala II (5)	Metals, BTEX, pesticides, PCB´s	Metals	Metals

The field centrifuge was adjusted to a flow rate of 4 L min^{-1} for optimal efficiency. Recovery efficiencies were obtained by comparing the turbidity of in-flowing and out-flowing water and were always higher than 90%. Sediment samples from the stainless steel centrifuge bowls were removed into glass containers and the bowls cleaned between each sampling.

Deposited sediment samples were obtained with a hand-operated Van Ween sediment grap-sampler and placed into glass containers. Water samples were collected from the end of the hose connected to the March submersible pump. Additionally, water samples collected for dissolved metals were filtered on site through 0.45 μm filters with an electrical vacuum pump connected to the generator.

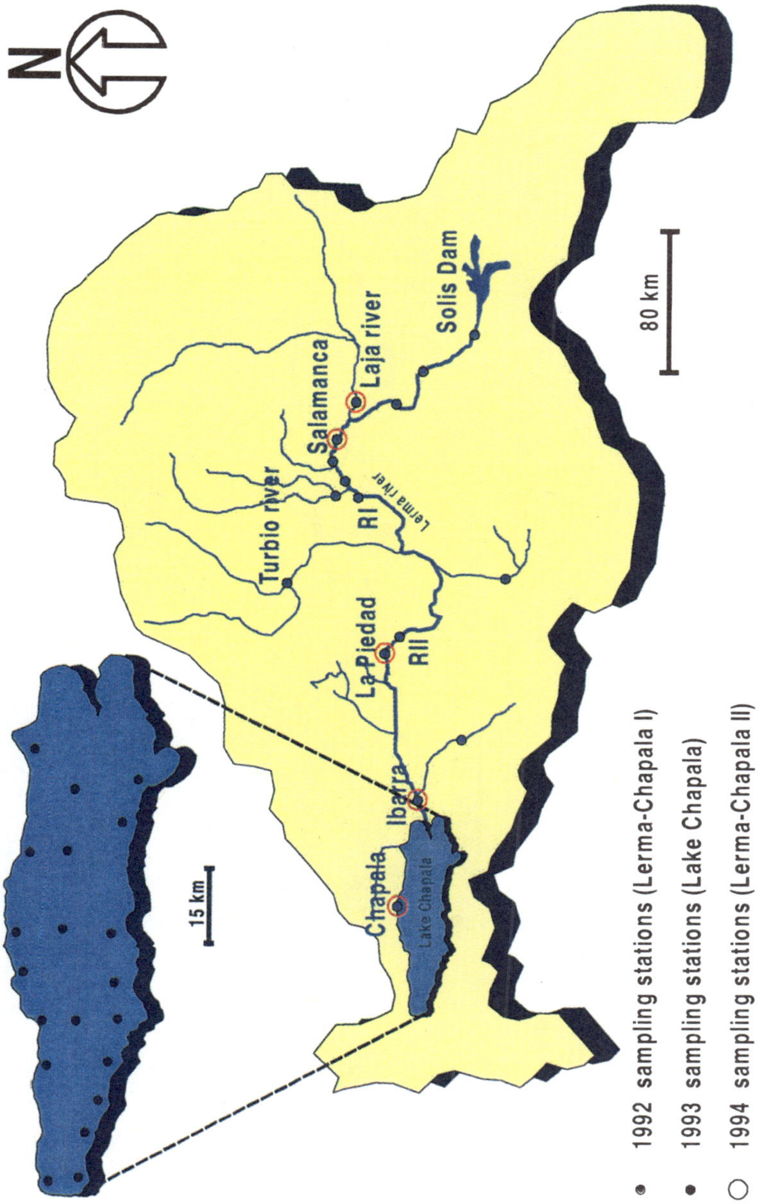

Figure 1 Lerma-Chapala watershed with sampling stations

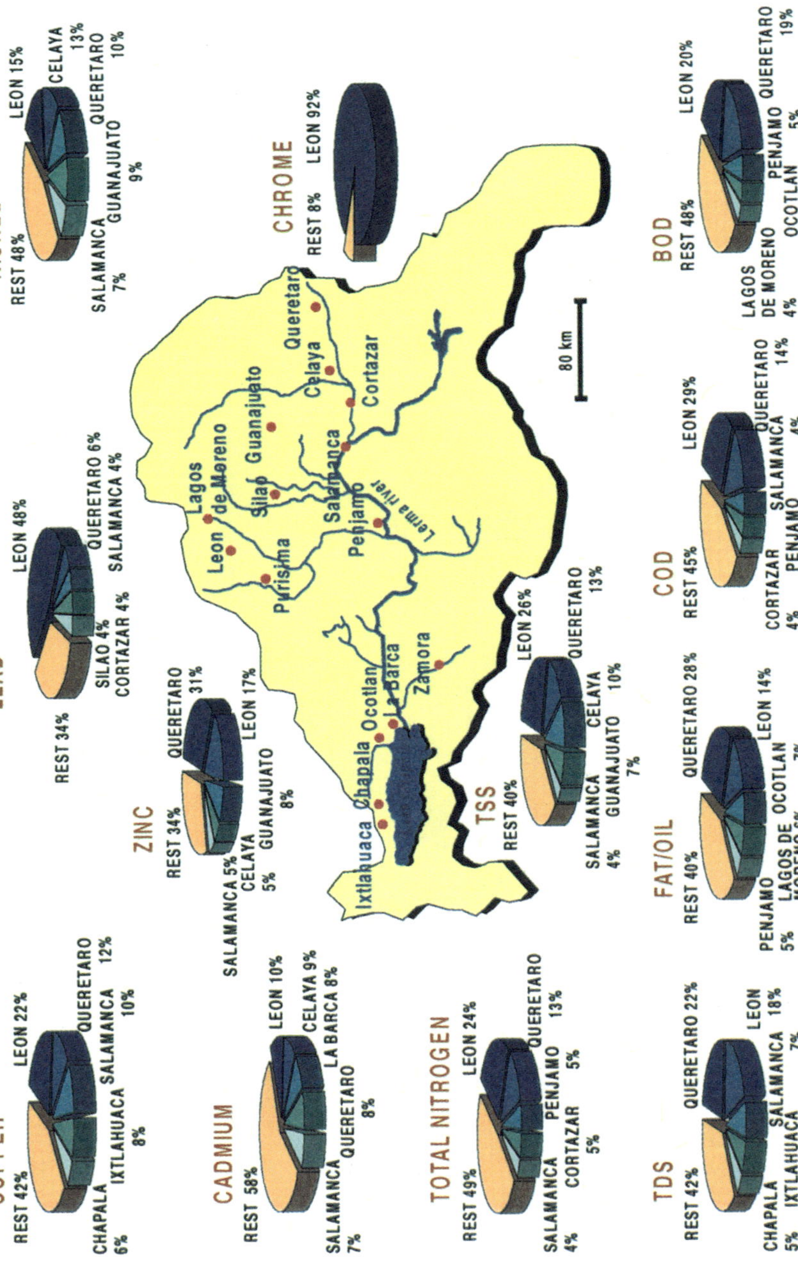

Figure 2 Estimated industrial discharges of contaminants and indicators in the Lerma-Chapala basin

Samples for metal analysis were preserved in thoroughly acid-cleaned polyethylene flasks; the water samples were acidified to pH below 2, and stored cool and dark during transportation. Once in the laboratory, sediment samples were freeze-dried and digested with high purity nitric and hydrochloric acids at temperatures lower than 85 °C according to USEPA method 200.2 (USEPA, 1991). Extracts were analyzed by flame atomic absorption spectrometry (Perkin Elmer) for Co, Cr, Cu, Mn, Ni, Pb, and Zn; hydride generation for Hg and As; and graphite furnace for Cd. Calibration curves were obtained with 5 data points for matrix-matched standards, covering the samples range of concentrations. Standard reference materials (river sediments) from the National Institute of Standards and Technology (NIST) were prepared as samples and analyzed for quality control purposes. Sampling, method, and spiked blanks were analyzed with each run. Duplicates were performed on at least 10% of the samples.

Samples for analysis of organic pollutants were preserved in glass flasks previously rinsed with pesticide-grade acetone, and dried. All samples were maintained frozen and in the darkness during transportation and storage. In the laboratory, freeze dried sediment samples and water samples were extracted with pesticide grade dichloromethane and analyzed on a Hewlett Packard 5890 series II gas chromatograph with EC detector. The presence of pesticides was confirmed by GC-MS (Hewlett Packard 5971) according to the procedure published by Keith *et al.* (1992).

Acidified samples were Soxhlet extracted with dichloromethane during 24 h for determination of PAH in GC-MS (see above). Compounds were quantified by comparison with calibration curves for 16 compounds according to USEPA (1990).

At least 10% of the samples for organic pollutants were extracted and analyzed twice. Extraction efficiencies were determined on samples previously spiked with standard compounds.

3.2 Inorganic river pollutants

Heavy metals in the Lerma river originate from either natural processes or human activities. The main natural processes are erosion and weathering of natural exposed metal-bearing ores that lead to the release of certain amounts of heavy metals.

When compared to the natural exposure of ore bodies through erosion within the last centuries, the exposure rate due to mining industry plays an important role. Since the Guanajuato district has been explored since 1548 and is primarily a producer of silver and gold (Wilson *et al.*, 1950) with minor amounts of base metals derived, leaching of tailings and drainage from these mining areas are suspected to introduce substantial amounts of metals into the Guanajuato river, which is a tributary to the Lerma river.

In previous studies, enrichment factors were calculated with respect to aluminum (Al) concentrations in suspended solids and deposited sediments of several sampling stations in the basin. (Hansen *et al.*, 1995; Posada-Ramos *et al.*, 1994). These factors were related to average values as reported for the continental crust (Lantzy and MacKenzie, 1979). Concerning these results, high enrichment factors were found for silver (Ag), Zn, Pb, Ni, Cu, Cd, and mercury (Hg) in sediments from several areas in the Lerma-Chapala basin. Metal enrichment factors were very high in sediments from the Guanajuato river for Cu (140), Ag (66) and Pb (39) and high for Zn (4), Hg (3), and Ni (3). This may be due to leaching of mine tailings in this area.

Additionally in recent times the industrial, municipal and rural developments of the river basin may contribute more importantly to the contaminant load of the river water and sediments.

During the monitoring period of 1992-1994, environmental samples were analyzed with the purpose of identifying risks concerning metals concentrations in sediments and water from the basin. In Table 2 the mean metal concentrations in river sediments are compared to international quality criteria for these solids.

Table 2 Mean metal concentrations in river sediment samples of the Lerma-Chapala basin

Metal	Limit ($mg\ kg^{-1}$)		1992 Deposited Sediments (15 Stations)		1994 Deposited Sediments (4 Stations)		1994 Suspended Sediments (4 Stations)	
	LEL[1]	SEL[2]	Concentration ($mg\ kg^{-1}$)	δ (%)	Concentration ($mg\ kg^{-1}$)	δ (%)	Concentration ($mg\ kg^{-1}$)	δ (%)
As	6	33	3.7	40.3	n.d.	n.d.	3.0	37.6
Cd	0.6	10	0.1	135.8	1.6	41.0	0.5	44.8
Co	-	-	n.d.	n.d.	27.5	6.3	22.7	55.6
Cr	26	110	62.2	74.4	52.5	11.5	29.5	45.3
Cu	16	110	23.9	82.7	145.3	148.6	53.5	64.1
Hg	0.2	2	0.07	49.5	n.d.	n.d.	0.3	101.1
Mn	460	1100	375.3	60.0	550.3	38.4	530.6	25.6
Ni	16	75	33.0	82.1	32.8	8.8	24.2	60.8
Pb	31	250	14.3	44.8	32.1	58.2	16.8	21.3
Zn	120	820	93.6	88.4	160.8	75.6	138.8	20.9

[1] = LEL: Lowest Effect Level. The majority of benthic organisms can tolerate this level. It is derived using field-based data on the co-occurrence of sediment concentrations and benthic species (Persaud *et al.*, 1993).

[2] = SEL: Severe Effect Level. Represents contaminant concentrations in sediment that could potentially eliminate most of the benthic organisms (Persaud *et al.*, 1993).

n.d.= not detected

Mean concentrations of Zn, Pb, cobalt (Co), Cu, Cd, Ni, Cr, arsenic (As), and Hg in suspended sediments from four river stations sampled in 1994 (Laja river, Salamanca, La Piedad, and Ibarra) did not exceed the quality criteria SEL for highly polluted sediments given in Table 2.

For deposited sediments, the mean concentrations of Zn, Pb, Co, Cd, Ni, Cr, As, and Hg from four river stations sampled in 1994 did not exceed the above limits. For Cu, the detected concentrations were higher than the SEL in the sampling station at La Piedad with a concentration of 468 mg kg^{-1}. Therefore, this sample can be classified as highly polluted in Cu. Also, more than a factor of two enriched the concentrations of Zn and Pb compared to the mean values measured in the other river stations.

The concentrations of metals in deposited sediments of the La Piedad station were significantly higher than those in suspended sediments. This may be due to the different organic carbon contents of the suspended and deposited sediment samples (0.9% and 7.1%, respectively). The high TOC-content of the bottom sediments in this sampling station can be explained by the predominance of pork-breeding industries in this area.

In 1992, mean concentrations of all metals in deposited sediments of the 15 river stations were below the SEL guidelines given in Table 2. Exceptions were found for Cr in three sampling stations with concentrations higher than the limit of 110 mg kg^{-1} established by the Great Lakes Water Quality Board (Persaud *et al.*, 1993). In one of these stations, the Guanajuato river near the confluence with the Lerma river, the Cu and Ni concentrations were also higher that the corresponding SEL values for highly contaminated sediments.

Average concentrations of 8 metals (Cd, Cr, Cu, Hg, Mn, Ni, Pb, and Zn) were higher than the lowest effect levels (LEL), which indicate concentrations of contaminants in sediment where further testing is recommended. This will require additional information and a thorough review of the site and may include both chemical and biological tests. Nevertheless, it has to be mentioned that risk assessment should include a comparison with non-polluted sediments in the study area. This is not available for the Lerma-Chapala watershed. Therefore, the observed enriched concentrations could be due to natural sources.

Sampling sites with the highest concentrations of metals in sediments are: the Guanajuato river, downstream from an important mining area; and the Lerma river downstream from La Piedad, due to the accumulation of metals in the organic-rich sediments resulting from the pork-breeding industry in this area.

3.3 Inorganic lake pollutants

In 1993, an evaluation of metal pollution of the sediments in the lake was carried out. All 20 sampling stations in the lake correspond to those where the Mexican National Water Commission (CNA) collects water samples on a three-monthly basis. Additionally, in 1994, water and sediment quality were evaluated in one station near the outflow of the lake, where the lake water is used as drinking water supply for the city of Guadalajara.

Mean concentrations and standard deviations of metal contents are given in Table 3. Detected concentrations of metals in the bottom sediments of all 20 lake stations were lower than the SEL limits given in this Table. The 1994 sampling confirmed these results, since measured metal concentration for lake sediments was below these limits.

Table 3 Metal concentrations in Lake Chapala sediments

Metal	Limit (mg kg^{-1})		1993 Deposited Sediments (20 Stations)		1994 Deposited Sediments (1 Station)	1994 Suspended Sediments (1 Station)
	LEL[1]	SEL[2]	Concentration (mg kg^{-1})	δ (%)	Concentration (mg kg^{-1})	Concentration (mg kg^{-1})
As	6	33	17.4	35.1	n.d.	2.3
Cd	0.6	10	0.1	32.9	0.7	1.3
Co	-	-	n.d.	n.d.	25.0	21.8
Cr	26	110	62.8	16.6	32.0	36.5
Cu	16	110	28.4	22.9	80.0	20.0
Hg	0.2	2	0.1	51.4	n.d.	0.7
Mn	460	1100	827.2	21.7	340.0	816.0
Ni	16	75	50.5	13.9	35.0	30.9
Pb	31	250	17.2	26.7	19.1	20.1
Zn	120	820	96.8	14.5	67.0	158.0

1 = LEL: Lowest Effect Level. The majority of benthic organisms can tolerate this level. It is derived using field-based data on the co-occurrence of sediment concentrations and benthic species (Persaud *et al.*, 1993).

2 = SEL: Severe Effect Level. Represents contaminant concentrations in sediment that could potentially eliminate most of the benthic organisms (Persaud *et al.*, 1993).

n.d. = not determined

Average concentrations of the 1993 sampling showed that five metals (As, Cr, Cu, Mn, and Ni) were higher than the lowest effect levels (LEL). It is observed that three of these metals are also enriched in the river sediments, (Table 2) indicating that these pollutants are being transported by the river and have already been deposited in the lake.

In general, the variability of metal concentrations in river deposited sediments is higher than for lake sediments (see standard deviation in Figure 3). This may be due to different point sources along the rivers in the watershed. Considering the lake's dimensions, this also indicates that the system is well mixed, since metal's concentrations in the sediments of different stations do not present considerable variations.

No significant difference between the mean concentrations of seven metals in deposited sediments in rivers and the lake was determined, as proven by t-distribution test at a significance level of 99%. By comparing 1992 river- and 1993 lake-sampling data there is, however, a tendency that the lake deposited sediments have slightly higher concentrations of metals than the river sediments. This would confirm the above-mentioned hypothesis that metals could accumulate in the lake with time.

Exceptions to this general statement were found for mean concentrations of Mn and As. With a significance of 99%, concentrations of both metals were higher in the lake than in the rivers (Figure 3). Results of Shine *et al.* (1998) as well indicate a direct source of As to the lake. This is probably due to an interaction of the lake water with hydrothermal springs in the lake bottom (Amezcua, 1993).

3.4 Organic pollutants

The Lerma river contains the largest and most intensive farmed region in the nation, El Bajío. In order to increase yields of agricultural crops, pesticides are used to protect them from weeds, insects, and other pests. The use of pesticides is of concern because of potential adverse effects on the quality and use of water resources. The most immediate concerns are for aquatic life and for the population in the basin, who rely on surface-water resources for drinking water.

Some of the pesticides used in the basin may be present in the Lerma river and its tributaries and be flushed into streams during storm runoff each year following application to cropland. Beside agricultural runoffs, industrial discharges may also influence water and sediment qualities in the watershed.

The five water samples obtained in 1994 were analyzed for representative industrial and agricultural contaminations and the results were compared to Mexican norms. When these did not exist, international criteria were used to identify potential risks.

The results indicated, that toluene, gasoline, phenols and PCB´s concentrations in river water were higher than the limits in the sampling stations at La Piedad and Salamanca while the other two river stations, at Ibarra and the Laja river, exceeded the limits for toluene, gasoline, and phenols. This indicates that the water samples of the river stations are affected by industrial discharges, whereas no significant levels of the measured agricultural pesticides were found (Table 4).

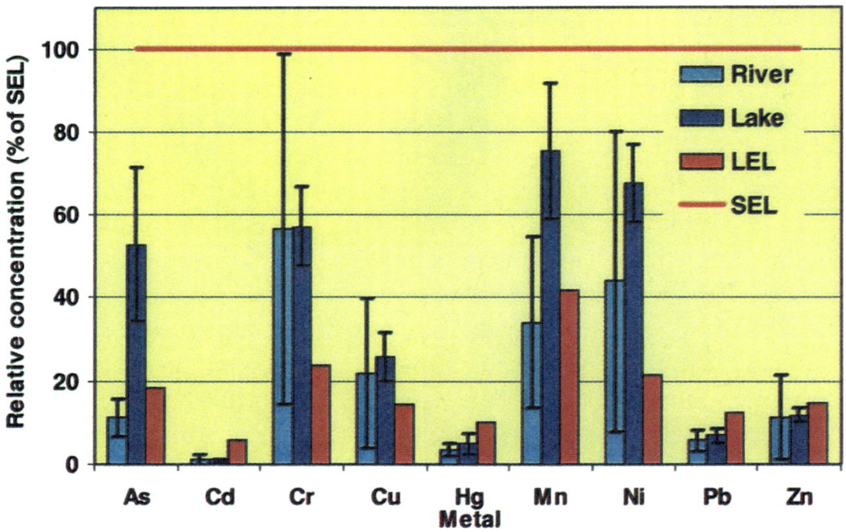

Figure 3 Comparison of metal concentrations in lake and river sediments with Lowest Effect Levels (LEL) and Severe Effect Levels (SEL) proposed by Persaud *et al.* (1993) for sediment risk assessment

In 1992, deposited sediments from 7 of the 16 sampled stations were analyzed for organic contaminants. The results are presented in Tables 5 and 6. Polycyclic aromatic hydrocarbons (PAH) and chlorinated pesticides concentrations in the organic fractions of the sediments were lower than those recommended by the SEL guidelines (Persaud *et al.*, 1993) for highly polluted sediments (Table 6). Nevertheless, total PAH concentrations were higher than the LEL sediment quality criteria published by the same authors in four river stations and total chlorinated pesticides in five out of six river stations (Table 5). The stations with the highest amounts of total chlorinated pesticides were Salamanca, and La Piedad with concentrations higher than the LEL guidelines by factors of over three.

The total PAH concentrations of were higher than these guidelines by one order of magnitude at the sampling station of the Turbio river downstream from the city of León and by a factor of six in station at La Piedad (Table 5).

A high variety of PAH´s was found, and concentrations were high for those of pyrolytic origin (incomplete combustion of fossil fuel) or petrochemical industry like phenanthrene, fluorantene, and benzo(g,h,i)-

pyrilene. Benzo(a)pyrene was practically absent, and only present in the sampling station at La Piedad.

Table 4 Concentrations of organic contaminants in lake and river water (1994 sampling)

Parameter	Limits ($mg\ L^{-1}$)	Sampling Stations (Concentrations in $mg\ L^{-1}$)				
		Ibarra	Chapala	La Piedad	Laja river	Salamanca
Benzene	0.01[3]	n.a.	<0.01	<0.01	<0.01	<0.01
Toluene	0.7[3]	n.a.	0.52	1.35	0.7	0.73
Ethylbenzene	0.3[3]	n.a.	0.008	0.014	0.081	0.01
Xylene	0.5[3]	n.a.	0.16	0.18	0.16	0.16
Phenols	0.001[1]	0.039	0.007	0.021	0.005	0.004
Gasoline	0.2[4]	3.26	0.02	0.47	17.82	26.72
PCB´s	0.0002[4]	<0.0002	<0.0002	0.0016	<0.0002	0.0007
Aldrin	0.00003[1]	<0.00003	<0.00003	<0.00003	<0.00003	<0.00003
Dieldrin	0.00003[1]	<0.00003	<0.00003	<0.00003	<0.00003	<0.00003
Endrin	0.001[2]	< 0.001	<0.001	<0.001	<0.001	<0.001
Heptachlor	0.00003[1]	<0.00003	<0.00003	<0.00003	<0.00003	<0.00003
Heptachlor epoxide	0.00003[1]	<0.00003	<0.00003	<0.00003	<0.00003	<0.00003
Hexachloro-benzene	0.00001[1]	<0.00001	<0.00001	<0.00001	<0.00001	<0.00001
Hexachloro cyclopentadiene	0.001[2]	<0.001	<0.001	<0.001	<0.001	<0.001
Lindane	0.002[3]	<0.002	<0.002	<0.002	<0.002	<0.002

[1] = NOM-127-SSA1-1994 (NOM, 1996; drinking water supply)
[2] = CNA 1989 (drinking water supply)
[3] = WHO 1996 and 1998 (drinking water)
[4] = The Netherlands-List 1994 (B-values for groundwater: when exceeding this value, further investigation is recommended)
n.a.= not analyzed

With exception of lake sediments and downstream from the Solis Dam, higher concentrations of organic pollutants than the lowest effect levels (LEL) were found in all sampling stations. This implies that further testing is recommended to decide if there exists an actual risk where intervention is necessary or future accumulation may occur. The occurrence of PAH may originate from several sources, industries, forest fires, open combustion of agricultural plant residues and solid waste. Contrary to this, the elevated levels of pesticides are probably only of agricultural origin. The registered level of organic contaminants suggests the necessity to gather additional information and to include these groups of substances in a future monitoring program.

Table 5 Concentrations of organic contaminants in lake and river sediments (1992 sampling) compared to Lowest Effect Levels (LEL)

Contaminant group	Limit[1] LEL	Sampling Stations (Concentrations in mg kg^{-1})						
		Solis Dam	Salamanca	RI	Turbio river	RII	La Piedad	Chapala
Total PAH[2] (mg kg^{-1})	4	1.8	5.7	4.4	46.8	1.4	24.0	2.3
Total chlorinated pesticides[3] (mg kg^{-1})	0.057	0.016	0.179	0.088	0.113	0.100	0.189	0.010

[1] = Persaud *et al.* (1993). Sediment Quality Criteria (Canada). The Lowest Effect Level. (LEL) is the level that can be tolerated by the majority of benthic organisms. It is derived using field-based data on the co-occurrence of sediment concentrations and benthic species.

[2] = PAH (total) is the sum of 16 PAH compounds: Acenaphthene, Acenaphthylene, Anthracene, Benzo[k]fluoranthene, Benzo[b]fluorene, Benzo[a]anthracene, Benzo[a]pyrene, Benzo[g,h,i]perylene, Chrysene, Dibenzo[a,h]anthracene, Fluoranthene, Fluorene, Indeno[1,2,3-cd]pyrene, Naphthalene, Phenanthrene, and Pyrene

[3] = Sum of Alpha-BHC, Beta-BHC, Gamma-BHC, BHC, Heptachlor epoxide, Aldrin, p,p´-DDE, p,p´-DDD, DDT, Dieldrin, and Endrin

4. CONTAMINANT ACCUMULATION CAUSED BY DECREASING WATER LEVEL

As discussed in the above sections, the water quality of the main rivers in the Lerma-Chapala basin and especially in Lake Chapala was of acceptable quality during the 1994 sampling.

Since transport of contaminants discharged into rivers mainly occurs with suspended sediments, this will eventually lead to an accumulation of metals in the lake sediments, which represent a natural sink for transported river sediments. This is especially valid for Lake Chapala due to the low water level in the lake during recent years, and the consequent reduced outflow through the Santiago river.

As has been reported by Aparicio (2001), between 1975 and 1999 the lake volume has reduced to about 40% of the storage capacity and the area to about 80% of its original size. This reduction is partly due to natural fluctuations of precipitation in the watershed. Nevertheless, available data support the possibility that within the basin, increasing evaporation due to the construction of dams, irrigation in agriculture, industrialization, and urbanization play a significant role in the reduction of water flow to the lake through the Lerma river.

Table 6 TOC-corrected concentrations of organic contaminants in lake and river sediments (1992 sampling) compared to Severe Effect Level (SEL)

Contaminant group	Limit[1] SEL	Sampling Stations (Concentrations in mg kg^{-1} TOC)						
		Solis Dam	Salamanca	RI	Turbio river	RII	La Piedad	Chapala
Total PAH[2] (mg (kg $TOC)^{-1}$)	10,000	150	2850	1467	2463	54	1000	209
Total chlorinated pesticides[3] (mg (kg $TOC)^{-1}$)	386.0	1.3	90.0	29.3	5.9	3.8	7.8	0.9
% TOC	100	1.2	0.2	0.3	1.9	2.6	2.4	1.1

[1] = Persaud *et al.* (1993). Sediment Quality Criteria (Canada) The Severe Effect Level (SEL). This level represents contaminant concentrations in sediment that could potentially eliminate most of the benthic organisms.

[2] = PAH (total) is the sum of 16 PAH compounds: Acenaphthene, Acenaphthylene, Anthracene, Benzo[k]fluoranthene, Benzo[b]fluorene, Benzo[a]anthracene, Benzo[a]pyrene, Benzo[g,h,i]perylene, Chrysene, Dibenzo[a,h]anthracene, Fluoranthene, Fluorene, Indeno[1,2,3-cd]pyrene, Naphthalene, Phenanthrene, and Pyrene

[3] = Sum of Alpha-BHC, Beta-BHC, Gamma-BHC, BHC, Heptachlor epoxide, Aldrin, p,p´-DDE, p,p´-DDD, DDT, Dieldrin, and Endrin

A decreasing water storage in the lake will lead to increasing pollutant, salt and nutrient concentrations, which will likely affect the quality of life of the surrounding population and the ecological integrity of the lake. Protection and water management of the whole watershed already include a water level control of Lake Chapala as a main action. Nevertheless, the question arises of the minimum acceptable water level that will both sustain the social and economical needs of the basin as well as ecological intactness of the lake and its surroundings.

As a first approach, and on the basis of available data, increase in salinity and accumulation of Cd are estimated, due to a possible continuous decrease in the water storage in the lake. These two parameters were chosen since salinity represents a general parameter of water quality and ecosystems classification, and Cd as an indicator of water quality due to its high toxicity and potential to bio-accumulate (Korte, 1992). Additionally, salinity influences the distribution of metals like Cd between aqueous and sediment phases (Hansen and Maya, 1997).

The projection of a decrease in storage of water is based on the recently published annual mean reduction of about -400 Mm^3 yr^{-1}, which was determined during the years 1994-1999 (CNA-IMTA, 2000).

The changes in lake water area and volume were estimated by using the data published by CNA-IMTA (2000). Figure 4 shows the relation between area and volume of the lake. Considering the geometric characteristics of the lake bottom topography, a reduction of volume to about 5,000 Mm^3 do not reduce the lake area significantly, where as a reduction to volumes below 5,000 Mm^3 will lead to a decrease in lake area as well.

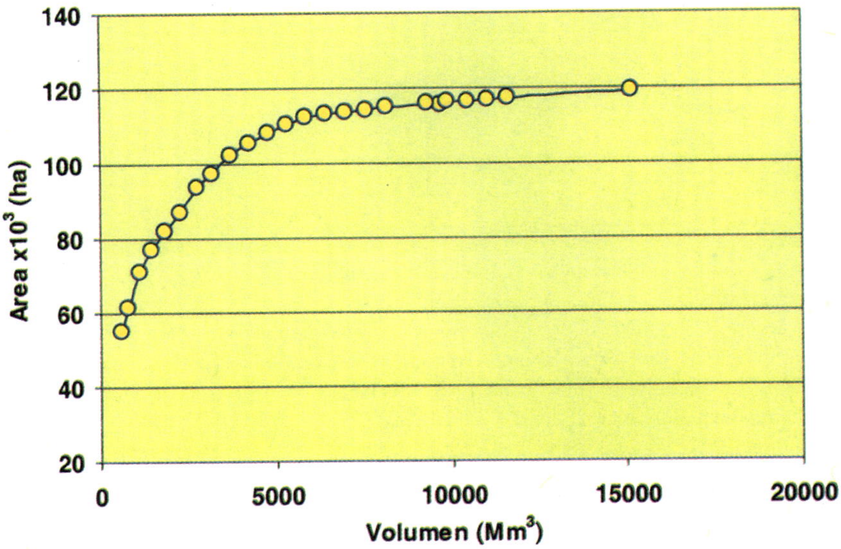

Figure 4 Relation between Lake Chapala area and volume

Aparicio (2001) reported a discrepancy in water inflow to the lake of about – 500 Mm^3 per year and that evaporation represents with about 84% the main water outlet in the lake water balance during the years 1994-1999. Decreasing evaporation due to the lake area reduction will therefore compensate the reduced water inflow into the lake and a new equilibrium in lake volume will be established with time.

Therefore the simulation includes annual evaporation rates proportional to a decreasing lake surface. The calculated time dependent adjustment of a new equilibrium of the lake volume is shown in Figure 5. The simulation of 30 years starts in the year 1995 with an annual evaporation for the lake of 1.4 Mm^3 for that year. Water inflow was assumed to be constant and the mean value of the years 1994-1999 was used as reported by Aparicio (2001).

Salt accumulation that occur during this adjustment process was calculated as reciprocally proportional to volume reduction, due to the fact that salinity is mainly defined by highly soluble salts like chlorides, and sulphates, were sorption effects are negligible.

The modeled adjustment to a new equilibrium of the lake volume and the resulting increase in salinity is shown in Figure 5. Real data shown in this Figure were reported in CNA-IMTA (2000). According to this simulation, in

2015 the lake volume will reach a stable water quantity of about 900 Mm^3 and salinity of 3000 mg L^{-1}.

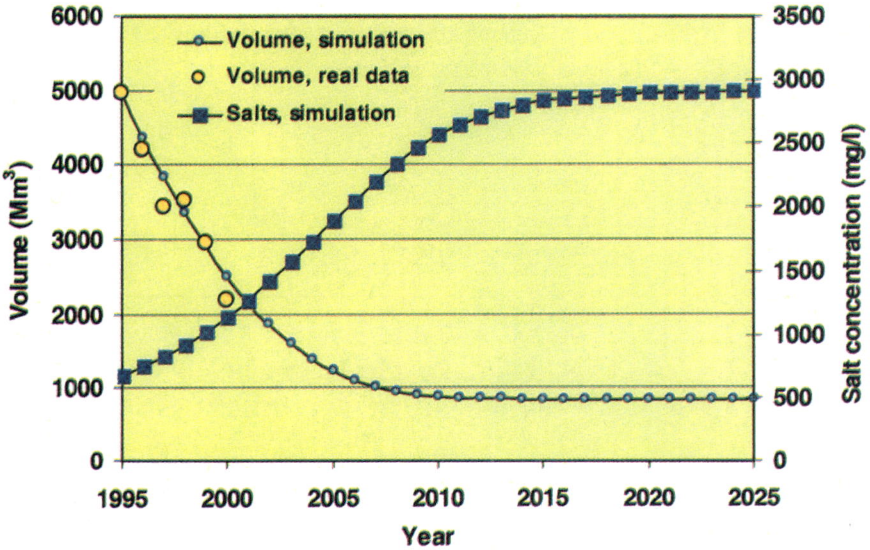

Figure 5 Simulation of lake volume and resulting increase in salinity in Lake Chapala, considering an initial deficit in the water balance of -500 Mm^3 yr^{-1}, and showing an adjustment to a new equilibrium

The accumulation of Cd in the lake water was modeled with a chemical equilibrium model, Hydraql (Papelis *et al.*, 1988) taking into account an increase in salinity and suspended solids due to a decrease in water level and volume. Adsorption of Cd on the sediment surfaces was simulated as triple layer surface complexation. The parameters and surface reactions used were obtained from adsorption experiments and modeling reported in previous studies (Hansen and Maya, 1997; Dzombak and Morel, 1990), and is presented in Table 7.

This exercise was based on the supposition that water in the lake is well mixed and oxidized conditions prevail in the water column. Since the mean water depth in the lake in 1995 varied between 4 and 5 m, these conditions are very probable and caused by a wind-driven mixing of the water column and by the low primary production of biomass which is due to the high suspended sediment concentration and corresponding low light penetration in the water column (Dávalos-Lind and Lind, 2001).

It should be mentioned that a forecast of the future changes in lake volume and area is a multifactor approach and depends on a number of unknown factors. The estimation presented here, does not include long-term cyclic variations, climate and temperature changes, and variations in water and land uses. Therefore, the presented scenario includes a significant error. Nevertheless, the focus of the presented work is directed on the accumulation of salts and contaminants (Cd) at different water levels and only secondarily on the time, when this will happen.

Table 7 Conditions used for simulation the accumulation of Cd in Lake Chapala

Parameter	Conditions	
Time of simulation	30 years (from 1995 to 2025)	
Lake volume	4,961 Mm^3 in 1995	
Lake area	111,110 ha in 1995	
Evaporation	1,400 $Mm^3\ yr^{-1}$ in 1995	
Salinity	Total salts 673 mg L^{-1} in 1995 (expressed as 0.012 mole NaCl L^{-1})	
Resuspended sediments	Resuspension of the top five centimeters of bottom sediments (density=1.2 g cm^{-3}) in the lake water volume (expressed as 13.4 g L^{-1} in 1995)	
Total Cd	Deposited sediments 0.7 mg kg^{-1} in 1995 (expressed as $0.8x10^{-7}$ mole L^{-1})	
Sediment surface acidity constants	$SOH = SO^- + H^+$ $SOH + H^+ = SOH_2^+$	log K = -9.01[1] log K = 5.01[1]
Interaction with electrolyte	$SOH + Na^+ = SO^- - Na^+ + H^+$ $SOH + H^+ + Cl^- = SOH_2^+ - Cl^-$	log K = -7.51[1] log K = 7.91[1]
Adsorption of Cd^{2+}	$SOH+Cd^{2+}+H_2O=SO^--Cd^{2+} + 2H^+$	log K = -2.92[2]

[1] = Hansen and Maya, 1997
[2] = Dzombak and Morel, 1990

During this simulation, the following assumptions were made:

- Total salt, sediment and contaminant concentrations were considered to be in "steady state" conditions. No further inputs of suspended and dissolved solids were considered in these simulations. The only mechanism of accumulation of salts, nutrients, and metals considered here is due to the water deficit in the lake.

- Mixing caused by the wind could re-suspend a sediment layer of an average of 5 cm in the lake.

The evolution of Cd during 1995-2025 under these assumptions is illustrated in Figure 6 for different pH-values of the water phase in equilibrium with suspended sediments. For slightly acidic conditions (pH 6), Cd concentrations would soon exceed the limit of 0.005 mg L^{-1} for for drinking water. Even at the most alkaline situation (pH 8), Cd concentrations should exceed this limit in 2010, albeit the water inflow to the lake will be of the same order of magnitude as has been observed during the last two decades (CNA-IMTA, 2000).

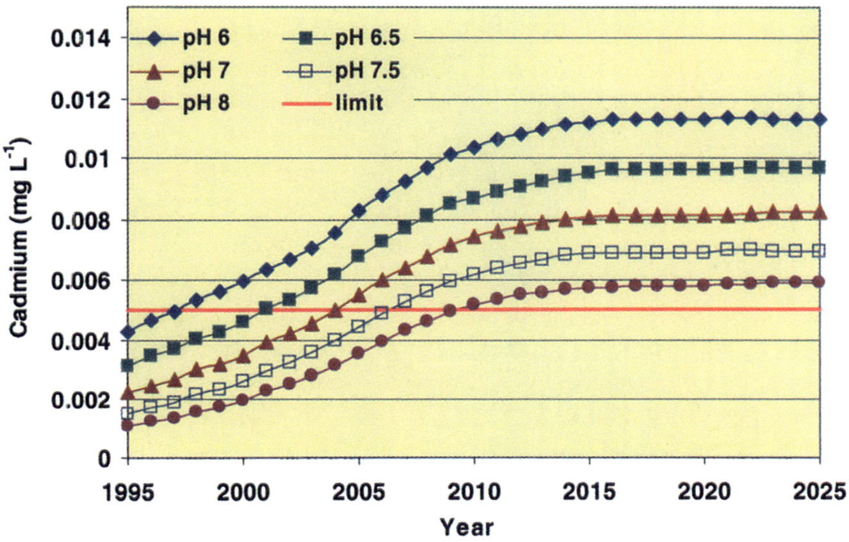

Figure 6 Prognosticated time-dependency of dissolved Cd in Lake Chapala at different water compositions

The simulation demonstrates a time dependent trend towards increasing Cd accumulation in the water-phase due to continuous deficit in the lake's water balance. According to this estimation, a new equilibrium would be reached in the year 2015 with Cd concentrations varying from 0.006 mg L^{-1} (pH 8) to 0.011 mg L^{-1} (pH 6). This means that concentrations may reach levels twice as high as those established by current drinking water limits.

Even if this simulation does not include future changes in rainfall, water consumption or climatic changes, the results clearly demonstrate that decreasing water levels in Lake Chapala represent a risk of dissolved heavy metals accumulation in the lake water, which in the near future may seriously affect the lake ecosystem and as a water supply for different uses.

5. SUMMARY

The water of the Lerma river carries dissolved and sediment-associated contaminants that originate from a variety of municipal, agricultural, and industrial sources. The distribution of these substances along the river depends on the nature and location of these sources, the stability of the contaminants in water and sediments, and the dilution by inflowing water. To evaluate the state of pollution of the Lerma-Chapala basin and make a first risk assessment approach, an industrial inventory of the watershed was elaborated, and priority contaminants in water and sediment samples from different sites in the study area, determined. Additionally, the probability of future contaminant accumulation in Lake Chapala was assessed, with the intention to make this information available for the inclusion in future monitoring, assessment, or management programs.

Information on industrial discharges to the Lerma-Chapala basin included over 3,500 enterprises from the food, chemical, pharmaceutical, petrochemical, mining, automotive, wood and derivatives, leather, textile, metallurgical, and agricultural chemical sectors. Total industrial contributions of metals in the watershed were estimated. Main discharges were found for Cr with 464 kg day^{-1}; Zn (46 kg day^{-1}); and Pb (13 kg day^{-1}). Elevated saline and organic loads were identified in several municipalities, which in total contributed with 275 t day^{-1} of TDS and 141 t day^{-1} of COD. Among the 65 industrialized municipalities, the five main sources for each pollutant were identified (Figure 1). In general the highly industrialized city of León was identified as the main producer of industrial discharges.

Since discharges may affect water and sediment qualities, samples were collected along the Lerma-Chapala basin to establish a first assessment of the ecological health concerning metals and toxic organic contaminants. The results indicated that the concentrations of the studied contaminants in almost all sediment samples were below the SEL-criteria that have been adopted by the Canadian government for highly polluted sediments. Nevertheless, it should be pointed out that a detailed comparative analysis of the results with ecotoxicological LEL criteria reflects a different situation. Whereas sediment samples of the 20 lake stations all exceed these criteria for 5 metals, the concentrations in 14 out of 15 river samples were higher in at least one parameter. Cr was enriched in all lake sediments by at least a factor of two with respect to the Canadian LEL guidelines for sediment pollution, which may affect benthic species (Persaud *et al.*, 1993). For this metal, there is evidence that it is released by industrial discharges due to the high loads as described above. From this point of view, Ni was enriched by a factor of three in lake sediments and of two in the river sediments. Cu was also higher than the above-mentioned LEL criteria. This indicates that the Lerma-Chapala sediments are affected by industrial, municipal, or agricultural discharges in the rivers.

As a consequence of decreasing water levels in Lake Chapala during the last decades, the subsequent risk of contaminant accumulation was assessed. This could contribute to define the minimum acceptable water level that will both sustain the social and economical needs of the basin as well as the ecological intactness of the lake and its surroundings. On the basis of available data, increasing salinity and accumulation of Cd were estimated, due to a possible continuous decrease in the storage of water in the lake.

The modeling results demonstrate a time-dependent tendency towards increasing accumulation of Cd in the water-phase due to a continuous deficit in the lake's water balance. A new equilibrium may be reached where Cd concentrations are two times higher than the limits recommended for drinking water. Even if this simulation only represents a first approach, the results clearly demonstrate that decreasing water levels in Lake Chapala represent a risk of accumulation of dissolved heavy metals in the lake water, which in the future may affect the ecological intactness and the quality of the water resource.

In the absence of adequate Mexican guidelines for sediments, the brief risk assessment presented in this chapter, was mainly based on Canadian regulations and quality criteria. It should be noted, that these criteria were elaborated for benthic organisms in freshwater from this country and thus include some uncertainty to whether the values are valid for tropical lake and river sediments as those from the study area. Another important step in risk assessment procedures is, beside of the information of chemical data, an evaluation and comparison with local background levels. This information was not available for metals and toxic organic chemicals in the Lerma-Chapala basin. Especially concerning the mining areas found in the basin, an interpretation of the presented results might be misleading and need further efforts in studying sediment contamination in selected areas identified here. For these reasons it was not possible to derivate or specify concrete action plans like remediation, restrictions in water and ecosystem uses, discharge reductions, etc., which normally should be a result of detailed risk assessment studies. Future water management plans should therefore include a more detailed risk assessment for the parts of the basin that show increased concentrations of toxic contaminants and represent sources that could affect the lake. Since remediation actions are difficult and expensive for lakes of this size, with the exception of spills, which should be cleaned up immediately, the most urgent need is to protect the ecosystem from further exposure. Thus, contaminant source control and water level management of Lake Chapala must be the main foundations of remedial actions in the basin.

Recent development and increase in land use practices in the Lerma-Chapala watershed have significantly affected the economical development and ecological quality of the basin. Therefore, in addition to the above-mentioned technical considerations, final decision as to the proper course of action must also be based on considerations of social and economic criteria (Dávila-Poblete and Treviño-Carrillo, 2001; Wester *et al.*, 2001).

ACKNOWLEDGEMENTS

The authors wish to thank the following persons and institutions for their support during the development of the work presented here: James O. Leckie (Stanford University), Jesús Amezcua and Vicente Ortega, Mexican National Water Commission (CNA); Silvia Gelover, Mexican Institute of Water Technology (IMTA); and Ian Droppo, Environment Canada

REFERENCES

Amezcua J (1993) CNA (Mexican National Water Commission) Personal communication.

Aparicio J (2001) Hydrology of the Lerma-Chapala Watershed *In: The Lerma-Chapala Watershed: Evaluation and Management* (Hansen AM and van Afferden M eds) Kluwer Academic/Plenum Publishers, London.

Becker RL, Grube AH and Toria R (1989) Pesticides: Surface Runoff, Leaching and Exposure Concerns. University of Minnesota, Minnesota Extension Service AG-BU-3911, 32 p.

Benoit G and Rozan TF (1998) Humic Substances and Sulfide Control Seasonal Variability of Free Metals in Four Connecticut Rivers. 214th National Meeting of the American Chemical Society.

CCME (1999) Canadian Sediment Quality Guidelines for the Protection of Aquatic Life *In* Canadian Environmental Quality Guidelines (Canadian Council of Ministers of the Environment), No1299, Winnipeg (ISBN 1-896997-34-1).

Cerniglia CE and MA Heitkamp (1989) Microbial Degradation of Polycyclic Aromatic Hydrocarbons in the Aquatic Environment *In* Metabolism of Polycyclic Aromatic Hydrocarbons in the Aquatic Environment (Varanasi U ed) Boca Ratón, FL: CRC Press. pp 41-68.

CNA (1989) Criterios Ecológicos de Calidad del Agua *In* Diario Oficial de la Federación, December 13, 1989.

CNA-IMTA (2000) Banco Nacional de Datos de Aguas Superficiales (BANDAS). Hidrometría y Sedimentos hasta 1999. Mexican Institute of Water Technology

Dávila-Poblete S and Treviño-Carrillo AH (2001) Social Characterization of the Watershed *In: The Lerma-Chapala Watershed, Evaluation and Management* (Hansen AM and van Afferden M eds) Kluwer Academic/Plenum Publishers, London.

Dávalos-Lind L and Lind OT (2001) Phytoplankton and Bacterioplankton Production and Trophic Relations in Lake Chapala: A Review *In: The Lerma-Chapala Watershed, Evaluation and Management* (Hansen AM and van Afferden M eds) Kluwer Academic/Plenum Publishers, London.

Dzombak DA and Morel FMM (1990) Surface Complexation Modeling. Hydrous Ferric Oxide. John Wiley & Sons, Inc, New York, 393 p.

Goolsby DA and Pereira WE (1996) Pesticides in the Mississippi River. Contaminants in the Mississippi River, 1987-92 (Meade RH ed) US Geological Survey, circular 1133, 140 p.

Hansen AM and Maya P (1997) Adsorption-desorption Behaviors of Cd and Pb in Lake Chapala, Mexico. Environment International, 23, 4, 553-564.

Hansen AM, León-Zavala A and Bravo-Inclán L (1995) Fuentes de Contaminación y Acumulación de Metales en la Cuenca Lerma-Chapala. Ingeniería Hidráulica en Mexico. Vol. X, 3, 55-69.

Hansen AM, Villalba G and Tiscareño-López M (1998) Adsorción-degradación de 2,4-D en Suelo Agrícola. VIII Congreso Nacional de Irrigación, Gómez Palacio, Dur., Mexico, p. 5.20-24.

Hellyer GM and Balog GE (1999) Derivation, Strengths and Limitations of Sediment Ecotoxicological Screening Benchmarks (ESB´s). USEPA-New England Regulatory Laboratory, Lexington, MA, 13 p.

Korte F (1992) Lehrbuch der Ökologischen Chemie. Georg Thieme Verlag, Stuttgart, pp 257-277.

Lantzy JR and MacKenzie TF (1979) Atmospheric Trace Metals: Global Cycles and Assessment of Man's Impact. Geochim Cosmochim, Acta, 43, 511-525.

Moore JW and Ramamoorthy S (1984) Organic Chemicals in Natural Waters – Applied Monitoring and Impact Assessment. Springer-Verlag, Inc, NY, 289 p.

Mueller JG, Cerniglia CE and Pritchard PH (1996) Bioremediation of Environments Contaminated by Polycyclic Aromatic Hydrocarbons *In* Bioremediation. Principles and Applications (Crawford RL and Crawford DL eds) Cambridge University Press, Cambridge, pp 125-194.

NOM-127-SSA1-1994 (1996) Norma Oficial Mexicana. Salud Ambiental, Agua para Uso y Consumo Humano – Limites Permisibles de Calidad y Tratamientos a que debe Someterse el Agua para su Potabilización *In* Diario Oficial de la Federación, January 18, 1996, First Section, pp. 41-46.

Papelis C, Hayes KF and Leckie JO (1988) Hydraql: A program for the Computation of Chemical Equilibrium Composition of Aqueous Batch Systems Including Surface-Complexation Modeling of Ion Adsorption at the Oxide/Solution Interface. Technical report No 306, Department of Civil Engineering, Stanford University, CA, 134 p.

Persaud D, Jaagumagi R and Hayton A (1993) Guidelines for the Protection and Management of Aquatic Sediment Quality in Ontario. Ontario Ministry of Environment and Energy Report, 30 p.

Posada-Ramos ME, Hansen AM and Ortega-Lara V (1994) Enriquecimiento de Metales en Sedimentos de la Cuenca Lerma-Chapala. Memorias. 4° Congreso Nacional de Geoquímica, Jiutepec, Mor, Mexico, ISBN 968-7417-01-3, 151-156.

Rozan TF and Benoit G (1999) Geochemical Factors Controlling Free Copper Ion Concentration in Rivers. Geochim Cosmochim, Acta 63, 3311-3319.

Shine JP, Ryan DK and Ford TE (1998) Annual Cycle of Heavy Metals in a Tropical Lake – Lake Chapala, Mexico. J Environ Sci Health, A33 (1), 23-43.

The Netherlands-List (1994) Interventiewaarden Bodemsanering, Staatscourant 95 v. 24. Mai, Nr DBO-07494013 v Directorat Generaal Milieubeheer.

Wester P, Melville R and Ramos S (2001) Institutional Arrangements for Water Management in the Lerma-Chapala Basin *In: The Lerma-Chapala Watershed, Evaluation and Management* (Hansen AM and van Afferden M eds) Kluwer Academic/Plenum Publishers, London.

WHO (1996) Guidelines for Drinking-water Quality. 2nd Ed Health criteria and other supporting information, Geneva, World Health Organization, pp. 940-949.

WHO (1998) Guidelines for Drinking-water Quality. 2nd Ed Addendum to Vol 2, Geneva, World Health Organization, pp. 281-283.

Wilson IF, Milton C and Houston JR (1950) A Mineralogical Study of the Guanajuato, Mexico, Silver-gold Ores. US Geological Survey Open-File Report, 15-70, 12 p.

Chapter 5

Water Concentrations, Bioaccumulation, and Human Health Implications of Heavy Metals in Lake Chapala

JENNY AYLA JAY[1] AND TIM E. FORD[2]
Harvard School of Public Health. 665 Huntington Avenue Boston MA 02115, USA. email: [1]*jennyayla@yahoo.com;* [2]*ford@endor.harvard.edu*

Keywords: Metals, drinking water, fish consumption, bioaccumulation, biomagnification

Abstract: Inorganic contaminants in Lake Chapala may pose risks to human health through exposure routes such as drinking water, recreation, and fish consumption. Accurately assessing these risks is difficult, as the fate and transport of metals in surface waters is governed by multiple processes. In this chapter, data for metals in Lake Chapala water, plants, and fish are reviewed, and risks to human health due to fish consumption and drinking water are evaluated. At times during the year, the concentrations of Cd and Pb were above USEPA Maximum Contaminant Levels for safe drinking water, indicating a need for monitoring. Levels of dissolved arsenic in water were consistently above recommended guidelines for safe consumption of water. Data on metal concentrations in fish point to a need for monitoring mercury in all edible species of fish from Lake Chapala, and suggest elevated Cu and Pb levels in fish tissue as well. Suspended clay particles have been shown to be important in metal accumulation in Lake Chapala, due to their high affinity for metals. To more accurately characterize the Lake, we need a more comprehensive monitoring program for toxic metals in the sediments, the water column (dissolved and particulate) and within the biota, and an assessment of population exposures.

1. INTRODUCTION

Historic and current waste inputs to Lake Chapala are suspected to have resulted in elevated levels of some contaminants in water, sediments, and aquatic life. Sources of toxic elements to the lake are both natural and anthropogenic, and include geological sources such as hot springs,

The Lerma-Chapala Watershed, Edited by Hansen and van Afferden,
Kluwer Academic/Plenum Publishers, New York, 2001

agricultural and urban run-off, domestic and industrial wastes, and atmospheric deposition. Specific potential anthropogenic metals sources in this watershed include battery manufacturing, copper products, paint and paint additives, electrical transformers, synthesis of petroleum products, iron and bronze foundries, and crude petroleum refining (Ford *et al.*, 2000).

Even if contaminant inputs can be quantified, accurate assessment of risk of metal exposure to humans and ecosystems is difficult, as the fate and transport of metals in surface waters is governed by multiple physical, biological, and chemical processes. Metals are subject to abiotic and biologically mediated transformations, and each form is involved in partitioning between solid, dissolved and gas phases. Contaminants may sorb to particles in the water column and settle to the sediment, only to be remobilized due to sediment resuspension or dissolution due to chemical changes in the overlying water (Hansen and van Afferden, 2001). Exchange with the atmosphere and deposition in precipitation are especially important in the case of mercury, which has a stable gaseous form. The cycling of a particular metal will determine its potential to adversely affect ecosystem and human health via various exposure routes.

Main pathways for human exposure to metals in Lake Chapala exist through drinking water, recreation, and the consumption of fish. Contaminants in the dissolved phase of the water column will affect the quality of drinking water. Both dissolved and particulate phase contaminants in the water column, as well as sediment contamination, may result in recreational exposures. Due to bioaccumulation of an element throughout an organism's lifetime, and biomagnification of the contaminant up the food chain when contaminated organisms are consumed, the concentrations of a toxic element in aquatic biota may be many times higher than levels in the surrounding water. For example, the level of mercury in fish can be as high as one million times more concentrated than that in the surrounding water (USEPA, 2001a). Thus, ingestion of fish may be a major exposure route for contaminants with a high tendency to biomagnify. In this chapter, we will review the data available for metals in Lake Chapala water, plants, and fish, and provide preliminary risk assessment calculations to evaluate risks to human health due to consumption of fish and drinking water.

2. METALS IN LAKE CHAPALA WATER AND IMPLICATIONS FOR HEALTH

In this section, we will describe the temporal and spatial variability observed in Lake Chapala surface water metal concentrations and possible health risks due to use of the water as a drinking water source. Figure 1 illustrates a simplified conceptual model for the cycling of metals in Lake Chapala. Because the lake is shallow, constant resuspension of small particles of clay, together with sorption of dissolved organic compounds and bacteria to the clay, results in clay-organic-bacteria aggregates (COBA) in the water column (Lind and Dávalos-Lind, 1999; Lind *et al.*, 1997), which can be important in the sorption of metals (discussion following). COBA may be consumed by macro-zooplankton (Arruda *et al.*, 1983) and some fish (Dávalos-Lind *et al.*, 1992), possibly leading to increased metal accumulation in fish. Dissolved metals may be taken up by algae and macrophytes, which may in turn be grazed by zooplankton or fish.

Precipitation during the rainy season (July through September) provides the major portion of freshwater to the lake. Evaporation extraction result in the largest outflow from the lake (Ford and Ryan, 1992). Wind stress over such a shallow water column maintains vertically well-mixed conditions throughout the year.

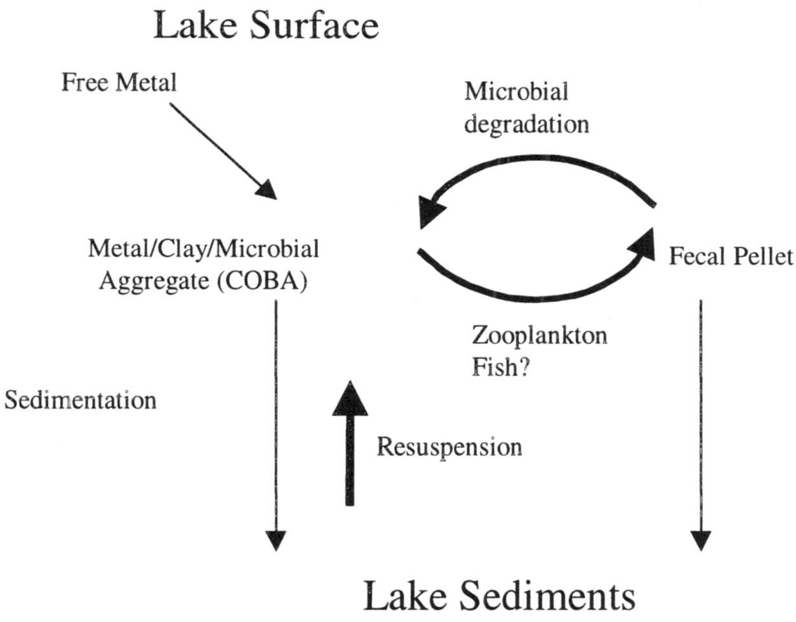

Figure 1 Simplified conceptual model of metal cycling in Lake Chapala

Ford and Ryan (Ford and Ryan, 1992 and 1995; Shine *et al.*, 1998b) studied the dissolved (< 0.4 μm) concentrations of seven metals: Arsenic (As), lead (Pb), cadmium (Cd), zinc (Zn), copper (Cu) and chromium (Cr)

every month over a one-year period, at six locations (Figure 2). Sampling sites were chosen to represent various areas within the lake. The easternmost site within the lake, Site 26, is located near the Santiago river outlet, representing the final stage of lake circulation (at the time these studies were conducted and prior to operation of the present water extraction site). Water is turbid at this site, and sediments are deep and flocculent. Water at Site 15, at the center of the lake, is much less turbid, and the sediments are sandier, because prevailing winds drive currents that minimize flocculent sediment at this region. Site 3, at the western end of the lake, has relatively less turbid water than Site 26, but also has deep, flocculent sediments.

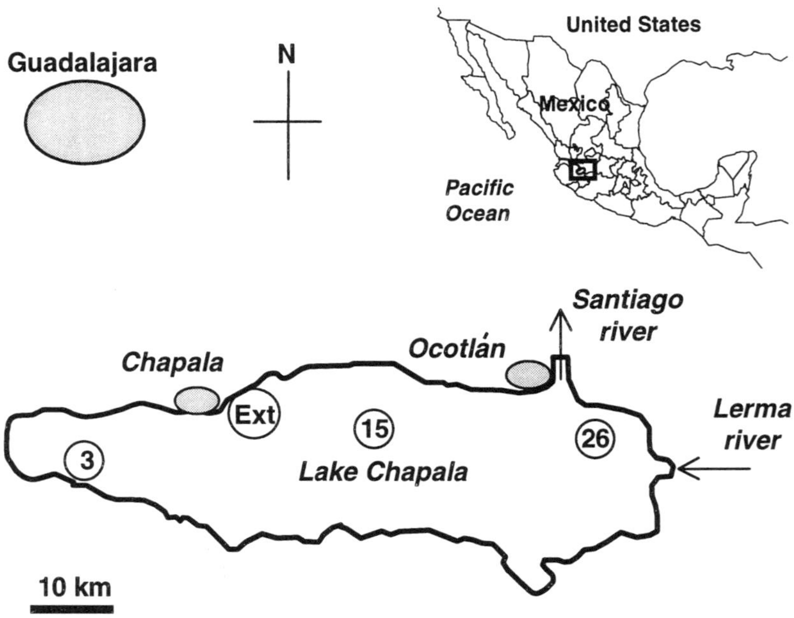

Figure 2 Lake Chapala sampling locations (3, 15, 26, and Ext). Freshwater flows into Lake Chapala through the Lerma river and out through the Santiago river at the time of sampling. The site labeled "Ext" is the municipal water extraction point supplying drinking water to the city of Guadalajara

A water extraction site (labeled Ext) in the lake, just East of the town of Chapala, was also sampled. Due to dredging near the time of sampling, the sediments here were compacted sand. The Lerma river, the major input to Lake Chapala, was sampled at a site that ensured mixing from all major tributaries of the river. The Santiago river, an outlet from the lake, was sampled upstream of Ocotlán.

Table 1 provides a summary of lake water dissolved metal data and the USEPA Freshwater Chronic Criteria for the protection of ecosystem health. Table 2 provides USEPA drinking water standards and toxicological information for the same elements. In general, most metals displayed a concentration peak in April, May, and June, when the lake level is at its lowest, which could possibly indicate evaporative concentration. In August and September, after the onset of the rainy season, some metal concentrations displayed second peaks, thought to be due to increased input from the watershed, or resuspension of metals within the lake. Metal concentrations were lowest from October to April, when the lake level was at its highest.

Table 1 Summary data for dissolved metals in Lake Chapala (Stations 3, 15, 26, and Ext.) and the surrounding rivers (Lerma and Santiago) in samples collected monthly between November 1990 and December 1991. The values shown are the means (standard deviation) in μg L^{-1}. USEPA Chronic Criteria are based on a mean water hardness of 148 mg L^{-1} (Shine *et al.*, 1998b; Limón *et al.*, 1992)

Me-tal	April-June		July		August-September		October-March		US EPA chronic criteria
	Rivers	Lake	Rivers	Lake	Rivers	Lake	Rivers	Lake	
As	29 (18)	52 (3.0)	8.3 (5.2)	27 (14)	9.0 (4.1)	34 (14)	21 (14)	26 (10)	190
Cr	2.3 (1.4)	2.1 (1.1)	4.7 (0.9)	6.0 (2.1)	1.7 (1.9)	2.1 (2.1)	1.3 (1.5)	2.6 (2.8)	250
Cd	4.3 (2.1)	5.7 (1.5)	0.4 (0.5)	1.8 (1.1)	1.0 (0.1)	1.0 (0.1)	0.6 (0.8)	0.8 (0.7)	1.4
Pb	17 (5.3)	28 (9.5)	9.0 (6.0)	13 (4.8)	9.2 (1.7)	14 (4.0)	3.1 (4.3)	6.8 (7.1)	3.8
Zn	34 (10)	39 (21)	9.3 (4.2)	12 (4.7)	19 (15)	28 (16)	30 (32)	23 (23)	120
Ni	5.3 (3.9)	6.2 (4.3)	0.3 (0.4)	4.4 (5.4)	35 (27)	39 (24)	20 (20)	12 (10)	220
Cu	0.7 (0.8)	0.6 (0.9)	3.2 (2.6)	<1.0	6.2 (1.0)	12 (7.3)	4.4 (5.8)	2.9 (2.3)	16

Arsenic. Concentrations of As peaked twice during the year, just before and after the onset of the rainy season. During the earlier peak, the dissolved As concentrations in the Lerma river were much lower than those in the lake or the Santiago river. Possible explanations for this could be that As was introduced to the lake from hot springs known to exist on the northern shore of the lake, or from reversal of the Santiago river. During the second peak, the As concentrations increase along a gradient from West to East, and are higher in the lake than in either of the river sites, supporting the existence of an internal source of As. Ford and Ryan (Ford and Ryan, 1992) measured As in hot spring-fed streams at the Northwestern end of the lake of 90 μg L^{-1}.

Table 2 Maximum Contaminant Level Goals (MCLGs), Maximum Contaminant Levels (MCLs), and toxicity information for inorganic contaminants in drinking water (USEPA, 2001b)

Metal	MCLG[1] (μg L^{-1})	MCL[2] (μg L^{-1})	Potential health effects from exposure above the MCL
As	NA	50	Cancer of bladder, lungs, skin, kidney, nasal passages, liver and prostate Non cancerous effects to cardiovascular, pulmonary, immunological, neurological and endocrine systems
Cr	100	100	Short-term: Allergic dermatitis Long-term: Kidney, liver, nerve tissue damage
Cd	5	5	Short-term: Gastrointestinal distress, liver injury Long-term: Kidney, liver, blood and bone damage
Pb	0	15	Children: Delays in physical or mental development. Adults: Stroke and kidney disease, cancer
Zn	NA	NA	Above 3 mg L^{-1}, taste and appearance of water is affected
Ni	100	100	Short-term: No known effects Long-term: Heart and liver damage, skin irritation
Cu	1,300	1,300	Short-term exposure: Gastrointestinal distress Long-term exposure: Liver or kidney damage

[1]MCLG: The level of a contaminant in drinking water below, which there is no known or expected risk to health. MCLGs allow for a margin of safety and are non-enforceable public health goals

[2]MCL: The highest level of a contaminant that is allowed in drinking water. MCLs are set as close to MCLGs as feasible using the best available treatment technology and taking cost into consideration. MCLs are enforceable standards

While the dissolved As concentrations in the lake and river are below the USEPA Freshwater Criteria for the protection of ecosystems (190 μg L^{-1}), they are above the WHO drinking water standard of 10 μg L^{-1}, and they approach or exceed the USEPA drinking water standard of 50 μg L^{-1}. In 1999, the U.S. National Research Council found that the USEPA's MCL for As of 50 μg L^{-1} does not provide adequate public health protection, and should be revised (NRC, 1999). The World Health Organization estimates the excess lifetime cancer risk associated with ingesting water with 10 μg L^{-1} As to be $6x10^{-4}$ (WHO, 1993). Thus, treatment of extracted water for As may by required for the lake to provide a safe drinking water supply.

A recent USEPA report describes a number of treatments to remove As in drinking water, including precipitative processes such as coagulation/filtration, adsorptive processes, ion exchange, membrane processes such as reverse osmosis, and alternative technologies including photo-oxidation (USEPA, 2000).

Cadmium and Lead. Similar to As concentrations, Cd and Pb concentrations showed peaks just before the rainy season. This is consistent with the hypothesis that evaporative concentration is responsible for the observed peaks. A second peak was not observed after the onset of the rainy season.

During the dry season, concentrations of Cd and Pb (5.7 and 28 $\mu g\ L^{-1}$) exceed the USEPA Freshwater Criteria (1.4 and 3.8 $\mu g\ L^{-1}$, respectively) and the Maximum Contaminant Levels (5 and 15 $\mu g\ L^{-1}$, respectively). Exposure to Cd results in short-term effects such as nausea, vomiting, and diarrhea, while long-term effects include kidney, liver, bone and blood damages. Water treatment methods that have been approved by the USEPA for removal of Cd include coagulation/filtration, ion exchange, lime softening, and reverse osmosis. The Risk Reference Dose for Cd established by the USEPA is 70 μg Cd day^{-1} (Dudka and Miller, 1999). Drinking 2 L day^{-1} of untreated Lake Chapala water during the dry season, would result in an intake of 12 μg Cd day^{-1}.

Pb exposure can lead to delays in normal physical and mental development in babies and young children. Over the long-term, Pb exposure can cause stroke and kidney disease, and cancer. Although concentrations of Pb should be monitored in drinking water, it should be noted that exposures to Pb through contaminated soils or Pb-based paint are probably more significant (USEPA, 2001c).

Chromium. Cr concentrations rose and fell throughout the wet and dry seasons, but were well below the USEPA Freshwater criteria of 250 $\mu g\ L^{-1}$ and drinking water MCL of 100 $\mu g\ L^{-1}$.

Zinc. Zn concentrations were quite variable, and may be a result of contamination (a major problem in the analysis of Zn); however, all measured concentrations were well below the USEPA freshwater chronic criteria of 120 $\mu g\ L^{-1}$. A drinking water MCL for Zn has not been developed.

Nickel and Copper. In contrast to the other metals, Ni and Cu concentrations did not peak during the months preceding the rainy season. Peaks occurred after the onset of the rainy season, likely due to increased input or resuspension due to higher flow in the Lerma river. All of the observed Ni concentrations were below the USEPA freshwater chronic criteria of 220 $\mu g\ L^{-1}$; however, the standard for Cu (16 $\mu g\ L^{-1}$) was exceeded during the peak. All measured values were well below the drinking water standards for Ni and Cu (100 and 1,300 $\mu g\ L^{-1}$, respectively).

3. METALS IN LAKE CHAPALA BIOTA AND IMPLICATIONS FOR HEALTH

3.1 Metal accumulation in plants

Metal accumulation in aquatic plants such as the water hyacinth can lead to magnification of metal in the food chain (Kay and Haller, 1986). Metal concentrations in water hyacinth (*Eichornia crassipes*) from Lake Chapala, collected in March and November 1990 are presented in Table 3 (Shine *et al.*, 1998b). Samples of both leaves and roots were collected from the Lerma river, Station 15, and the Santiago river and measured for As, Cr, Cd, Pb, Zn, Ni, and Cu. In all cases, the plant roots contained a higher concentration of metals than the leaves. Plants collected in November had higher metal concentrations than those collected in March, presumably due to the age of the plant. Metal concentration in plants occurred to a high degree. For example, in the November Santiago river sample, Cr was concentrated 22,200 times in *Eichornia* roots compared to the water column (μg g^{-1} root (μg mL^{-1} water)$^{-1}$).

Table 3 The concentration (dry weight) of metals in the leaves and roots of water hyacinth plants (Eichornia crassipes) collected in the Lerma river in March 1990 and in the Lerma and Santiago rivers, and Lake Chapala (Station 15) from November, 1990 (Shine *et al.*, 1998b)

Metal	Lerma river March 1990		Lerma river November 1990		Station 15 November 1990		Santiago river November 1990	
	Leaves (μg g^{-1})	Root (μg g^{-1})	Leaves (μg g^{-1})	Root (μg g^{-1})	Leaves (μg g^{-1})	Root (μg g^{-1})	Leaves (μg g^{-1})	Root (μg g^{-1})
As	<13	<10	<10	23	<10	<10	<11	<10
Cr	1.4	11	1.3	25	<1	8.5	1.7	33
Cd	<1	<1	<1	6.4	<1	1.1	<1	3.1
Pb	4.0	13	3.9	39	<3	7.7	7.0	15
Zn	21	33	24	110	26	44	48	84
Ni	<2	9.0	<2	28	1.5	9.2	<2	17
Cu	4.3	11	2.3	28	6.2	14	9.1	25

Recent analyses of bulrushes (*Typha latifolia*) from Lake Chapala have shown mercury (Hg) concentration in this species of up to 15 μg g^{-1} of plant material (dry weight) (Burton, 1997).

3.2 Metal accumulation in fish

The Lake Chapala fishery is important both economically and as a food source for surrounding communities (Ford *et al.*, 2000). Three fishes in this lake are important food sources: the primarily planktivorous charal (*Chirostoma* spp.), and the omnivorous tilapia (*Oreochromis mozambique*, Family Cichlidae) and common carp (*Cyprinus carpio*, Family Cyprinidae) (Lind *et al.*, 2000).

In 1990, metal concentrations (As, Pb, Cd, Zn, Ni, Cu, and Cr) in tilapia and carp muscle and liver tissue collected from the city of Chapala and from the Santiago river were determined (Table 4). Most fish collected had fin erosion and parasitic worms, suggesting a stressed environment. Cu concentrations in tilapia livers were approximately 2240 $\mu g\ g^{-1}$, much higher than the action level of 30 $\mu g\ g^{-1}$ recommended by the Australian National Health and Medical Research Council for fish tissue (Ahmed, 1991). Pb concentrations in liver and muscle tissue of both tilapia and carp ranged from 0.9 to 74 $\mu g\ g^{-1}$, greatly exceeding action levels suggested by the Canadian Health Protection Board (0.5 $\mu g\ g^{-1}$) and the Australian National Health and Medical Research Council (2.0 $\mu g\ g^{-1}$) (Ahmed, 1991).

Table 4 The concentration (dry weight) of metals Tilapia and Carp liver and muscle tissue collected in the city of Chapala and from the Santiago river in November 1990. The results shown are the mean of three individuals of each species (Shine *et al.*, 1998b)

Metal	Tilapia Chapala		Carp Chapala		Tilapia Santiago river		Carp Santiago river	
	Liver ($\mu g\ g^{-1}$)	Muscle ($\mu g\ g^{-1}$)	Liver ($\mu g\ g^{-1}$)	Muscle ($\mu g\ g^{-1}$)	Liver ($\mu g\ g^{-1}$)	Muscle ($\mu g\ g^{-1}$)	Liver ($\mu g\ g^{-1}$)	Muscle ($\mu g\ g^{-1}$)
As	<1	NA	<1	NA	<1	NA	<1	NA
Cr	7.9	0.28	3.6	<0.1	6.0	0.51	0.81	0.19
Cd	0.5	NA	1.2	NA	1.1	NA	0.21	NA
Pb	7.8	4.1	12	4.8	33	6.1	2.5	3.2
Zn	105	36	1090	180	131	30	309	59
Ni	1.8	0.26	4.9	1.9	10	0.34	1.3	0.19
Cu	2260	8.0	131	3.9	2220	3.0	13	2.2

NA = Not analyzed

A study conducted in 1996 determined the concentrations of Cd, Cu, Ni, Pb, Zn, and Hg in charal (a mix of species of the Genus *Chirostoma*) from three locations in Lake Chapala. The goals of the study were to measure and compare metal concentrations in 1) freshly caught charal from the East end of the lake, Ocotlán (the region with the most highly contaminated sediments), 2) freshly caught charal from the West end of the lake, Ajijic, 3) fresh charal purchased at a Chapala fish market (obtained from western and central portions of the lake), and 4) fried fish from the same market (Ford *et al.*, 2000).

All metals with the exception of Hg were present at levels below US and other international guidelines (Table 5). Fish tissue concentrations of Hg ranged from 0.2 to 8 μg g^{-1} (dry weight), corresponding to between 0.05 and 1.84 μg g^{-1} (wet weight). Ocotlán fish had the highest levels of Hg; the average concentration in fish from this site was 1.1 μg g^{-1} (wet weight, Ford *et al.*, 2000). The USEPA recommends a tissue residue criterion of 0.3 μg g^{-1} (USEPA, 2001d). However, recent evidence points toward the necessity of even lower levels to ensure the protection of public health (Dickman *et al.*, 1998).

Table 5 Concentrations of whole fish samples of *Chirostoma* sp. from different regions of Lake Chapala, Mexico (Ford *et al.*, 2000)

Metal	Concentration in fish at each collection site μg g^{-1} dry weight (standard deviation)			
	Ajijic, n = 5	Ocotlán, n = 6	Market, n = 10	Fried, n = 5
Cd	0.01 (0.001)	0.021 (0.03)	0.059 (0.059)	0.01 (0.004)
Cu	0.92 (0.2)	1.2 (0.21)	0.83 (0.12)	0.76 (0.02)
Ni	0.82 (0.4)	0.65 (0.17)	0.60 (0.12)	0.21 (0.03)
Pb	0.067 (0.04)	0.13 (0.03)	0.47 (0.17)	0.55 (0.27)
Zn	123 (36)	159 (31)	96 (18)	50 (8.2)
Hg	0.74 (0.8)	4.9 (1.7)	1.4 (0.3)	0.70 (0.21)

In general, fried fish contained lower concentrations of metals than did fresh fish; this difference was significant for Cd, Ni, Zn, and Hg. Possible contributing factors include dilution from the batter, loss of metals to the frying oils, or volatilization.

3.3 Importance of clay in metal accumulation

COBA have been shown to be important in metal speciation and accumulation in Lake Chapala, due to their high affinity for metals (Lind *et al.*, 2000). For example, Cr concentrations in the COBA were 35,000 times that dissolved in the water, equivalent to 79% of the Cr per volume of water Ford and Ryan, 1995). Macro-zooplankton or small fish may then graze the COBA, possibly resulting in magnification of the metal concentration in the food chain Lind *et al.*, 2000).

Mesocosm experiments in which Cu and Zn accumulation by fish was compared in the presence and absence of clay showed different results for tilapia and charal (Table 6). Cu and Zn concentrations in tilapia tissue were decreased in the presence of clay, indicating that sorption of metals to COBA (which are not consumed by tilapia) could have a protective effect on metal accumulation in this fish. Zn concentrations in charal tissue were not decreased in the presence of clay, presumably because these small fish consume COBA. In this experiment, it appeared that zooplankton did not provide a pathway for metal accumulation in fish, possibly because they were not consuming COBA Lind *et al.*, 2000).

Table 6 Concentration (μg g^{-1} dry weight) of Zn in the tissues of Tilapia and Charal after six weeks exposure in the presence and absence of suspended clay Lind *et al.*, 2000)

Exposure concentration (μg L^{-1})	Tilapia		Charal	
	Clay	No Clay	Clay	No Clay
300	249 (37)	405 (82)	28 (6)	29 (10)
600	396 (78)	546 (181)	32 (19)	26 (6)

3.4 Risk calculations due to fish consumption

The WHO has set the tolerable intake level for Hg to be 33 μg day^{-1}. This amount of Hg can easily be consumed in one meal based on conservative estimates. A plateful of fried charal consisting of 10 to 15 fish, each with a dry weight of 3 g, could contain between 20 and 30 μg Hg, if the mean Hg content of 0.7 μg g^{-1} dry weight for market fried fish is assumed. In Ocotlán, where Hg levels were highest, fresh fish contained an average of 5 μg Hg g^{-1} dry weight. A 50% reduction in concentration through frying would correspond to 75-113 μg Hg in a plateful of fried charal, well above the WHO tolerable daily level.

If a person weighing 65 kg consumed three platefuls of fried charal from Ocotlán per week, the average daily dose would be 0.75 μg Hg kg^{-1} day^{-1} due to charal consumption alone. This would result in a hazard quotient of 7.5, as the USEPA reference dose for chronic exposure to methyl mercury (the predominant form of Hg in fish) is 0.1 μg Hg kg^{-1} day^{-1} (USEPA, 1997). A hazard quotient > 1 indicates risk of adverse health effects over a lifetime of exposure. In the urban area of Alta Floresta in the Amazon Basin, hazard quotients from fish consumption were calculated to be 2 and 22 for the general public and adult fishermen, respectively (Hacon *et al.*, 1997).

Hg content in a particular fish species would be expected to depend on the feeding behavior of the species. Because Hg concentrations magnify up the aquatic food chain, piscivorous fish generally have higher Hg levels. Lacerda and co-workers (Lacerda *et al.*, 1994) measured Hg concentrations in fish muscle in the Carajas mining region of up to 0.14, 0.31, and 2.3 μg g^{-1} wet weight, in detritivores, herbivores, and carnivores, respectively. Because charal are herbivores/detritivores, higher Hg levels would be expected in carp and tilapia, which are omnivorous species.

4. CONCLUSIONS AND RECOMMENDATIONS FOR FUTURE RESEARCH

Available metal data from Lake Chapala indicate the need for monitoring levels of As, Cd, and Pb in drinking water extracted from the lake. Levels of dissolved As in water were consistently above recommended guidelines for safe consumption of water. Data on metal concentrations in fish point to a need for monitoring Hg, Pb, and Cu in all edible species of fish from Lake Chapala.

The above assumptions are based on relatively limited data sets obtained from only three sites (for water samples). However, as mentioned in the introduction, fate of contaminated metals in aquatic ecosystems is highly complex. In order to effectively provide a human and ecological risk assessment for Lake Chapala and surrounding populations, including the Guadalajara municipal drinking water supply, we need far more extensive quantification of contaminant concentrations, their potential bioavailability and population exposure routes.

The fate of a contaminant metal discharged to the Chapala ecosystem will depend on a wide range of biotic and abiotic effects. Limited data has shown similar metals concentrations in both sediment samples and COBA samples (unpublished data). To more accurately characterize the lake, we need a more comprehensive monitoring program for toxic metals in the sediments, the water column (dissolved and particulate) and within the biota.

It is likely that sinks exist for metals within sediments at specific locations within the lake, particularly if relatively protected from prevailing winds (embayments, for example). The fate of sediment bound metals will then depend on redox conditions and degree of microbial transformation to more bio-available forms that can then accumulate in the benthos, and, subsequently, in food fishes (Shine *et al.*, 1998b; Ford and Mitchell, 1992). In addition, the fate of these contaminants is closely tied to sediment carbon concentrations. In Lake Chapala, Ford (unpublished data) has measured extremely high concentrations of dissolved organic carbon (50-70 $\mu g\ g^{-1}$) in the Santiago river. It is likely that sediment carbon concentrations are also very high. Mineralisation of that carbon by microbial activity is thought to be important in releasing sediment bound metals (Ford, 1994; Shine *et al.*, 1998a).

The fate of metals such as As and Hg are well known to be closely associated with microbial activity, including transformation to more bio-available forms (e.g., methylation of Hg (USEPA, 2001d). In terms of fundamental research, Chapala presents a unique opportunity to examine metal cycling in a Lake that is dominated by suspended particles, the COBA. These COBA have been shown to be composed of high numbers of microorganisms (Lind and Dávalos-Lind, 1991) and the high organic content should result in formation of anaerobic micro-niches. It is primarily the anaerobic community of microorganisms that methylate Hg to the highly lipid soluble form that rapidly bio-concentrates in consumer species. To our knowledge no attempts have been made to speciate any of the metals associated with COBA, but research in this area would provide an opportunity to improve our understanding of the dynamics of metal cycling in Lake Chapala, and by extrapolation, other turbid lake ecosystems.

Although much remains to be done to improve our knowledge of the fate of toxic metals in lake Chapala, even less research appears to have been conducted to assess population exposures. For drinking water, this may be relatively straightforward, requiring point-of-use sampling and accurate metals analysis, combined with assessment of drinking water consumption patterns and other exposure routes within the home. For fish consumption, recreational and occupational exposures, biomarkers (metals or metabolites produced in the blood, breast milk or urine; metals accumulating in hair or nails, etc.) are often a first approximation of exposure. However, these measurements need to be interpreted cautiously with information carefully collected on fish consumption patterns, or exposure times to lake water, and compared with alternative exposure routes, e.g., drinking water, produce, vehicle exhaust, smoking habits, etc. As a first step, we suggest a population study using non-invasive metal accumulation markers (e.g. hair samples) would be useful in determining temporal trends in exposure to Hg in lake communities. This would be combined with a questionnaire-based study of the many exposure routes.

REFERENCES

Ahmed F (1991) Chemical Health Risk Assessment: Critique of Existing Practices and Suggestions for Improvement In Seafood Safety, National Academy Press, Washington DC, pp 172-266.

Arruda J, Marzolf R and Faulk R (1983) The Role of Suspended Sediments in the Nutrition of Zooplankton in Turbid Reservoirs. Ecology, 64, 1225-1235.

Burton T (1997) Can Mexico's largest lake be saved? An international rehabilitation effort gets under way. Ecodecision, 23.

Dávalos-Lind L, Sada R, Guerra A, Velarde G, Orozco L, Chrzanowski T and Lind O (1992) La Producción Bacteriana y su Importancia en la Cadena Trófica en el Lago de Chapala. Ingeniería Hidráulica en México, 7, 30-37.

Dickman MD, Leung CKM and Leong MKH (1998) Hong Kong Male Subfertility Links to Mercury in Human Hair and Fish. Sci Tot Environ, 214, 165-174.

Dudka S and Miller WP (1999) Accumulation of Potentially Toxic Elements in Plants and their Transfer to the Human Food Chain. Environ Sci Health, Part B, B34, 681-708.

Ford T (1994) Pollutant Effects on the Microbial Ecosystem. Environ Health Perspect, 102 (suppl. 12), 45-48.

Ford TE, Ika R, Shine J, Dávalos-Lind L and Lind OJ (2000) Trace Metal Concentrations in *Chirostoma* sp from Lake Chapala, Mexico: Elevated Concentrations of Mercury and Public Health Implications. Environ Sci Health, A35, 313-325.

Ford T and Mitchell R (1992) Microbial Transport of Toxic Metals *In* Environmental Microbiology, Wiley-Liss, Inc. p. 83-101.

Ford T and Ryan DK (1992) Evaluation of Heavy Metals in Water, Sediment and Organisms in Lake Chapala. Final Report to the State of Jalisco, Harvard University, 32 pp.

Ford T and Ryan D (1995) Toxic Metals in Aquatic Ecosystems: A Microbiological Perspective. Environ Health Perspect, 103, Suppl 1, 25-28.

Hacon S, Rochedo ER, Campos RRR and Lacerda LD (1997) Mercury Exposure through Fish Consumption in the Urban Area of Alta Floresta in the Amazon Basin. J Geochem Explor, 58, 209-216.

Hansen AM and van Afferden M (2001) Toxic Substances: Sources, Accumulation, and Dynamics *In: The Lerma-Chapala Watershed: Evaluation and Management* (Hansen AM and van Afferden M eds) Kluwer Academic/Plenum Publishers, London.

Kay SH and Haller WT (1986) Heavy Metal Bioaccumulation and Effects on Water Hyacinth Weevils, *Neochetina eichhorniae*, Feeding on the Water Hyacinth, *Eichhornia crassipes*. Bull Environ Contam Toxicol, 37, 239-245.

Lacerda LD, Bidone ED, Guimaraes AF and Pfeiffer WC (1994) Mercury Concentrations in Fish from the Itacaiunas-Parauapebas River system, Carajas region, Amazon. An Acad Bras Ci, 66, 373-379.

Limon JG, Lind OT, Vodopich DS, Doyle R and Trotter BG (1989) Long- and Short-term Variation in the Physical and Chemical Limnology of a Large, Shallow, Turbid Tropical Lake (Lake Chapala, Mexico). Arch Hydrobiol Suppl, 83, 57-83.

Lind O and Dávalos-Lind L (1991) Association of Turbidity and Organic Carbon with Bacterial Abundance and Cell Size in a Large, Turbid, Tropical Lake. Limnol Oceanog 1991, 36, 1200-1208.

Lind O and Dávalos-Lind L (1999) Suspended Clay: Its Role in Reservoir Productivity *In* Theoretical Reservoir Ecology and Its Applications. (Tundisi J and Straskraba M eds) Backhus, Leiden.

Lind OT, Chrzanowski TH and Dávalos-Lind L (1997) Clay Turbidity and the Relative Production of Bacterioplankton and Phytoplankton. Hydrobiologia 353, 1-18.

Lind OT, Dávalos-Lind L and Ford TEJ (2000) Clay and the Movement of Metals into Food Chains. Environ Sci Health, A35, 1171-1182.

NRC (National Research Council) (1999) Arsenic in Drinking Water, National Academy Press, Washington DC.

Shine JP, Ika R, Ford TE (1998a) Relationship Between Oxygen Consumption and Sediment-Water Fluxes of Heavy Metals in Coastal Marine Sediments. Environ Toxicol Chem, 17, 2325-2337.

Shine JP, Ryan DK and Ford TE (1998b) Annual Cycle of Heavy Metals in a Tropical Lake-Lake Chapala, Mexico. J Environ Sci Health, A33, 23-43.

USEPA (1997) Methylmercury (MeHg). Integrated Risk Information System. CASRN 22967-92-6. Washington DC

USEPA (2000) www.epa.gov/safewater

USEPA (2001a) www.epa.gov/waterscience/criteria/methylmercury/mercappa.pdf

USEPA (2001b) www.epa.gov/safewater

USEPA (2001c) www.epa.gov/opptintr/lead/index.html

USEPA (2001d) www.epa.gov/waterscience/criteria/methylmercury/merc68.pdf

WHO (World Health Organization) (1993) Guidelines for drinking-water quality Vol. 1, Recommendations, Second ed, Geneva

SECTION III

LAKE CHAPALA

Chapter 6

An Introduction to the Limnology of Lake Chapala, Jalisco, Mexico

OWEN T. LIND[1] AND LAURA DÁVALOS-LIND[2]
[1]*Limnology Laboratory, Department of Biology, Baylor University. Waco, Texas, 76798-7388, USA. email: owen_lind@baylor.edu.*
[2]*Chapala Ecology Station (Univ. Autónoma de Guadalajara and Baylor University), PO Box 97388, Waco, Texas 76798-7388, USA. email: laura_davalos-lind@baylor.edu*

Key words: Limnology, Lake Chapala, physical properties, chemical properties, biotic properties

Abstract: This chapter describes some of the physical, chemical, and biological properties of Lake Chapala. It brings together descriptive data from several sources and provides a reference for the more specific papers, which follow. Lake Chapala, located at the western edge of the neo-volcanic zone of Mexico, is a tectonic lake of great age and is one of the oldest lakes in North America. At approximately 1000 km^2 it has the greatest area of any Mexican lake. Never the less, because of its' age and because of water extraction, it is extremely shallow and, with a relative depth of < 0.02%, it is one of the most shallow of the world's large lakes. The water budget consists of inputs principally from two sources, the Lerma river and direct surface rainfall. The outputs are to the city of Guadalajara's domestic water supply, to the Santiago river and evaporation. The proportions of these input and output sources has changed dramatically in the past 25 years. The Lerma river's input is reduced from approximately two-thirds of the total to approximately one-half. Output to the Santiago river has stopped as water demands by Guadalajara increased. The lake is elongate lying in a rift valley with an east to west orientation. Consequently wind action is principally along the long axis with the generation of large waves and complete mixing. The prevailing East to West winds generates currents capable of transporting materials entering via the Lerma river along the south shore with a return in the mid-lake. The east end, near the river delta, has extensive areas less than one-meter depth for much of the year (the annual rainy season – dry season depth change averages approximately 0.8 m). In attempts to lessen water loss to evaporation, significant areas have be isolated by dykes and converted to agricultural land. Sediments resuspended from the shallow area are transported along the south shore with return gyres evident in space images and photography. These suspended clays are a major forcing factor in determining the lake communities' functions. The mean annual lake-wide Secchi depth is 0.5 m. This varies from 20 cm at the shallow east end to 70 cm at mid-lake. The mean coefficient of PAR extinction averages 2.3 m^{-1}. Transparency is rarely, and only in the shallowest regions, influenced by algae mass. Lake wide, inorganic suspended solids average 18 mg L^{-1}. Because of the shallowness and wind mixing, the lake does not seasonally stratify, but like other tropical lakes, can stratify on a daily basis with appropriate climatic conditions. The mean annual water temperature is approximately 22 °C with an annual seasonal range of only 21 to 23 °C. The water is moderately hard (TH = 150 mg L^{-1}) and alkaline (TA = 185 mg L^{-1}, pH = 8.7). Because of the soil properties of this volcanic region and the inputs of anthropogenic materials from the Lerma river and villages surrounding the lake, the total phosphorus concentrations are high (0.4-0.5 mg L^{-1}). Most of this (≈90%) is SRP. Total nitrogen is of a similar concentration (0.5-0.8 mg L^{-1}) and total inorganic nitrogen is 0.1-0.4 mg L^{-1}. With the exception of nitrogen, chemical elements follow a pattern of dilution with the rainy season. Nitrogen, with much supplied in rainwater, tends to increase at that time. The nature of the lake's biota is less well known than that of the chemistry. Species lists are incomplete for most taxa. And the functional ecology of most communities is poorly understood. Undoubtedly the very low transparency is a major determinant of the structure and function of the biota. For example, macrophyte communities are limited to either floating or emergent forms. Cattail (*Typha angustifolia*) is abundant in the shallow waters of the eastern region. Water hyacinth (*Eichornia crassipes*) can be abundant with a cycle associated with water level fluctuations. Recently and in 1989, scientists from SIAPA studied phytoplankton taxonomy in relation to water treatment problems. The occurrence of cyanobacteria appears to have increased in the past decade. Perhaps because of the high nitrogen content of the water, cyanobacterial nitrogen fixation does not occur. Phytoplankton biomass as chlorophyll is generally low (<10 µg L^{-1}) over most of the lake but can be high (>25 µg L^{-1}) in shallow regions with a favorable light climate. Phytoplankton photosynthesis in the water column is low. Because of the light-limited phytoplankton photosynthesis, bacterioplankton production is unusually important is the formation of small grazable particles as the base of the food web. The lake's zooplankton has been studied taxonomically, but poorly with regard to community function. The benthos is less well known either taxonomically or functionally. Little of the taxonomic data on zooplankton and benthos has been published in peer-reviewed literature. The lake's fish community is rather well known taxonomically and from a fishery perspective. The importance of the fishery to the local economy has necessitated serious study of the abundance and distribution of harvested fish species. The unusual occurrence of the *Chirostoma* species flock has attracted evolutionary biologists for study of this taxon. The functional role of the fishes in the lake ecosystem has not been investigated, but study of niche partitioning in the *Chirostoma* flock is now underway. Opportunities abound for both descriptive and functional investigations of many lake communities.

1. PHYSICAL PROPERTIES

Lake Chapala between 102° 42' 00"-103° 25' 30" West and 20° 6' 08" - 20° 18' 08" North, is located at the western edge of Mexico's neo-volcanic zone is one of the oldest tectonic lakes in North America. Garduño-Monroy *et al.* (1993) concluded that the Chapala Rift dates from the late Miocene. The present lake dates from the middle Pleistocene although Clements (1963) suggests the possibility of the late Pliocene. The present lake elevation is approximately 1520 m above sea level but terraces on the surrounding mountains provide evidence of a very deep lake, which had considerable climatic variation (de Anda *et al.*, 1998; Barbour 1973). At approximately 1000 km^2 it has the greatest area of any Mexican lake. Nevertheless, because of its age and the water extraction, it is extremely shallow an, with a relative depth of < 0.02%, it is among the shallowest of the world's large lakes.

The water budget consists of inputs principally from two sources– the Lerma river and direct surface rainfall (Limón *et al.*, 1989). The catchment area is 54,400 km^2 of which 43,800 km^2 are in the Lerma river basin. The present outputs are to Guadalajara's domestic water supply and evaporation. The proportions of these input and output sources have changed dramatically in the past 25 years. This has been associated with over-exploitation both up-river and by Guadalajara (de Anda *et al.*, 1998; Limón and Lind, 1990). With the exception of 1993, there has been little or no outflow via the natural drain, the Santiago river, since the 1980s (Filonov *et al.*, 1998). The Lerma river's input is reduced from approximately two-thirds of the total to approximately one-half. For such a large lake the water residence time is short ranging between 1.5 and 2.5 years for extreme wet and dry years respectively (Limón *et al.*, 1989). The climatic processes in the region are extremely complex and variable due to their space and time scales (Filonov *et al.*, 1998). Their data show little, if any change in precipitation and evaporation since 1971 when the lake was at "normal" capacity.

The lake is elongate and lies in a rift valley with an East to West orientation. Because of such fetch and shallowness, mean depth of about 3 m (Table 1), winds, principally along the long axis, generate large waves and complete mixing.

The shallowness is the principal physical feature determining the function of this ecosystem. There is no persistent thermal or chemical (vertical) stratification. Most importantly, the shallowness permits great inorganic turbidity through continuous fine-clay resuspension (Lind *et al.*, 1997). Lake Chapala's morphometric variables at the historical normal (pre-1977) 1524 m, declining (1982-83) 1521 m, and very low 1519 m water level of 1988-89. The East end, near the Lerma river delta, has extensive areas less than one meter deep for much of the year (the annual rainy – dry seasons depth change averages approximately 0.8 m) and consequently is the most turbid. Sediments resuspended from the East shallow area are transported along the South shore with mid-lake return gyres evident in satellite images and photography (Figure 1). The prevailing East to West winds generate currents capable of transporting dissolved and sediment-bound materials entering via the Lerma river along the South shore with a return in mid-lake (Simons, 1980). The water intake for the city of Guadalajara located somewhat West of mid-lake on the North shore significantly affects currents (Desentis, personal communication).

Table 1 Lake Chapala's morphometric variables at the historical normal (pre-1977) 1524 m, declining (1982-83) 1521 m and very low 1519 m water level of 1988-89. (From Lind *et al.*, 1997)

Variable	Surface Elevation (mamsl)		
	1524	1521	1519
Volume (Mm^3)	7962	4667	2700
Area (km^2)	1112	1039	930
Maximum length (km)	77	75	72
Maximum width (km)	22.5	22.5	19
Maximum depth (m)	10.5	7.5	5.5
Mean depth (m)	7.2	4.5	2.9
Relative depth	0.028	0.021	0.016

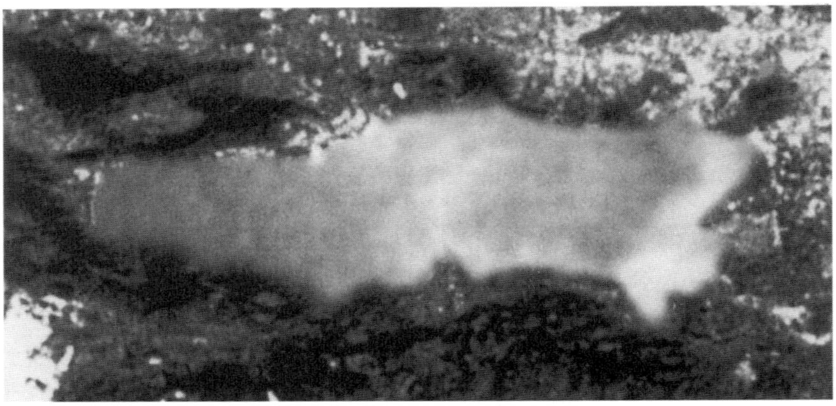

Figure 1 Computer enhanced image of Lake Chapala from Ektachrome photograph taken from the space shuttle showing mid-lake turbid water return gyre originating in the shallow East and southeastern lake regions

Suspended clays are a major forcing factor in determining the functions of communities in the lake (Lind *et al.*, 1992; Lind *et al.*, 1997). Various optical properties for the major lake regions are presented in Table 2. Of these, the most significant relative to community function is the ratio of mixing (i.e. lake depth to photic depth). Lind *et al.* (1992) showed that the critical ratio to enable net phytoplankton accumulation in this lake was two. This ratio only occurs during the dry season at the East end of the lake and a few other shallow shoreline regions. Transparency is rarely influenced by algae mass, and then only in the shallowest regions. The East-end has the greatest variability in optical properties primarily due to the occasional blooms of phytoplankton (see CV, Table 2). Ultraviolet radiation can be significant to the processing of dissolved organic matter. Velarde (1995) measured penetration of UV-B radiation. The mean UV-B radiation attenuation coefficient was great at 10.1 m^{-1}.

Table 2 Mean ± SD, annual range (in parentheses), and CV (%) of optical and physical features of major regions of Lake Chapala. PAR = photosynthetically active radiation, ISS = inorganic suspended solids, Z_{mix}:Z_{eu} = ratio of mixing and euphotic depths (Modified from Lind *et al.*, 1992; Lind *et al.*, 1997)

Secchi depth (m)	PAR attenuation coefficient (m^{-1})	Photic Depth (m)	Turbidity (NTU)	ISS (mg L^{-1})	Z_{mix}:Z_{eu}
Western					
0.6 ± 0.2 (0.4 - 0.9) 33	2.5 ± 0.4 (1.6 - 3.8) 16	1.9 ± 0.4 (1.0 - 2.4) 21	47 ± 48 (24 - 220) 102	38 ± 40 4 - 175 105	2.4 ± 0.6 (1.8 - 4.8) 25
West central					
0.7 ± 0.2 (0.4 - 1.1) 28	2.3 ± 0.5 (1.2 - 3.0) 22	2.1 ± 0.4 (1.4 - 3.4) 19			3.4 ± 0.7 (2.0 - 5.3) 20
Center					
0.6 ± 0.1 (0.4 - 0.9) 17	2.6 ± 0.3 (1.8 - 3.1) 12	1.8 ± 0.3 (1.0 - 2.8) 17	36 ± 10 (25 - 59) 28	25 ± 15 4 - 55 60	4.0 ± 0.8 (2.3 - 6.7) 20
Northeast					
0.5 ± 0.2 (0.2 - 1.0) 40	2.9 ± 0.8 (1.9 - 6.2) 28	1.8 ± 0.4 (1.0 - 2.4) 22			3.2 ± 0.9 (2.2 - 5.2) 28
Eastern					
0.2 ± 0.1 (0.03 - 0.7) 50	9.7 ± 3.2 (2.8 - 16.8) 33	0.6 ± 0.3 (0.2 - 1.6) 50	121 ± 74 (42 - 285) 61	114 ± 71 12 - 237 62	3.7 ± 2.0 (1.5 - 9.4 54

The mean annual water temperature is about 22°C with an annual seasonal range of only 21 to 23°C (Limón *et al.*, 1989). Transient thermal stratification occurs on non-windy days. At such times surface temperatures up to 24°C with bottom temperatures of 21°C occur (Doyle, 1985).

2. CHEMICAL PROPERTIES

The water is moderately hard (TH ≈ 150 mg L^{-1}) and alkaline (TA ≈ 185-mg L^{-1}, pH ≈ 8.7) however the most significant chemical feature is the high concentration of the nutrients nitrogen and phosphorus (Table 3). Because of the soil properties of this volcanic region and the input of anthropogenic materials from the Lerma river and surrounding villages, the total phosphorus concentrations are high. Most of this (≈ 50 to 90%) is soluble reactive phosphorus (SRP). Total inorganic nitrogen is also high relative to many other lakes but low relative to total phosphorus. The TIN:TP is approximately one, indicating of nitrogen limitation of phytoplankton production. With the exception of nitrogen, chemical elements follow a pattern of dilution with the rainy season. Nitrogen, with much supplied in rainwater, tends to increase at that time (Limón *et al.*, 1989).

Table 3 Mean or mean ± SD and range of physical and chemical variables by major region. IDS = inorganic dissolved solids, POC = particulate organic carbon, DOC = dissolved organic carbon. (Modified from Limón *et al.*, 1989 and Lind *et al.*, 1997)

Variables (units)	Eastern	Center	Western
Total Hardness (mg L^{-1})	156	167	154
Total Alkalinity (mg L^{-1})	181	191	191
IDS (mg L^{-1})	305	279	282
Total Phos. (µg L^{-1})	755	1060	764
NH_3-N (µg L^{-1})	113 ± 126 17 – 491	68 ± 41 27 – 166	65 ± 35 13 – 139
NO_3-N (µg L^{-1})	235 ± 187 62 – 770	188 ± 138 47 – 610	235 ± 136 53 – 463
POC (mg L^{-1})	19 ± 9 10 – 36	12 ± 3 6 – 18	12 ± 5 6 – 19
DOC (mg L^{-1})	28 ± 10 12 – 50	25 ± 7 15 – 40	23 ± 6 14 – 35

Shine *et al.* (1998) reported heavy metal concentrations in the water, sediments, and biota. Maximum concentrations in the water occurred at the end of the dry season (presumably by evaporative concentration) and again after river inflow during the rainy season. Cadmium and lead reached concentrations greater than those allowed for the protection of aquatic life as recommended by the U. S. Environmental Protection Agency (USEPA, 1995). In the biota, they reported the highest concentrations in the roots of *Eichornia crassipes* (water hyacinth) and in the livers of carp and tilapia. Hansen (1992) compared the metals concentrations of the lake sediments with those of other large American and European lakes and found that concentrations of most metals were not markedly different. Nevertheless, concentrations of mercury and cadmium were elevated in Lake Chapala.

3. BIOTIC PROPERTIES

The nature of the lake's biota is less well known than that of its physics and chemistry. Species lists are incomplete for most taxa. Moreover, the functional ecology of most communities is poorly understood. Undoubtedly the very low transparency is a major determinant of the biota's structure and function. For example, macrophyte communities are limited to either floating or emergent forms. Cattail (*Typha angustifolia*) is abundant in the shallow waters of the eastern region. Water hyacinth (*Eichornia crassipes*) can be abundant with a cycle apparently associated with water level fluctuations. Most recently, during the early 1990s, it ultimately covered approximately 25% of the lake's surface.

3.1 Phytoplankton

Regularly, scientists from the Sistema Intermunicipal de Agua Potable y Alcantarillado (SIAPA) monitor phytoplankton taxonomy in relation to water treatment problems (Ledesma, personal communication). Water samples for this purpose are taken from the shallow region near the intake pumps located in the region of the Isla de las Alacranes. These data, although probably the most extensive of recent information, are generally not available in published form. Cortés-Altamirano and Guzmán (unpublished) reported 41 phytoplankton taxa for the same lake region based on spring 1988 samples. *Anabaena limnetic*, presently abundant, was not found at that time. Pim and Aguirre (unpublished) sampled the phytoplankton of the central and western regions in July and September of 1989 and identified 57 taxa. One unidentified species of *Anabaena* was listed as infrequent. They commented upon the low abundance of cyanobacteria in their samples.

Pelagic cyanobacterial nitrogen fixation does not occur (Glass, 1987). Perhaps this failure in a nitrogen-limited lake is due to the lack of solar energy penetration sufficient to support significant heterocystic cyanobacteria populations. When *Anabaena* from near-by shallow ponds was inoculated into Lake Chapala water under artificial illumination, nitrogen fixation occurred demonstrating the absence of chemical or feedback inhibition. Presumed nitrogen-fixing taxa are found in shallow shoreline areas. (Stoermer and Andresen, personal communication). Sampling in 1999 along the North shore near the Chapala Ecology Station at Ajijic they found abundant heterocystic filaments in both the surface psammon and the shallow benthos.

Phytoplankton biomass as chlorophyll is generally low ($<10\ \mu g\ L^{-1}$) over most of the lake but can be high ($66\ \mu g\ L^{-1}$) in shallow regions with a favorable light climate. Phytoplankton chlorophyll concentrations increased at all lake regions during the 1980s (Table 4). Phytoplankton photosynthesis in the water column is low. Annual water column photosynthesis (particle and exudates) ranged between 125 and 139 g C m^{-2} yr^{-1} for various lake regions in 1989-90 (Lind *et al.*, 1997). This was an increase over the lake-wide average particulate primary production of 80 g C m^{-2} yr^{-1} measured in 1983-84 (Lind *et al.*, 1992). These low (oligotrophic) rates are not due to nutrient deficiencies – the nutrient content of the water is high (see above).

Laboratory algal-assays conducted periodically since 1983 consistently show that, when provided with adequate illumination, nitrogen is the chemical element limiting algal growth (Dávalos *et al.*, 1989, Hernández *et al.*, in press). Nevertheless, phytoplankton production *in situ* is not nutrient limited.

Table 4 Mean ± SD, annual range (in parentheses), and CV (%) of net daily phytoplankton chlorophyll (μg L^{-1}) at each region of Lake Chapala, Mexico, June 1983 - June 1984 and June 1989 - July 1990. For 1983-84, N = 23 except Station 15 where N = 46. For 1989-90, N = 16. Modified from Lind *et al.*, 1992 and Lind *et al.*, 1997

Region	1983-84	1989-90
West	5.0 ± 2.5 (1.6 - 11.0) 50	7 ± 2 (5 - 12) 28
West central	4.5 ± 2.1 (0.9 - 9.8) 47	----
Center	4.2 ± 1.4 (1.5 - 8.0) 33	7 ± 2 (4 - 11) 28
Northeast	4.3 ± 1.6 (1.9 - 8.1) 37	----
East	9.3 ± 3.4 (2.2 - 14.1) 37	19 ± 14 (2 - 66) 74

Assuming stoichiometric relationships of phytoplankton components, the quantity of nitrogen available – given the abundant phosphorus – is capable of producing hyper-eutrophy. Phytoplankton production is low because of the unfavorable light climate described above. The ratio of mixing depth to photic depth describes the fraction of time in an algal cell's life that the cell is in the dark and incapable of net growth.

For Lake Chapala, Lind *et al.* (1992) determined that the ratio favoring algal biomass accumulation is two or less – phytoplankton at least 50% of the time in the photic zone. A ratio of two or less occurs only in the shallow areas (see above).

3.2 Bacterioplankton

Lind *et al.* (1994) calculated that the quantity of photosynthesis was insufficient to support the commercial fishery sustained in the lake and suggested a large role for micro-heterotrophy. Because of the light-limited phytoplankton photosynthesis, bacterioplankton production is unusually important in the formation of small grazable particles as the base of the food web. Lind *et al.* (1997) reported a bacterioplankton production of 9.3×10^7 kg C yr^{-1}, which accounted for 40% of the small organic particle production. Because of the low phytoplankton production and the absence of any known large sources of allochthonous organic carbon inputs to the lake, it was postulated that such bacterioplankton heterotrophy could only occur by increased efficiency in the use of ambient organic matter. The role of adsorptive concentration of organic carbon on the abundant clays was proposed as an efficiency-increasing mechanism. This is supported by the fact that approximately 90% of the lake's bacterioplankton are associated with clay particles and that those cells are of significantly greater volume than those free in the water (Lind and Dávalos-Lind, 1991). Dávalos-Lind *et al.* (1992) demonstrated the possibility of direct ingestion of these clay-bacteria particles by the native atherinid fishes, *Chirostoma* spp.

The ingestion of clay-bacteria particles by plankton and fish has other consequences. These particles concentrate ionic and polar water contaminants. Heavy metal tissue burdens in fish are large (Shine *et al.*, 1998). Wind-driven resuspension of contaminated sediments can be a significant source of contamination to the water column's biota. Dávalos-Lind (1996) demonstrated this effect with elutriate bioassays using both bacterioplankton and native phytoplankton under aerobic and anaerobic conditions. Inhibition of phytoplankton growth was elutriate-dose dependent while bacterioplankton metabolism was not inhibited even at high elutriate doses. She postulated bacteria community adaptation to the contaminants not developed in the phytoplankton community.

3.3 Zooplankton and Benthos

The lake's zooplankton and benthos have been studied sporadically, and principally taxonomically (e.g. Flores unpublished keys to the plankton and benthic crustacea). Unfortunately these studies have not been published and exist only in the "gray literature" of theses, society proceedings, and in-house government reports. One functional study of the zooplankton was that of Trotter (1988) who found that one-third of the nitrogen used by phytoplankton was regenerated zooplankton nitrogenous wastes while the zooplankton contribution to the total lake nitrogen pool was only 15%.

3.4 Fish

The lake's fish community is rather well known taxonomically and from a fishery management perspective. The importance of the fishery to the local economy has necessitated serious study of the abundance and distribution of harvested fish species. Most fisheries data, as that of the invertebrates, exist only as "gray literature". Government reports showed a decline in the fishery during the 1980s (unpublished report, Delegación de la Secretaría de Pesca en Jalisco) but it is not clear whether that was due to fewer fishes and catch per unit effort or just to less effort as fishers sought alternate means of income. Lyons *et al.* (1998) concluded that the fishery is not in decline with catch of exotic species more than compensating for the loss of native species. Pomeroy (1993, 1994) described the sociological implications of the changed fishery. Soto-Galera *et al.* (1998) reported a decline in the number of fish taxa for the Lower Lerma river basin including Lake Chapala.

The unusual occurrence of the *Chirostoma* species flock has attracted evolutionary biologists for study of this taxon (Barbour, 1973). The functional role of the fishes in the lake ecosystem has not been investigated, but study of niche partitioning in the *Chirostoma* flock is now underway (J. Kelly, personal communication). Seventy percent of the native fish species in the Lerma river basin are endemic. Three species have become extinct, but none from the lake (Lyons *et al.*, 1998). Opportunities abound for both descriptive and functional investigations of many lake communities.

REFERENCES

Barbour C (1973) A biogeographical History of Chirostoma (Pisces: Atherinidae): A Species Flock from the Mexican Plateau. Copeia 3: 533- 556.

Clements T (1963) Pleistocene History of Lake Chapala Jalisco, Mexico. Essays in Marine Geology in Honor of Emery KO. Univ Calif Press, Los Angeles.

Dávalos L, Lind O and Doyle R (1989) Evaluation of Phytoplankton-limiting Factors in Lake Chapala, Mexico: Turbidity and Spatial and Temporal Variation in Algal Assay Response. Lake Reservoir Mgt, 5: 99-104.

Dávalos-Lind L (1996) Phytoplankton and Bacterioplankton Stress by Sediment-borne Pollutants. J Aquat Ecosys, Health 5:99-105.

Dávalos-Lind LO, Sada R, Guerra A, Velarde G, Orozco L, Chrzanowski T and Lind O (1992) La Producción Bacteriana y su Importancia en la Cadena Trófica en el Lago de Chapala. Ingeniería Hidráulica en Mexico, Mayo-Diciembre: 30-36.

de Anda J, Quiñones-Cisneros S, French R and Guzmán M (1998) Hydrologic Balance of Lake Chapala, Mexico. J Amer Water Resources Assoc, 34: 1319-1331.

Doyle R (1985) Phytoplankton Productivity of Tropical Lake Chapala, Mexico. MS Thesis, Baylor University, 190 p.

Filonov A, Tereschenko I and Monzón C (1998) Oscillations of the Hydro-meteorological Characteristics in the Region of Lake Chapala for Intervals of Days to Decades. Geofísica Internacional, 37: 293-307.

Garduño-Monroy V, Spinnler J and Ceragioli E (1993) Geological and Structural Study of the Chapala Rift, State of Jalisco, Mexico. Geofísica Internacional, 32:487-499.

Glass J (1987) Biological Nitrogen Fixation in a Nitrogen Limited Tropical Lake, Lake Chapala, Mexico. MS Thesis, Baylor University, 90 p.

Hansen AM (1992) Metales Pesados en el Sistema Lerma-Chapala: Distribución y Migración. Ingeniería Hidráulica en México, Mayo-Diciembre: 92-98.

Hernández S, Bernal F, Velarde G, Ortiz D, Lind O and Dávalos-Lind L. In Press. The Algal Growth Potential and Limiting Nutrient of 30 Lakes and Reservoirs of the Mesa Central of Mexico. Verh Internat Verein Limnol.

Limón J and Lind O (1990) The Management of Lake Chapala, Mexico. Lake Reser Manage, 6: 61-70.

Limón JG, Lind OT, Vodopich DS, Doyle RD and Trotter BG (1989) Long- and Short-Term Variation in the Physical and Chemical Limnology of a Large, Shallow, Turbid Tropical Lake (Lake Chapala, Mexico) . Arch Hydrobiol Suppl 83 (Monogr Beit), 1: 57-83.

Lind O and Dávalos-Lind L (1991) Association of Turbidity and Organic Carbon with Bacterial Abundance and Sell Size in a Large, Turbid, Tropical Lake. Limnol Oceanogr, 36: 1200-1208.

Lind O, Dávalos-Lind L, Chrzanowski T and Limón J (1994) Inorganic Turbidity and the Failure of Fishery Models. Int Revue ges Hydrobiol, 79: 7-16.

Lind O, Doyle R, Vodopich D, Trotter B, Limón J and Dávalos-Lind L (1992) Clay Turbidity: Regulation of Phytoplankton Production in a Large, Nutrient-rich Tropical Lake. Limnol Oceanogr, 37: 549-565.

Lind O, Chrzanowski T and Dávalos-Lind L (1997) Clay Turbidity and the Relative Production of Bacterioplankton and Phytoplankton. Hydrobiologia, 353: 1-18.

Lyons J, González G, Soto E and Guzmán M (1998) Decline of Freshwater Fishes and Fisheries in Selected Drainages of West-central Mexico. Fisheries, 23: 10-18.

Pomeroy C (1993) Organized Fishers' Responses to Social Dilemmas of Common Pool Resource Use. Ph D dissertation, Texas A & M Univ, 175 p.

Pomeroy C (1994) Obstacles to Institutional Development in the Fishery of Lake Chapala, Mexico. p 17-41 *In* Dyer C and McGoodwin J (eds.) Folk Management in the World's Fisheries. Lessons for Modern Fisheries Management. U Colorado Press.

Shine J, Ryan D and Ford T (1998) Annual Cycle of Heavy Metals in a Tropical Lake – Lake Chapala, Mexico. J Environ Sci Health, Part A33: 23-43.

Simons T (1980) Circulation Models of Lakes and Inland Seas. Can J Fish Aquatic Sci, 203: 1-146.

Soto-Galera E, Díaz-Pardo E, López-López E and Lyons J (1998) Fish as Indicators of Environmental Quality in the Lerma River Basin, Mexico. Aquat Ecosys Health Mgt, 1: 267-276.

Trotter B (1988) The Role of Zooplankton-Phytoplankton Community Interactions in the Nutrient Dynamics of Lake Chapala, Mexico. MS Thesis, Baylor University, 108 p.

USEPA (1995) Federal Register 60(86): 22229-22237.

Velarde G (1995) Bacterial Growth on UV-B Photolytically Produced Dissolved Organic Matter. MS Thesis, Baylor University, 33 p.

Chapter 7

Hydro-meteorology of Lake Chapala

ANATOLIY E. FILONOV[1], IRINA E. TERESHCHENKO, AND CÉSAR O. MONZÓN

Physics Department, University of Guadalajara Apdo.Postal 4-040, Guadalajara 44421, Jal., Mexico. email: [1]afilonov@ccip.udg.mx

Key words: Lake Chapala, hydro-meteorological characteristics, temperature and current measurements, thermal fronts

Abstract: The purpose of this work is to examine fluctuations of the hydro-meteorological characteristics of Lake Chapala, such as air temperature, precipitation, evaporation, atmospheric pressure, wind, water level in the lake and volume of flow of the rivers in the Lake Chapala region. Linear and nonlinear (maximum likelihood) methods of spectral analysis were used to investigate these phenomena. The main periods were identified in the range from hours to decades, obtaining estimates of the mean square amplitudes. The thermodynamic mode of Lake Chapala was also investigated, particularly the vertical-horizontal distribution of temperature of the water and currents. For this purpose, measurements of the distributions of currents and temperatures were performed. Measurements show that Lake Chapala has a quite complex thermodynamic regime and a significant vertical stratification. The differences in temperatures detected between the surface and bottom in the center of the lake were between 0.5 and 1°C and in the eastern part, from 2 to 3°C. The warmest water was located in the lake's eastern shallow part. The temperature of the water in the central and southern parts of the lake was 1-2°C lower than that of the water at the northern shore. In all the cross-sections, ordinary spatial variations of 3°C on a distance of just 100 to 500 m were registered with a towed device. A buoy station registered the movements of the internal thermal front with a temperature fluctuation of 2°C. The advanced part of the front was accompanied by intense internal waves. The measurements show that the currents in the lake depend strongly on the intensity of the daytime breeze, which enhances wind velocities up to 10 cm s^{-1}, resulting in increases of the water levels of up to 10-30 mm near the shoreline.

The Lerma-Chapala Watershed, Edited by Hansen and van Afferden, Kluwer Academic/Plenum Publishers, New York, 2001

1. INTRODUCTION

Among shallow lakes in the world, Lake Chapala is the largest (Figure 1). Due to its size and to the settlement of fast growing industrial and cultural centers within its watershed, the lake plays an important role in the economy of Mexico (Sandoval, 1994).

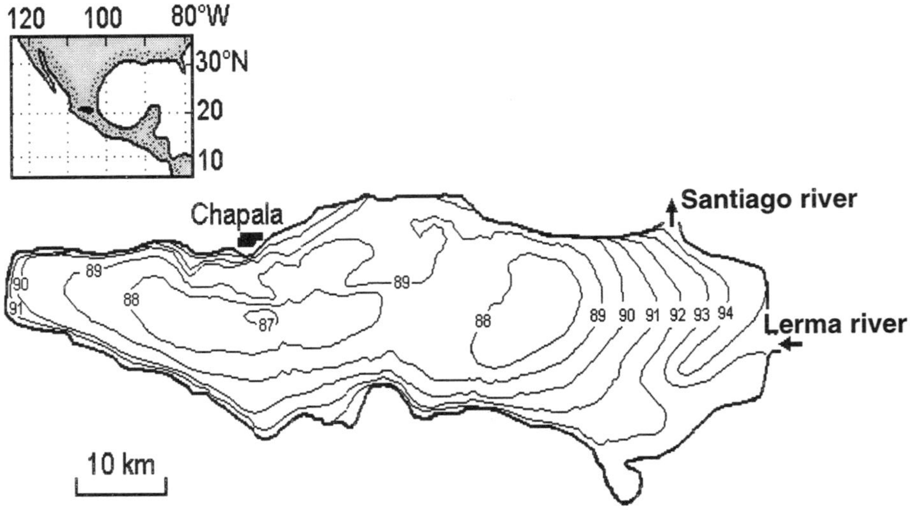

Figure 1 Bathymetrical map of Lake Chapala. Depth is given in meters in relation to the 87 m depth-mark

Average precipitation in the Lerma-Chapala watershed is about 735 mm per year, varying from 300 mm in dry years to 1,200 mm in wet years. Surface evaporation ranges from 1,400 to 1,600 mm per year and accounts for almost twice the average rainfall (Jauregui, 1995; Filonov *et al.*, 1998).

Precipitation has long-term temporary fluctuations related to the periods of El Niño, (i.e. 2 and up to 7 years) and of solar activity (i.e. 11, 22, 45 and more years) (Filonov and Tereshchenko, 1997). The variability of inter-annual precipitation causes significant temporary fluctuations in water level, area and volume of the lake.

Local winds and breeze determine the dynamic processes in the lake. They develop on a daily basis as a result of large temperature differences between surface water and the surrounding land with poor vegetation coverage. During the day, the land is strongly heated and in the night it is cooled, causing breeze circulation. The breeze directed towards the land in the daytime, is more intense than in the night. The speed of the breeze wind during the day can reach values of 8-10 m s^{-1} and it tends to strongly increase evaporation from the lake surface (Filonov, 1998).

A review of previous research about Lake Chapala includes 120 publications since 1987 (Guzmán and Morelos, 1992). A group of U.S. and Mexican biologists studied the factors that control the production of phytoplankton in the lake (Lind *et al.*, 1992). Some studies on water circulation and on the dynamics of pollution in the lake have also been published (Escalante, 1992; Hansen and van Afferden, 2001; León, 1994; Rodríguez, 1994).

Among these studies, there are no references on the variability of the hydro-meteorological characteristics of Lake Chapala and its surroundings. Due to large deviations of the means, the variation magnitudes can be a determining factor.

The variability of the hydro-meteorological and thermodynamic regime of the lake is quite useful to construct a complete hydrodynamic circulation model that responds to the variability of the atmospheric processes in a wide range of time scales. Our objective is to discover the physical significance of internal relations between all the processes that form the hydro-climate of the lake.

In this chapter we will discuss the variability of the hydro-meteorological characteristics of the Lake Chapala region in a range of different scales and periods, and using spectral analysis methods. We also refer to some features of the lake's thermodynamic regime.

2. VARIABILITY OF THE HYDRO-CLIMATE IN THE LAKE REGION

2.1 Information used and the processing method

In this work, time series of hydro-meteorological fluctuations from the records of the Comisión Nacional del Agua (CNA) were used. The records were made at the Chapala climatic station. Data were recorded with automatic meteorological stations of the CNA and the University of Guadalajara (UG). Annual means of air temperature and rainfall in Guadalajara were obtained from the UG station. The time series data are summarized in Table 1. Using a special program, minimum and maximum values, as well as maximum possible differences between neighboring data were controlled. Uncertain information and missing data were replaced with cubic spline interpolation. Depending on data intervals the time series were grouped into four sections (Table 1). The amplitudes of the spectra were estimated for each group of time series (Jenkins and Watts, 1969; Konyaev, 1990):

$$C_x(\omega) = \int_0^T x(t) \cdot \exp - i2\pi\omega t dt, \tag{1}$$

where $x\,(t)$ is the time series; T is the total length of the series; and ω is the frequency. The auto-periodogram $S_{xx}\,(\omega)$ and cross-periodogram $S_{xy}\,(\omega)$ were defined as

$$S_{xx}(\omega) = \frac{1}{T} C_x(\omega) \cdot C_x^*(\omega), \tag{2}$$

$$S_{xy}(\omega) = \frac{1}{T} C_x(\omega) \cdot C_y^*(\omega) = P_{xy}(\omega) - iQ_{xy}(\omega). \tag{3}$$

Here P_{xy} (ω), Q_{xy} (ω) are the real and imaginary parts of the crossed periodogram; * is the complex conjugate.

The spectral estimate was obtained by smoothing the frequencies of the period-grams:

$$\hat{S}_{xx}(\omega) = \int_{-\infty}^{\infty} S_{xx}(\omega')Z(\omega - \omega')d\omega', \tag{4}$$

where $Z(\omega)$ is a smoothing function.

The estimates of the coherence function were also calculated for pairs of time series at the limits of each group:

$$C_{o\,xy}^2(\omega) = \frac{\left|\hat{S}_{xy}(\omega)\right|^2}{\hat{S}_{xx}(\omega) \cdot \hat{S}_{yy}(\omega)} \tag{5}$$

and the phase difference

$$\Delta\varphi_{xy}(\omega) = arctg\left(\hat{Q}_{xy}(\omega) / \hat{P}_{xy}(\omega)\right) \tag{6}$$

Table 1 Characteristics of time series data

Time series	Interval	Characteristics	Dimensions	Start date	End date	Number of data
I	15 min	Air temperature	°C	5/1/1994	6/271994	5472
		Humidity	%	5/1/1994	6/27/1994	5472
		Wind speed	$m\ s^{-1}$	5/1/1994	6/27/1994	5472
		Wind direction	degree	5/1/1994	6/27/1994	5472
	1 hr	Wind speed	$m\ s^{-1}$	12/3/1996	1/2/1997	720
		Lake Chapala water level	m	12/3/1996	1/2/1997	720
II	1 day	Air temperature	°C	1/1/1990	8/31/1994	1703
		Evaporation	mm	1/1/1990	8/31/1994	1703
		Precipitation	mm	1/1/1990	8/31/1994	1703
		Atmospheric pressure	mb	1/1/1990	8/31/1994	1703
		Wind speed	$m\ s^{-1}$	1/1/1990	8/31/1994	1703
		Wind direction	degree	1/1/1990	8/31/1994	1703
		Lake Chapala water level	m	1/1/1990	8/31/1994	1703
		Lerma river flow volume	$m^3\ s^{-1}$	1/1/1990	8/31/1994	1703
III	1 month	Air temperature	°C	1/1/1971	12/31/1998	336
		Evaporation	mm	1/1/1971	12/31/1998	336
		Precipitation	mm	1/1/1971	12/31/1998	336
		Atmospheric pressure	mb	1/1/1971	12/31/1998	336
		Wind speed	$m\ s^{-1}$	1/1/1971	12/31/1998	336
		Wind direction	degree	1/1/1971	12/31/1998	336
		Lake Chapala water level	m	1/1/1971	12/31/1998	336
		Lerma river flow volume	$m^3\ s^{-1}$	1/1/1971	12/31/1998	336
		Santiago river flow volume	$m^3\ s^{-1}$	1/1/1971	12/31/1998	336
IV	1 yr	Air temperature	°C	1900	1998	99
		Precipitation	mm	1900	1998	99
		Lake Chapala water level	m	1900	1998	99
		Lerma river flow volume	$m^3\ s^{-1}$	1932	1998	67

For the structure of the wind velocity and direction time series for groups I and II, the method of rotational components was used (Gonella, 1972; Mooers, 1973). At this point the estimates of the "clockwise" spectra are found,

$$S_{-}(\omega)=\frac{1}{8}\left[S_{uu}(\omega)+S_{vv}(\omega)-2Q_{uv}(\omega)\right] \tag{7}$$

and for the counterclockwise spectra,

$$S_{+}(\omega)=\frac{1}{8}\left[S_{uu}(\omega)+S_{vv}(\omega)+2Q_{uv}(\omega)\right] \tag{8}$$

where $S_{uu}(\omega)$, $S_{vv}(\omega)$ are the autospectra, and $Q_{uv}(\omega)$ is the spectral square of the zonal and meridional components, $u(t)$ and $v(t)$, respectively, of the wind velocity vector. The orientation of the major axes of the elliptic hodographs

$$\Theta(\omega)=\mathrm{arctg}\left\{2\mathrm{P}_{\mathrm{uv}}(\omega)/(\mathrm{S}_{\mathrm{uu}}(\omega)-\mathrm{S}_{\mathrm{vv}}(\omega))\right\} \tag{9}$$

and their stability

$$\mathrm{E}^2(\omega)=\frac{(\mathrm{S}_{\mathrm{uu}}(\omega)+\mathrm{S}_{\mathrm{vv}}(\omega))^2-4(\mathrm{S}_{\mathrm{uu}}(\omega)\cdot\mathrm{S}_{\mathrm{vv}}(\omega)-\mathrm{P}^2{}_{\mathrm{uv}}(\omega))}{(\mathrm{S}_{\mathrm{uu}}(\omega)+\mathrm{S}_{\mathrm{vv}}(\omega))^2-4\mathrm{Q}^2{}_{\mathrm{uv}}(\omega)} \tag{10}$$

were also calculated, where $P_{uv}(\omega)$ is the co-spectrum of $u(t)$ and $v(t)$.

The mean annual data series in group IV were used to find the energy peaks of the spectra, using nonlinear methods of maximum likelihood (ML) for relatively short time series, together with the traditional spectral analysis method (Burg, 1967, 1972; Akaike, 1969; Konyaev, 1990). Spectral estimates from the maximum likelihood method were calculated with

$$\hat{S}(\omega)=\Delta t/\left|A_{ML}(\omega)\right|^2, \tag{11}$$

where the averaging filter was found by

$$A_{ML}(\omega)=\sum_{k,j}^{N}a_{k,j}\exp-i2\pi\omega(k-j)\Delta t, \tag{12}$$

and the coefficients $a_{k,j}$ were obtained from the inverse of the initial time series correlation matrix:

$$\begin{bmatrix} B(0) & B(1) & \ldots & B(N-1) \\ B(1) & B(0) & \ldots & B(N-2) \\ \ldots & \ldots & \ldots & \ldots \\ B(N-1) & B(N-2) & \ldots & B(0) \end{bmatrix}^{-1} = \begin{bmatrix} a_{11} & a_{12} & \ldots & a_{1N} \\ a_{21} & a_{22} & \ldots & a_{2N} \\ \ldots & \ldots & \ldots & \ldots \\ a_{N1} & a_{N2} & \ldots & a_{NN} \end{bmatrix} \tag{13}$$

The mean square amplitudes of the harmonics of the dominant peaks in the spectra were found from the values of spectral density. Confidence intervals were calculated at each level for all spectral estimates, by using standard algorithms described in the literature (Jenkins and Watts, 1969; Konyaev, 1990). The number of degrees of freedom was $v = 2\alpha(2F + 1)$; where α is the number of independent realization segments, in which spectral estimates were averaged and F is the half-width of the filter used to average the period-grams.

2.2 Variability from one hour to one day

It is possible to relate fluctuations with periods of between ten minutes and one day to a scale of variability of the hydro-meteorological characteristics in any area of the lake. Its basic energy source is the daily course of temperatures caused by heat inflow change to a terrestrial surface or to the atmosphere during one day. Energy interactions of the soil-lake-atmosphere system are not simple. The area surrounding the lake is mountainous with various spatial orientation valleys. Acting by portions, with daily periodicity, thermal energy does not remain in a fixed site of a spectrum but is redistributed in a complex way into fluctuations with a wide range of frequencies (Scorer, 1978).

Figure 2 shows the fluctuations in the temperature, relative humidity and wind time series (time series group I), measured by using an automatic meteorological station in May and June, 1994. It was noticed, that in all the series, daily periodicity remains. The curves have not the accurate "sine wave" form. When significantly modulated, lower frequencies of the curves are visible at the expense of fluctuations. In every analysis, the boundary between dry and wet seasons is visible. In 1994 the wet season began on June 10. Mean air temperatures decreased sharply from 22 to 18 °C, and the amplitude range of its daily fluctuations fell from 14-15 to 7-8 °C. The relative humidity in the same time period changed from 50-70 to almost 100%. During the dry season, daily average wind speed was about 3 m s^{-1}, with a maximum of 6 m s^{-1}. The fluctuations of wind direction had a more or less ordered character. As the seasonal rains approached, fluctuations of the irregular components of high frequencies wind were amplified. In the northern hemisphere, such processes are related to the displacement of the equatorial low-pressure trough. Circulating processes in powerful cloudy structures accompanying the equatorial trough can generate large vertical and horizontal local pressure gradients. This can cause unexpected strengthening of wind, and strongly break its daily course (Palmén and Newton, 1969; Riehl, 1979).

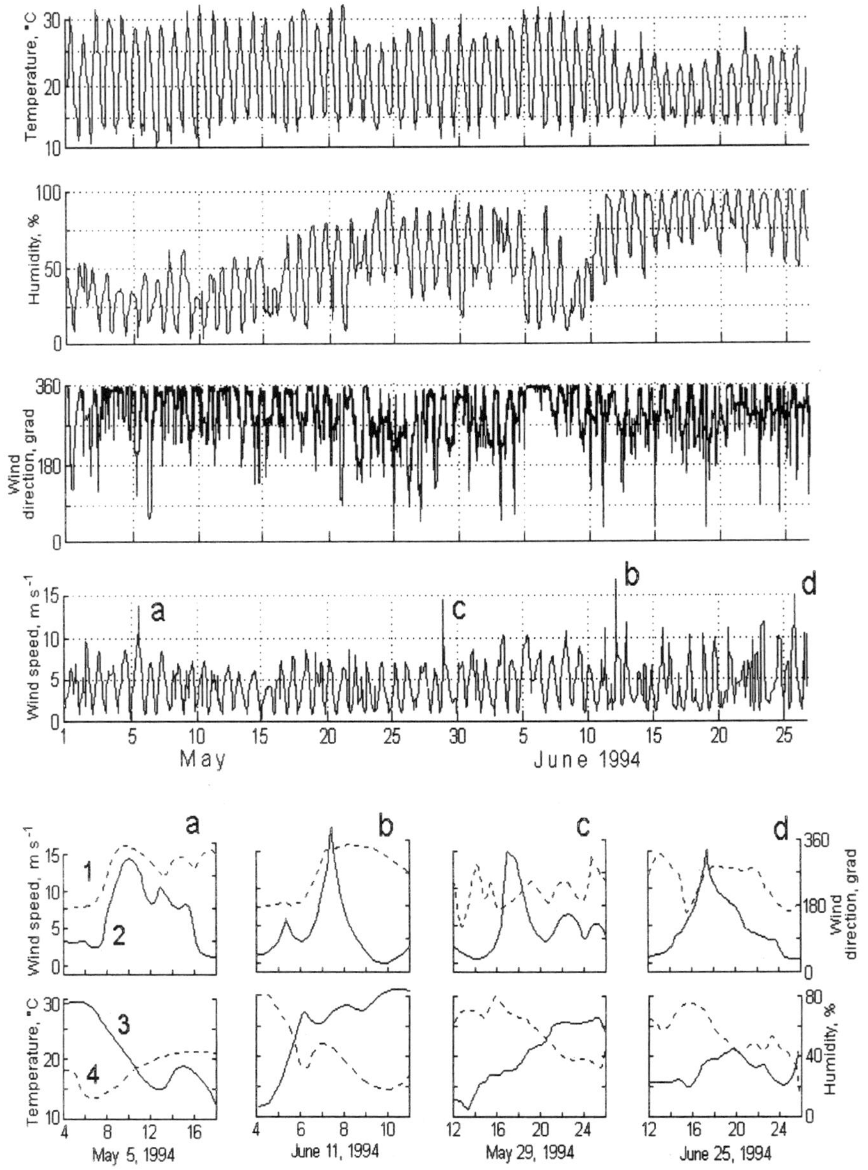

Figure 2 Time series fluctuations of meteorological characteristics measured by the automatic meteorological station at Chapala (group I). Letters a-d indicate sharp wind strengthening. The lower part of the Figure shows the temporary course of meteorological characteristics, for time intervals appropriate to the moments of wind strengthening. Numbers 1 and 2 correspond to wind direction and speed, respectively; 3 to the air temperature and 4 to relative humidity

In the lake area, where the Chapala climatic station is located, cases of not seasonally dependent sharp wind strengthening were registered in separate time intervals. In a period of 2-3 hours its speed was increased from 2-3 up to 10-17 m s^{-1}. Then the wind smoothly abated within 4-6 hours to one m s^{-1} or to complete calm. At the moment of the strengthening, wind sharply turned clockwise in almost 180 degrees from North to South.

In Figure 2, a and b indicate two cases of wind strengthening and change in direction, the air temperature was sharply decreased from 26-29 to 14 °C and the relative humidity raised from 30 to 40%. During the morning, wind speed reached its maximum between 9 and 10 am. These cases were undoubtedly caused by intense morning breeze, directed from the lake to the land.

Letters c and d in Figure 2 indicate two cases where air temperature increased from 14 to 16 °C, and relative humidity reduced to 40-50%, when sharp wind strengthened and changed its direction. Both cases were observed in the evening (about 8 pm) and they were related to the strengthening of the evening breeze, directed from the shoreline to the lake.

To analyze the detailed structure of the variability of meteorological characteristics, auto and cross-spectra were calculated for all time series. A preliminary analysis shows that this does not contain large amplitude fluctuations if periods of less than one hour are considered. Therefore, to reduce the number of calculations, less than one hour fluctuation periods were removed from time series by a low-frequency filter.

The normalized spectrum of the analyzed time series (in percentage, relative to the maximum) is specified in Figure 3. Daily peaks prevail in the temperature and relative humidity spectra. Within this spectrum, statistically authentic periodic peaks of 12, 8, 6 and 4 hours are identified. They demonstrate the "cascade" mechanism for energy transfer from the sun acting with a daily periodicity and generating fluctuations with small amplitudes and periods (and apparently with small linear sizes).

Daily peaks of wind fluctuations could be adequately expressed only in the clockwise spectrum. This confirms the assumption, that in the area of Lake Chapala, daily wind fluctuations are caused basically by daytime breeze and as the turn of its speed vector occurs in a clockwise direction, typical of the northern hemisphere. The coherence analysis showed that for all pairs of time series, only daily and semi-daily periods exceeded the 95% confidence interval. The difference of phases, the square coherence matrixes and the mean square of the harmonics' amplitudes of these periods are given in Table 2.

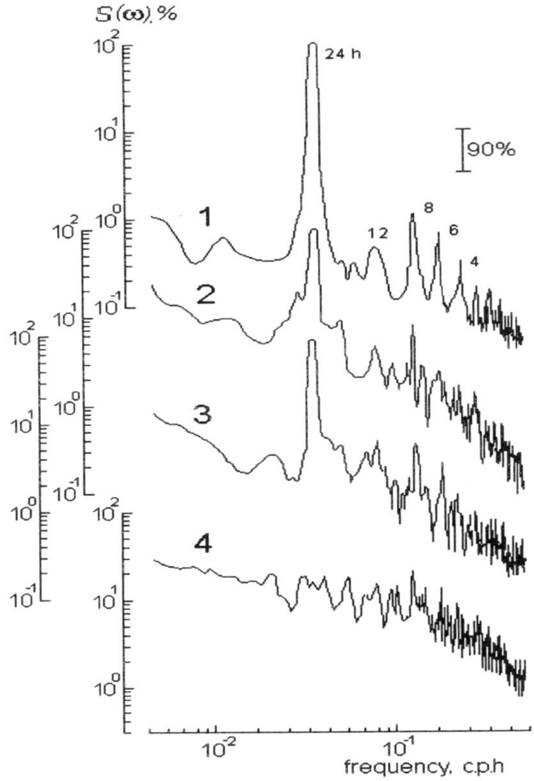

Figure 3 Normalized power spectra of fluctuations of the meteorological characteristics in group I: air temperature (1), relative humidity (2), "clockwise" (3) and "counterclockwise" (4) wind speed components. The vertical line in the spectrum indicates a 80% confidence limit

It is observed that daily fluctuations of meteorological characteristics are related almost linearly to each other but with different phase shifts, and an easily explainable physical meaning. As a rule, change of wind direction causes immediate alteration of relative humidity ($\Delta\varphi$ = 0 degrees). On the contrary, temperature and humidity fluctuations, as well as the speed and direction of wind, always occur in opposite phase ($\Delta\varphi$ = 180 degrees).

The results of the analysis have shown, that thermodynamic processes in Lake Chapala during periods from one hour to about one day are caused by local circulation processes (breeze and angled winds) whose primary source is the daily temperature fluctuation.

The mean square amplitudes of a daily harmonic and more amplitudes for semi-daily fluctuation were analyzed for all meteorological characteristics. There is a close connection between temporary temperature gradients and local circulation intensity. In the night, the wind speed decreases to close to zero, creating thermal stability, since the exchange of quantity of movement in a vertical direction is very small.

Table 2 Square coherence function values, phase differences and mean square amplitudes of daily (a) and semi-daily (b) harmonic of the hydro-meteorological characteristics of the Lake Chapala region (group I time series). Values of the phase differences (in parenthesis) are given in hours. Coherence values exceeding 95% of the confidence interval are in bold

		Air temperature	Humidity	Wind speed	Wind direction	Mean square amplitude
a	Air temperature		182 (12.1)	113 (7.5)	201 (13.4)	5.7
	Humidity	**0.901**		270 (18)	6 (0.4)	15
	Wind speed	**0.740**	**0.718**		176 (11.7)	2.5
	Wind direction	**0.979**	**0.862**	**0.684**		24
b	Air temperature		266 (8.9)	110 (3.7)	85 (2.8)	0.4
	Humidity	0.581		--	--	3
	Wind speed	0.427	0.041		156 (5.2)	0.5
	Wind direction	**0.475**	0.215	**0.688**		8

As a result of the weak evaporation during the night, the moisture is kept at the lake's surface and on the land. In the morning the steady layer of the atmosphere near the ground collapses under the influence of the sun and the wind speed quickly increases (Riehl, 1979). The classical mechanism of daily circulation in the atmosphere above the lake should be strongly broken as a result of the influence of breeze and angled winds. A mountain chain bounding North and South of Lake Chapala as well as the flat surface at its eastern and western shoreline, maintain this area dry enough to create the conditions that keep breeze circulation on the main axial line of the lake. On that line, the various circulating processes of small intensity created by angled winds are imposed.

Stability factors, **E,** and orientation, **Θ,** of the elliptic orbit's long axes were calculated for daily and semi-daily harmonics, resulting in the following data $\mathbf{E}_{24}$=0.987, Θ_{24}=89 degrees, and $\mathbf{E}_{12}$=0.864, Θ_{12}=82 degrees, respectively. This means that the average of daily and semi-daily elliptic hodographs of wind speed vector fluctuations have a stable form and are extended in the direction of an axial line of the lake.

The statistical reliability of short fluctuations of the meteorological characteristics available in the spectra of Figure 3 (for periods of 8, 6, and 4 hours) shows little coherence. It evidences the presence of nonlinear relations between these fluctuations in the atmosphere above the lake.

The synchronous fluctuations of the wind speed and anomalies at the lake level, measured during December 1996 at the Chapala climatic station, show the presence of well-defined seiches (Figure 4). During one month the wind speed fluctuations of daily periodicity were observed having a force of up to 3 m s^{-1} (average per hour values). Only on December 8 and 19 (1996) and on January 2 (1997), the wind speed went up to 6-7 m s^{-1}. This was probably caused by the influence of synoptic variability. Average anomalous fluctuations of the lake level, did not exceed $\pm$10 mm however in separate days they reached $\pm$30 mm.

The time series of the power spectra are illustrated in Figure 4b. Their forms are considerably different. Daily spectral peaks prevail in the wind speed spectrum, although there are also smaller peaks with periods of 12 hours and of 2, 3 and 7 days that are appropriate for synoptic fluctuations.

In the spectrum that shows water level fluctuations, peaks of spectral density in the periods of 24, 5.7 and 4.9 hours are observed. The first is apparently related to wind force caused by daytime breeze. The two peaks on periods 5.7 and 4.9 are undoubtedly connected to fluctuations of the lake water volume. These fluctuations have been described in the scientific literature (Bergamasco and Gacic, 1996; Parsmar and Stigebrandt, 1997) and occur always in lakes and semi-closed sea gulfs. Lake Chapala is extended from West to East and it can almost be described as having a rectangle form. Therefore, to estimate the periods of the first two seiches modes, the known formula "Meriann" can be used (LeBlond and Mysak, 1978):

$$\tau_s = 2l/(n\sqrt{gH}) \tag{14}$$

where l is the length of lake, n is the mode number of level fluctuations, H, the lake's average depth, and g , the acceleration due to gravity. These calculations show, that the period of the first seiches mode is equal to

5.4 hours and the second mode to 2.7 hours. In Lake Chapala the first mode of seiches prevails. Since the form of the lake differs from a rectangular one, there can be simultaneous free waves of the first mode of two close frequencies. The first seiches mode should cause significant fluctuations in levels in the western and eastern parts of the lake and intensive horizontal currents in its central part. This should be taken into account when constructing models and describing the dynamics of pollutants in the lake.

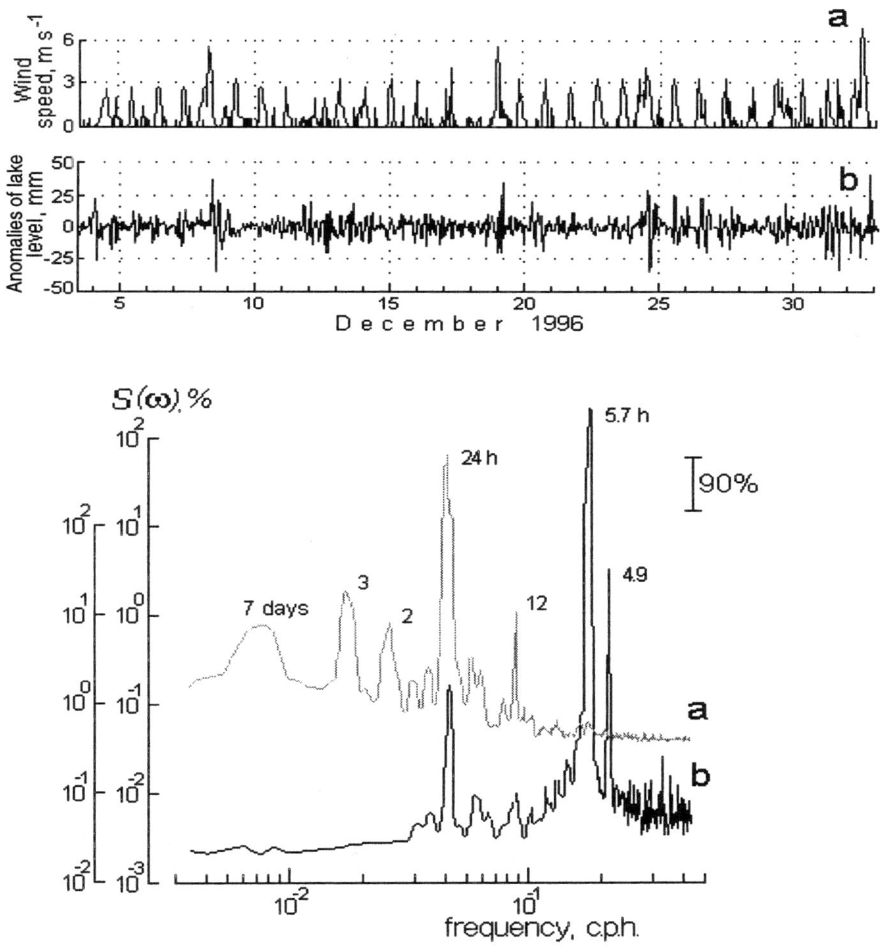

Figure 4 Time series of synchronous fluctuations of wind speed (a) and anomalies in the lake water level (b) during December 1996. The lower part of the Figure shows the power spectra

2.3 Synoptic variability

In the tropics, synoptic weather oscillations are weaker than at moderate latitudes (with the exception of tropical cyclones). Some significant deviations in the daily behavior of almost all of the meteorological characteristics are difficult to explain as consequences of local causes. The synoptic-scale in circulation systems presents variable forms, intensities and duration.

Lake Chapala is located in the southwestern part of the Mexican Plateau, at an average elevation of 1510 m above sea level. An anti-cyclonic circulation influences the central and southern part of this plateau with small gradients in the baric field and average pressure oscillations from 1008 to 1020 hPa (Riehl, 1979). However, due to the non-periodic strikes of cold air from North America and to the movements of atmospheric fronts, the variability of temperature and pressure near the earth's surface shows a synoptic periodicity of 5-9 days (Mosiño and García, 1973). For this reason the weather is typically dry with low humidity in winter.

For the synoptic variations, it was only possible to estimate the time scales and mean square amplitudes based on the daily data collected at the Chapala climatic station during almost 5 years (group II). A high-frequency filter was applied to the time series with a cosine filter at a pass width of 30 days.

Using the filtered series of the hydro-meteorological characteristics, the frequency-time spectra $S(\omega,t)$ were calculated. A time sequence of segments was formed for each series, some in front and some behind, overlapping by 50% (Rozhskov, 1979; Pourahmadi and Salehi, 1984). This procedure allowed us to study the seasonal dependency of the oscillation on the synoptic range of frequencies.

The appropriate length of the segments and their coverage magnitude were experimentally chosen. When excessive lengths were obtained, a consequence of non-stationary processes, the estimates of the frequency-time spectra $S(\omega,t)$ fitted poorly in time. Too short segments worsened the ability of analysis. Tests suggest an optimum length of two months (60 data points) when the objective is to cover one month. The period-gram of each segment was smoothened with three data points.

Spectral estimates $S(\omega,t)$ of the synoptic range of frequencies of the hydro-meteorological characteristics are shown in Figure 5.

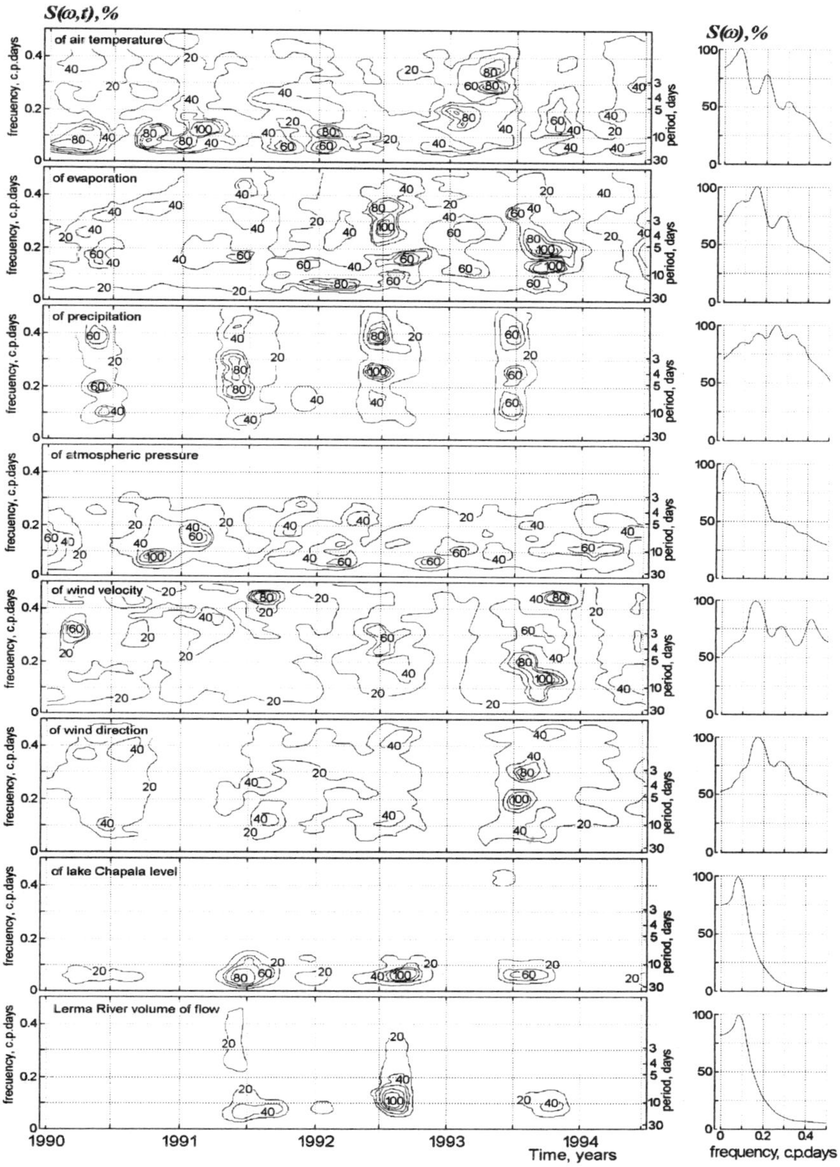

Figure 5 Time-space spectra of the synoptic oscillations, based on time series group II data. Relieves in spectral density are shown as percentages of the maximum value. Average spectra are shown in the right side of the Figure

Table 3 shows the oscillations' periods of the spectral peaks and their mean-square amplitudes. A seasonal cyclic behavior of the synoptic oscillations' intensity is clearly observed in all frequency-time spectra; this is a characteristic of correlated periodic random processes.

Table 3 Parameters of the synoptic variability in Lake Chapala (time series from group II)

Hydro-meteorological characteristics	Mean periods (days) and mean square amplitudes of the synoptic oscillations*
Air temperature	3.5(0.7), 5(0.9), 8(1.1), 10(0.9)
Evaporation	2.5(0.6), 4(0.8), 6(0.6), 9(0.7)
Precipitation	2.5(2.3), 4(9.5), 6(8.7), 9(7.8)
Atmospheric pressure	4.5(4), 6(5), 8(7), 10(6)
Wind module	2.5(56), 4(64), 6(48), 9(52)
Wind direction	2.5(56), 4(64), 6(48), 9(52)
Lake Chapala water level	Nonexistent synoptic variability
Lerma river flow volume	9(51), 15(32)

*Mean square amplitudes from the oscillations of the corresponding periods are shown in parenthesis.

Synoptic processes in the Lake Chapala region present periods from 2 to 15 days, which are in agreement with previous results (Mosiño and García, 1973). Air temperature and atmospheric pressure fields increase during the winter (dry season), and wind, evaporation, precipitation and flow volume of the Lerma river increase in the wet season.

We find considerable high mean-square amplitudes of the synoptic oscillations in the neighborhood of Lake Chapala. They can reach 20-40% of the annual harmonic amplitude; the variation in precipitation can be even greater. The synoptic oscillations of all meteorological characteristics show a similar periodicity. In the Lerma river flow there are only two synoptic periods: 9 and 15 days. Apparently, synoptic processes of shorter periods could not be identified in flow oscillations since they are smoothened by the large area of the basin and the time that the river requires to rise.

Frequently, short-period rain structures in the tropics provide more precipitation than the long-period section of the spectrum. This also confirms our results (see Table 3), and explains the increase in the Lerma river flow by a factor of almost one and a half in a 9-day compared to a 15-day period.

Cross-spectral analysis for meteorological characteristics' pairs were also conducted for the synoptic periods: temperature-atmospheric pressure in the winter and the other characteristics in the summer (see Table 1). All coherence estimates were greater than the confidence limit. Phase displacement for characteristics' pairs varied little in all periods. We may therefore assume that the phase lag of synoptic processes in the Lake Chapala region originates from the same physical regularities as the characteristics of its microclimate, and also from the physical-geographical situation of the lake.

2.4 Seasonal variability

The amplitude of seasonal variability, and its influence on the hydro-meteorological parameters allows us to understand the climatic peculiarities of the studied area. Monthly mean time series of the hydro-meteorological characteristics of group III were used to study seasonal variability. Not all the series have the same quality and number of data. Non-qualitative observations of the wind and atmospheric pressure series are particularly "contaminated". There are segments, which present constant pressure and wind for an entire year. However, these were the only data available.

An estimate of the influence of the "contaminated" parts of the records was made before the spectra estimates. The spectra of the 10-year segments were analyzed in addition to those of 5 years. It was found that segments with a great amount of non-qualitative data did not significantly distort the spectral density of seasonal variability frequencies, but there was a significant deformation of the high and low frequencies.

Figure 6 shows the spectra of the time series. It can be noticed that seasonal variability of the meteorological characteristics of the lake region is completely determined by two overlapping harmonics: annual and semi-annual, as well as by the contribution of the low-frequency tendency and of the high frequency random noise. The contribution of higher intra-annual harmonics with periods of 4, 3, etc. months is almost two orders of magnitude smaller than the main harmonic, even though these oscillations are clearly separated in almost all the spectra.

Thus, the low-frequency behavior of temperature, for example, in the lake region can be described as:

$$T(t) = T_0 + A_0 t + A_1 \sin [\, 2\pi(\omega_1 t + \varphi_1)] + A_2 \sin [\, 2\pi(\omega_2 t + \varphi_2)] + \gamma \quad (15)$$

where T_0 is the mean multi-annual air temperature in the lake region; A_0 is the angular coefficient of the linear tendency; A_1, A_2, ω_1, ω_2, φ_1, and φ_2, are the amplitudes, frequencies, and initial phases of the annual and semiannual harmonic; and γ is the high-frequency random complement. The unknown variables in the equation are easily obtained from the amplitude and energy spectra of the time series. It is clear that not all of the time series can be adequately described with this model; however, the approximation is quite reasonable in most of the cases.

Table 4 gives the values of square coherence, phase differences and mean square amplitudes for hydrological and meteorological elements' pairs. With the exception of air components, the variation of the annual harmonic amplitude of the meteorological elements presents a coherence that almost always exceeds the confidence level.

The pulsations of coherence estimates between the Santiago river flow and the other characteristics may be explained by the hydrological regime of this river that flows out of the lake. Mean square seasonal oscillations in the lake's level are small (21 cm for the annual harmonic), and cannot change significantly the transverse section of the river. There may be other non-periodic factors that produce irregular flow variations in this river. Phase differences, given in Table 4 for the principal seasonal harmonics, have a clear physical significance and reflect the inertial character of the reactions of the different hydro-meteorological characteristics in the seasonal energy balance to the land's surface.

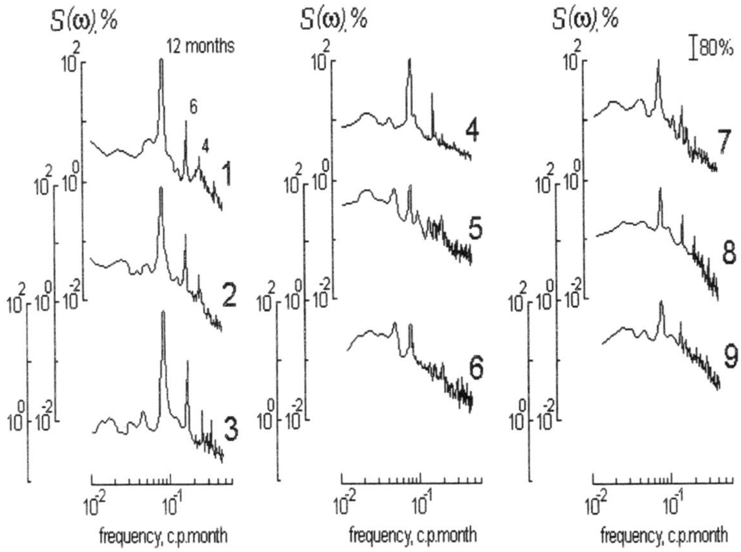

Figure 6 Normalized energy spectra of average monthly hydro-meteorological characteristics registered at the Chapala station during 1971-1998 (group III): (1) air temperature; (2) evaporation; (3) precipitation; (4) atmospheric pressure; zonal component (5) and southern component (6) of the wind speed; (7) lake level oscillation; (8) volume of flow from the Lerma river; (3) volume of flow from the Santiago river

Table 4 Square coherence function values, phase differences and mean square amplitudes of annual harmonic from the hydro-meteorological characteristics of Lake Chapala (time series from group III). Phase differences values in parenthesis are given in months

	a				b			
Hydro-meteorological characteristics	Evaporation	Precipitation	Atmospheric pressure	Mean square amplitude	Evaporation	Precipitation	Atmospheric pressure	Mean square amplitude
Evaporation		71 (2.2)	39 (1.4)	42		—	—	21.0
Precipitation	0.79		318 (10.2)	91	0.72		—	15.1
Atmospheric pressure	0.89	0.97		3.6	0.33	0.64		3.6

2.5 Inter-annual variability

During the present century, Lake Chapala has suffered two nearly catastrophic water level reductions. Since 1945, the annual mean water level decreased in almost four meters during the subsequent ten years; in 1955 and the four years that followed, it increased by almost five meters. From 1977 to 1998 the level decreased again by almost five meters. If the first catastrophic decrease was caused by climatic changes, the second was partly due to an increase in water extraction through the Chapala-Guadalajara and Calderón aqueducts and the Atequiza channel.

Low frequency level fluctuations with periods of 20-50 years are observed in many lakes. They can be attributed to climate fluctuations in the watershed, caused by cyclic changes in solar activity and by period changes in the atmospheric circulation. Within each cycle a regressive one replaces the multi-water phase. The rise and fall of the lake's water level have been traced in these phases as well as the increase or decrease in air temperature, evaporation, rivers' flow, etc. However they are highly dependent on cyclic precipitation fluctuations. Climatic periodic variations, with which nature regulates the lake's water level, have been confirmed during centuries, and is a complex process (Mason *et al.*, 1994; Lappo and Reva, 1997; Gertman, 1997; Meleshko *et al.*, 1998).

It is evident that the extraction of water for agriculture and other uses has not exceeded critical levels. In some way, the lake and its watershed have still a compensation mode from these anthropogenic losses. Figure 7 shows the spectral estimates for time series from group IV. Two types of spectral estimates were calculated: traditional, maximum entropy (ME) and maximum likelihood. The estimates of the ME method were calculated for the maximum length of the white filter P =36, with the relation:

$$P = 2N \cdot \Delta t / \lg(2n \cdot \Delta t)$$

where N is the series' length and Δt, the discrete time period of the data (Konyaev, 1990).

It is shown that ME outlines the details of the spectral density better than the traditional estimate of the spectrum (Figure 7). Table 5 shows the periods of the peaks in the spectral variations, based on both estimates. The inter-annual variations of all parameters present a wide range of periods, from three years to decades.

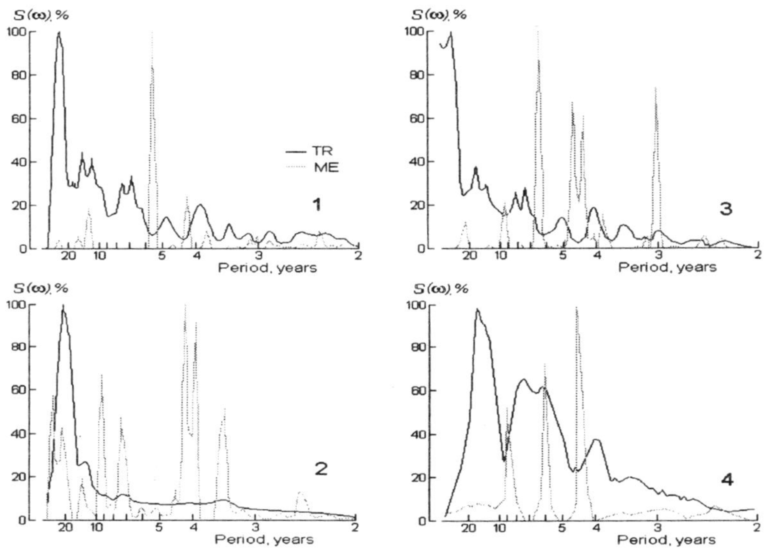

Figure 7 Normalized power spectra of average annual fluctuations in the hydro-meteorological characteristics of group IV: air temperature (1); precipitation in Guadalajara (2); Lake Chapala water level (3); Lerma river flow volume (4). Traditional linear estimations of the spectrum (TR) are shown and so are those using Maximum Entropy (ME) method

The characteristics of the cross-spectra were calculated for oscillations based on a 7-year period, which is present in all spectra. These estimates are shown in Table 6. It is observed that temperature fluctuations and other hydro-meteorological characteristics present coherence at a low confidence level. On the contrary, rainfall fluctuations, the lake water level and the Lerma river flow volume, are well correlated.

There is an urgent need to predict the water level fluctuation in Lake Chapala for a period of one or more years to come. Hansen and van Afferden (2001) simulated water level fluctuations during a 30-year period considering a permanent Lerma river inflow to the lake. These authors also estimated changes in water quality related to fluctuations in the lake's water levels. Another physically proved model that allows the execution of such forecast, is the level fluctuations model on the basis of the water balance equation of the lake (Musilev *et al.*, 1982; Privalsky, 1985).

Table 5 Parameters of variability between years in Lake Chapala (time series from group IV)

Hydro-meteorological characteristics	Mean periods (years) and mean square amplitudes of inter-annual oscillations	
	Traditional estimation of a spectrum	Estimation of a spectrum with ME
Air temperature in Guadalajara	22, 17, 7, 5, 3.8	16, 5.6, 4.3
Precipitation in Guadalajara	23, 16, 7, 5, 4	11, 6.3, 4.8, 4.5, 3
Lake Chapala level	28, 17, 8, 7, 5, 4	25, 9, 6.7, 4.9, 4.3, 3
Lerma river flow volume	28, 7, 4.3, 3	9, 6, 4.6

Table 6 Values of the square coherence function, phase differences and mean square amplitude of the harmonics of the hydro-meteorological characteristics within a 7-year period (time series from group IV)

	Temperature anomalies	Precipitation anomalies	Water level anomalies	Lerma river flow volume	Mean square amplitude.
Temperature anomalies		—	—	176 (4.4)	0.28
Precipitation anomalies	0.24		217 (4.6)	16 (0.4)	14
Water level anomalies	0.28	**0.46**		8 (0.2)	24
Lerma river flow volume	0.30	**0.43**	**0.48**		14

Values of phase differences for Lake Chapala (in parenthesis) are given in years

It is possible to obtain theoretical representations of fluctuations in water level and their connection to statistical characteristics of decisive factors, such as drainage of rivers, evaporation, fluctuations in precipitation, etc. However, this problem is not simple, because the time series of climatic variability are short, i.e., no more than 100 years. Therefore, methods that give effective results in short time series, are necessary to estimate the above-mentioned variables. With no doubt, parametrical (auto-regressive) methods for the estimation of natural processes spectra show some advantage, allowing the creation of stochastic models of the optimum linear level's forecast (Privalsky, 1985).

3. THERMAL AND DYNAMIC MODES

3.1 Experimental details

In September 1996, an experiment was conducted an experiment in Lake Chapala to measure the vertical and horizontal mixing and horizontal circulation processes (Filonov and Tereshchenko, 1997). The experiment was conducted in three stages. In the first stage, a CTD SBE-19 profiler was used to measure vertical temperature profiles and currents at 16 sampling points homogeneously distributed in the lake (Figures 8 and 9).

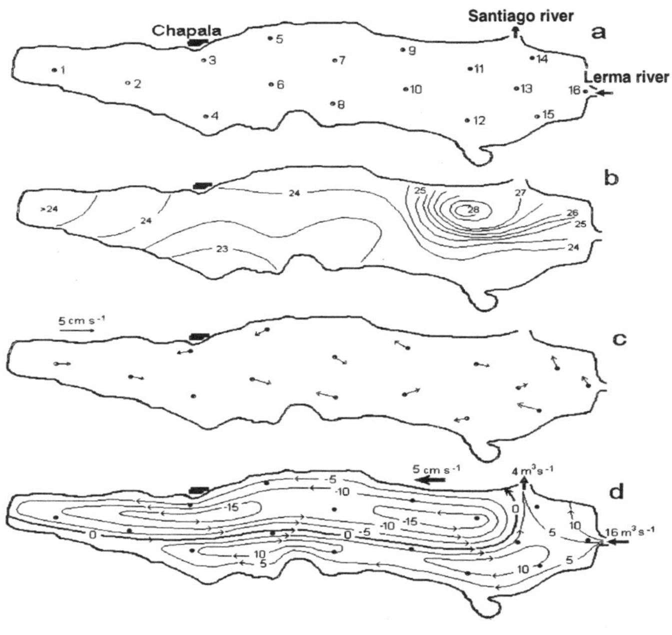

Figure 8 a) Standard hydro-meteorological stations in Lake Chapala. b) Surface temperature distribution (September 10-12, 1996). c) Velocity and direction of currents, measured on a 0.5 m horizon. d) Stationary integrated circulation, with eastern wind 5 m s^{-1}. Iso – streamlines in m^3 s^{-1}

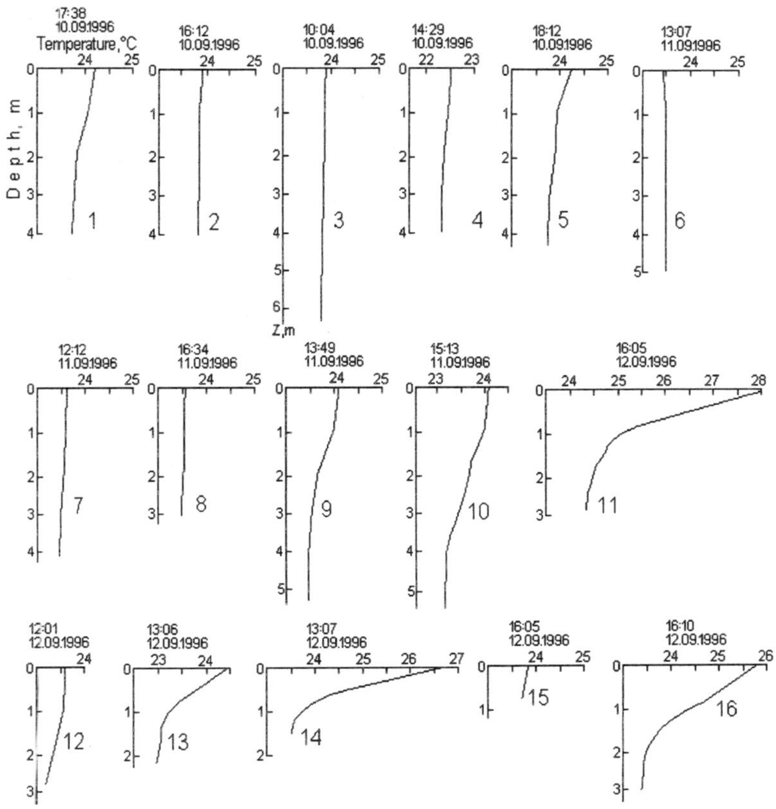

Figure 9 Temperature vertical structure in Lake Chapala on September 10-12, 1996, measured at standard hydrological stations using a SBE-19 SEACAT profiler

Measurements of the vertical temperature profiles and currents were conducted for three days from the 10th to the 12th of September 1996. Temperature profiles were recorded with a vertical resolution of 5 cm. Currents were measured with a current-meter type Ekman at 0.5, 3, 6 m and at the bottom of the lake. At the transects, surface temperatures were measured during two days, from September 18-19; during the first half of the day, from 8:00 am to 13:00 pm, when the early morning breeze develops.
The CTD was towed at a speed of 20 km h^{-1} and temperature data were recorded every second (every 5.6 m). To suppress possible influences of the boat and its outboard motor on the temperature sensors, the anchor instrument was moved 1 m off side of the boat with a special bar. This also provoked that water entering the sensor was mixed in the upper 40 cm layer.

In the third stage, an anchor with two vertical automatic CTD instruments, one at the thermocline and the other close to the bottom was installed in the eastern part of the lake. Several temperature and current profiles were measured around the anchor. Measurements at the anchor were made on September 24 at a depth of 5.8 m, using automatic SBE-16 instruments placed at 1.3 and 5.1 m. The records were made during eight hours with sampling time intervals of 15 seconds. At the same time, were measured from the anchored boat every 5 min with a resolution of 5 cm around the anchor, and the horizontal currents were measured at every meter from the surface to the bottom.

3.2 The circulation model

For the interpretation of instrumentally measured flows, the lake's stationary circulation was estimated when weak eastern wind as registered in most of the stations, as it can be seen in the series of hydrological measurements on September 10-12, 1996. The following shallow-water equations were used (Volzinger *et al.*, 1989; Simons, 1984):

$$gH\frac{\partial \xi}{\partial x} - fv = \frac{1}{\rho}(\tau_{wx} - \tau_{bx}) \tag{16}$$

$$gH\frac{\partial \xi}{\partial y} + fu = \frac{1}{\rho}(\tau_{wy} - \tau_{by}) \tag{17}$$

$$\frac{\partial u}{\partial x} + \frac{\partial v}{\partial y} = 0 \tag{18}$$

where x and y are the horizontal coordinates; u and v, the components corresponding to the vertically-integrated current; $H(x,y)$, the water depth; $\xi(x,y)$, the free surface in perturbation; g, the gravity; f , the Coriolis parameter; ρ , the water density; τ_w , the wind stress, and τ_b , the bottom stress. The wind had a constant speed (5 m s^{-1}) and direction over the whole lake. The wind stress components were calculated using equations $\tau_{wx} = \gamma|W|W_x$ and $\tau_{wy} = \gamma|W|W_y$, where W is the wind speed, W_x and W_y , the components of the wind speed, $\gamma = 3.25 \cdot 10^{-6}$ g cm^{-2} (Volzinger *et al.*, 1989). The bottom stress components were expressed by the equations:

$$\tau_{bx} = \frac{b\rho}{H^2}u \quad \text{and} \quad \tau_{by} = \frac{b\rho}{H^2}v$$

where b = 75 cm^2 s^{-1} is the constant bottom friction coefficient (Simons,1984). Equations (16-18) were transformed into one equation of the stream function $\psi(x,y)$, related to components of the vertically averaged currents by the following expressions:

$$u = -\frac{\partial \psi}{\partial y},\ v = \frac{\partial \psi}{\partial x}$$

$$\Delta^2 \psi - A(x, y)\frac{\partial \psi}{\partial x} - B(x, y)\frac{\partial \psi}{\partial y} - C(x, y) = 0\,, \qquad (19)$$

where

$$A(x, y) = (\frac{fH}{b}\frac{\partial H}{\partial y} + \frac{3}{H}\frac{\partial H}{\partial x}) \qquad B(x, y) = (\frac{3}{H}\frac{\partial H}{\partial y} - \frac{fH}{b}\frac{\partial H}{\partial x})$$

$$C(x, y) = \frac{H}{\rho b}(\tau_{wx}\frac{\partial H}{\partial y} - \tau_{wy}\frac{\partial H}{\partial x})$$

The following boundary conditions were applied:

$$(\psi)_L = 0\,,\quad (\Delta^2 \psi)_L = 0 \qquad (20)$$

Input values for the Lerma river inflow and the Santiago river outflow in September were 16 $m^3\ s^{-1}$ and 4 $m^3\ s^{-1}$, respectively. Equation (19) with boundary conditions (20) were solved for the stream function. For the initial system of equations approximated by a "chessboard" grid-points mesh, the solution was obtained numerically with a grid spacing of 0.5 km.

3.3 Results and discussion

The analysis showed that Lake Chapala (in spite of its shallowness) has a quite complex thermodynamic regime and a significant vertical stratification. The difference in temperature between the surface and the bottom in the center of the lake was of 0.5-1 °C and in the eastern part, of 2-3 °C.

The warmest water was located in the lake's eastern shallow part (0.5-2 m depth), where turbid water flowing from the Lerma river is rapidly heated and, which during calm periods, forms a sharp thermocline with gradients up to 1-2 °C m^{-1}.

Water in the central and southern parts of the lake is 1-2 °C cooler than that of the northern shore (Figure 8b, Figure 9). In the central, western and southern parts of the lake the vertical gradients of temperature are rather weak and do not exceed 0.05-0.2 °C m^{-1}. However, in the eastern shallow part the temperature gradients in the uppermost layer of water of 1-2 m thickness unusually reach high magnitudes, down to 3 °C m^{-1}.

Measurements conducted in early morning hours with no wind demonstrated the existence of cyclonic circulation in the lake with velocities of 2-3 cm s^{-1} at the surface and 1-2 cm s^{-1} in deeper layers (Figure 8c). The resulting calculations of the lake's stationary integrated circulation are shown in Figure 8d. The weak eastern wind causes the formation of wind-oriented currents and a much stronger east-oriented compensation flows in the central part of the lake. Together they form a cyclone water circulation in the North part of the lake and anticyclone circulation in the South. It also leads to the formation of currents that move the warm surface of water of the lake's eastern region to the West along the northern shore. Thus causing down-welling in the North and upwelling in the South, where the center of cold water was detected (see Figure 8b).

Velocities of the integral currents were estimated by the equation

$$|\vec{u}| = (\psi_2 - \psi_1)/Hd$$

where H is the mean depth, and d is the distance between the closest ψ_1 and ψ_2 streamlines. They change from 0 to 1 cm s^{-1}, depending on the region of the lake. Results of instrumental measurements turned out to be almost equal to the model's results. The differences in velocities occur due to the fact that generally, the speed of currents in the subsurface layer are higher than the vertically integrated velocity. Close to noon, there is usually breeze circulation over the lake, which presents a noteworthy, generally cyclonic, component along the shoreline. During these periods, the velocity of the movement of the thin subsurface water layer, influenced by the wind, was significant. Its magnitude was estimated to be 20-30 cm s^{-1}, based on the transport velocity of water lilies with submerged roots of 40-50 cm.

Measurements made by CTD towing showed special features of surface temperature distribution that had not been previously observed (Figure 10). In the eastern part of the lake, an abnormally high temperature spot was frequently observed, and it is marked in Figure 10b as a stain of fine points. In the eastern part of the lake, particularly at transects 12, 13 and 14, spatial variations of 1-2 °C were recorded. The largest temperature oscillations reached 3 °C in the form of cold water patches (Figure 11) of only 50-100 m width. The most intense group of patches was recorded at transect 13, and it is shown in the spatial spectrum of Figure 8c, where temperature patches of this group were observed with a spatial scale of 100 to 300-400 m.

Figure 12 shows the temporal structure of the temperature variation recorded at the anchor in the region of the lake with intense spatial temperature patches. As it was observed, it was possible to instrumentally record the transport of the internal thermal water layer from a device that generates intense internal waves. At the beginning of the observation period (6:35, 24-09-96), the water layer of the anchor area was weakly stratified and had a decrease in temperature between the surface and the bottom of only 0.3 °C. Starting at 10:00, the instruments recorded the onset of a warm water lens surface no thicker than 1.5 m. It was group I of internal waves at the front part of the lens, the group of precursory waves that generally precedes the main group of internal waves generated by the internal bore that moves in shallow water (Turner, 1973; Ostrovsky and Stepanyants, 1989).

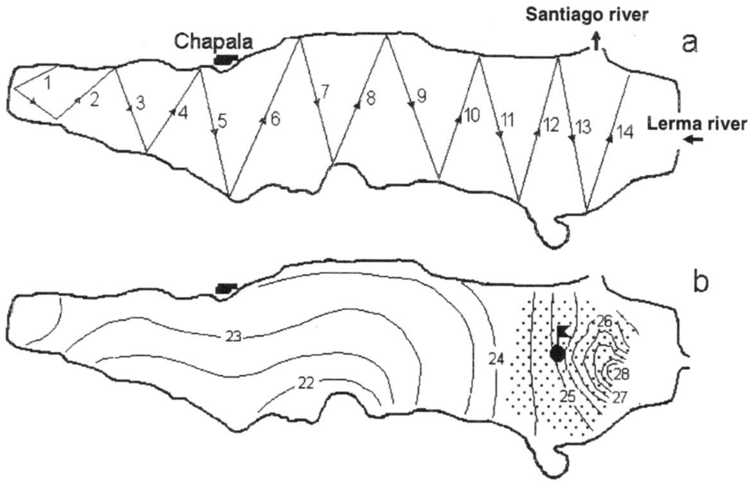

Figure 10 a) Direction and number of transects of the tow. b) Lake's surface temperature distribution based on the data from September 18-19, 1996. The right side of the figure shows the points where intense spatial variations in the surface temperature occurred. Larger points indicate anchor position

Their amplitudes did not exceed 0.5 m, as it was estimated with the formula:

$$\Delta\xi_z(t) = \xi_z(t_1) - \xi_z(t_0) = [T(t_1) - T(t_0)]/(\overline{dT/dz}) \tag{21}$$

where $\xi_z(t_0)$, $\xi_z\, t_1)$ are the vertical displacements of the water layers at level z at times t_0, t_1, which cause the fluctuations in temperature $T(t_0)$, $T(t_1)$; dT/dz is the average vertical temperature gradient during the time period $t_1 - t_0$.

At 11:10, a single nonlinear internal wave (oscillation II in Figure 12b) of 1.5 m amplitude passed through the anchor point. It resembled a Korteveg-de Vries soliton, and lasted only 4 min. Approximately 10 min later, an abrupt forward front of a warm water lens began to develop, accompanied by long series of internal waves (group III) with amplitudes of 0.5-1.3 m, which lasted almost 1 h. As a result of the onset of the warm lens, the water temperature at 1.3 m rose by 0.3 °C and by almost 2 °C at the surface. The internal bore was caused by the onset of the water lenses (patches) close to the anchor site, moving from the East with a velocity of 20-30 cm s^{-1} from a region with significantly lower depths, caused by the early morning breeze, which began to develop at 9:00 a.m. By 10:00, the wind speed close to the anchor reached 5-8 m s^{-1} and had a stable easterly direction. The structure of the variation patches of surface temperature, recorded with the towed CTD in the eastern part of Lake Chapala, was evidently caused by the following reasons:

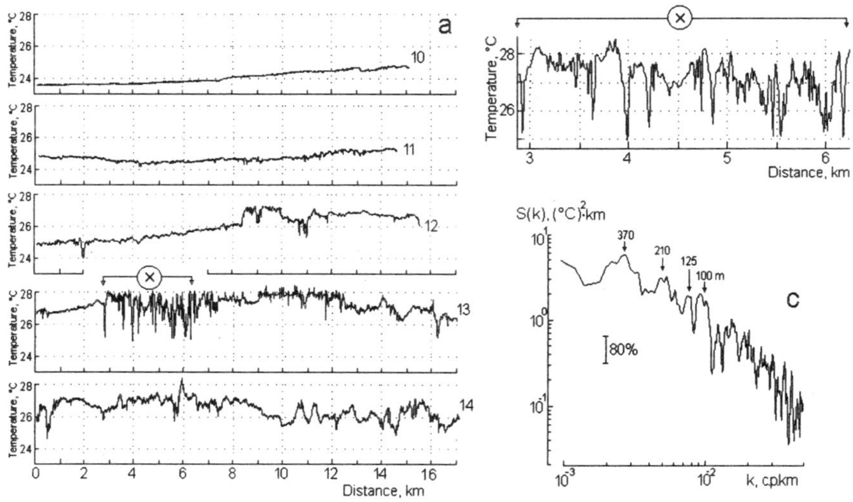

Figure 11 a) Surface temperature at transects 10 to 14. b) A section of the record with especially intense fluctuations in the temperature and spatial spectrum (c) corresponding to section 13

The predominantly easterly early morning breeze that blows from the shore causing, in this part of the lake, the displacement of the thin (only 1-1.5 m) warm water surface layer from the wide shallow region of the lake, which borders the Lerma river inlet towards the colder central part. It also generated a horizontal and vertical Langmuir-type circulation system of converging and diverging movements in the water layer. This initially "breaks" the warm water layer and continues in the lens space (patches), it does so separated and extended by the wind, which due to the instability of the displacement, generates groups of intense internal waves in their periphery, that transport energy from the wind towards the interior of the lake in the form of heat and movement (kinetic). Since breeze circulation occurs over the entire lake throughout the year, it can be concluded that it plays an important role in the mechanisms of vertical and horizontal mixing of water masses.

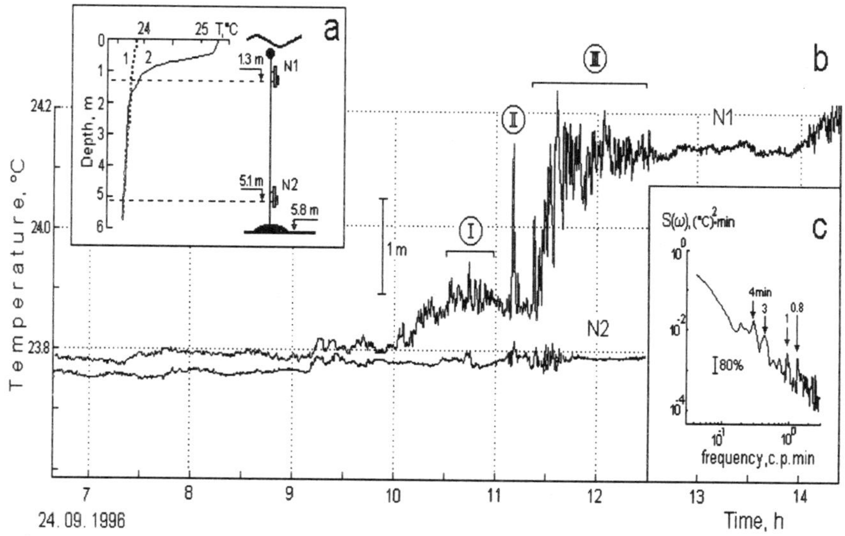

Figure 12 Temperature fluctuations at the anchor. a) Plot of the SBE-16 position and vertical temperature profiles recorded at 8:00 (curve 1) and at 13:00 (curve 2). b) Temperature's temporal behavior at levels 1.3 and 5.1 m. Roman numbers indicate the groups of internal waves and the thin vertical line indicates their scale. c) Frequency spectrum of the short-period internal waves of group III

4. CONCLUSIONS

Climatic processes in the Lake Chapala region and nearby areas are exceptionally complex and variable due to their space-time scales. In this work the spectral analysis of long time series of hydro-meteorological characteristics, such as: air temperature, precipitation, evaporation, atmospheric pressure, wind, water level of the lake, and the rivers' flow volume in the Lake Chapala region were studied. Analysis shows that the spectrum of fluctuations in the Lake Chapala area is wide. Daily, synoptical, annual and semi-annual fluctuations can be adequately expressed. Variations due to the El Niño (2-5 years) and longer periods (from 7 years up to several decades) are satisfactorily defined, especially the strong influences on the fluctuations of the lake's water level. The nature of these fluctuations is analyzed in detail, their mean square amplitudes, coherency and phase shifts between fluctuations of the different hydro-meteorological characteristics on identical frequencies, can be appreciated.

The main dynamic process occurring in the lake is the lake breeze circulation. Daytime breeze does not usually exceed 4 m s^{-1}. Beyond any doubt the lake breeze increases lake surface evaporation. Lake breeze and pressure variations generate free waves-seiches. The spectral analysis of lake level fluctuations from the limnograph show that the periods of free seiches are equal to 5.7 hours, and their amplitude is of 7.5 mm. Seiches generate periodic currents, which can reach an approximately 1 cm s^{-1} nodal line.

The CTD measurements suggest a complex thermodynamic regime in Lake Chapala, and a significant vertical stratification. Temperature differences between the surface and bottom waters ranged from 0.5 to 1.0 °C at the lake's center, and from 2.0 to 3.0 °C at its eastern part. In fact, the warmest water occurred in the eastern section where the lake is shallow and the water is turbid due to the discharge of the Lerma river (depth here ranges from 0.5 to 2.0 m). During calm weather, the water rapidly heats up developing sharp thermocline gradients of 1-2 °C m^{-1}. The water of the central and southern parts of the lake is 1 to 2 °C cooler than that at the northern shore. The CTD towing yielded particular features of surface temperature not described before. The instrument recorded spatial temperature variations of 1 to 2 °C. The largest variation recorded a value of 3 °C and occurred in the form of cold-water patches 50 to 100 m wide. A number of patches was recorded at transect 13 with spatial variation of 100 to 400 m. Larger variations of about 4.7 km were also recorded, and they were oriented along the wind direction.

A buoy station registered temperature fluctuations of 2 °C that belonged to the internal thermal front movements. A well-defined group of internal waves were also recorded at the leading edge of the front with the form of KdeV solitons and amplitudes of 1.5 m. It is deduced that the front by the buoy station was generated by the movement of warm water traveling from the eastern shallow part of the lake. The main force, which impelled the warm water movement, was the morning breeze.

A full picture of the lake's thermodynamic processes would have been possible if a wide range of meteorological stations within the area were established to monitor the parameters at different seasons of the year. It is also important to study the character and interaction of breeze circulation with valley winds in different parts of the lake. Particular attention should be paid to breeze penetration depth when it is landward and then travels back to the lake. How the lake breeze influences the water mass dynamics is another phenomena to be studied. Finally it would be particularly interesting to study local peculiarities of energy transfer mechanisms from the main diurnal harmonic to high-frequency oscillations, as well as the role of non-linear energy transformations.

REFERENCES

Akaike H (1969) Power Spectrum Estimation Through Autoregressive Model Fitting. Ann Inst Statist Math 21, pp 407-419.

Bergamasco A and Gacic M (1996) Baroclinic Response of the Adriatic Sea to an Episode of Bora Wind. JPO, 26(6), pp 1354-1369.

Burg, JP (1972) The Relationship Between Maximum Entropy Spectra and Maximum Likelihood Spectra. Geophysics, 37, pp 417-428.

Burg JP (1967) Maximum Entropy Spectral Analysis. Paper presented at the 37th Annual International Soc Exsplor Geophys Meeting, Oklahoma City, pp 111-123.

Escalante ME (1992) Transporte de Contaminación en el Lago de Chapala (etapa inicial). Informe técnico, CH-9206/07, pp 43-52.

Filonov AE, Monzon, CO and Tereshchenko IE (1996) A Technique for Fastconductivity-temperature-depth Oceanographic Surveys. Geofísica Internacional, 35(4), pp 415-420.

Filonov AE and Tereshchenko IE (1997) Preliminary Results on the Thermic Regime of Chapala Lake, Mexico. Suppl to EOS Transact, AGU, 78(46).

Filonov AE, Tereshchenko IE, Monzon CO (1998) On the Oscillations of the Hydro-Meteorological Characteristics in the Region of Lake Chapala in Time Frames of the Days to Tens of Years. Geofísica Internacional 37 (4), 293-307.

Filonov AE (1998). ¿Morirá el lago de Chapala? Existe la Posibilidad de Salvarlo. Teorema 3, 16-18.

Gertman IF (1997) An Anticipated Height of the Caspian Sea Level in the First Half of the 21st Century as Inferred from the Analysis of Secular Cycles of Solar Activity. Russian Meteorology and Hydrology, 12, pp 72-76.

Gonella J (1972) A Rotary-components Method for Analysis Meteorological and Oceanographic Vector Time Series. Deep-Sea Res, 19, pp 833-846.

Guzmán MA and Morelos ML (1992) La Bibliografía del Lago de Chapala (Análisis hasta 1987). Tiempo de Ciencias, 28, pp 1-22.

Hansen AM and van Afferden M (2001) Toxic Substances: Sources, Accumulation and Dynamics *In: The Lerma-Chapala Watershed: Evaluation and Management* (Hansen AM and van Afferden M eds) Kluwer Academic/Plenum Publishers, London.

Jauregui E (1995) Rainfall Fluctuations and Tropical Storm Activity in Mexico. Erdrunde, Band 49, pp 39-48.

Jenkins GM and Watts DG (1969) Spectral Analysis and Its Applications. Holden-Day. San Francisco-Cambridge-London-Amsterdam, 345 p.

Konyaev KV (1990) Spectral Analysis of Physical Oceanographic Data AA BALKEMA/ROTTERDAM, 200 p.

Lappo CC and Reva YA (1997) A Comparative Analysis of Long-term Variability of the Black and Caspian Sea Levels. Russian Meteorology and Hydrology, 12, pp 41-50.

LeBlond PH and Mysak LA (1978) Waves in the Ocean. Elsevier Scientific Publications Co, Amsterdam, 602 p.

León LF (1994) Modelado del Transporte de Contaminantes. Caso de Estudio: Lago de Chapala. Gaceta del Lerma, Número especial, pp 94-99.

Lind OT, Doyle R, Vodopich DS, Trotter BT, Glass J, Dávalos-Lind L and Limón G (1992) Lago de Chapala: Factores que Controlan la Producción de Fitoplancton. Ing Hidrául en México, pp 17-29.

Mason M, Guzkovska CG and Sreet-Perrot FA (1994) The Response of Lake Levels and Areas to Climate Change. Climate change, 27(2), pp 124-136.

Meleshko VP, Golitsin GS, Volodin EM, Galin VY, Govorkova VA, Meshcherskaya AV, Mokhov II, Pavlova TV, Sporyshev PV (1998) Calculation of Water Components over the Caspian Sea Watershed with a Set of Atmospheric General Circulation Models. Izvestiya, Atmospheric and Oceanic Physics, 34(4), pp 534-542.

Mooers CNK (1973) A Technique for the Cross-spectrum Analysis of Pairs of Complex Valued Times Series with Emphasis on Properties of Polarized Components and Rotational Invariance. Deep-Sea Res., 20, pp 1129-1141.

Mosiño P and García E (1973) The Climate of Mexico. In the climates of North America. Elsevier Scientific Publishing Company, Amsterdam, 345-404 p.

Musilev SV, Privalsky VE and Ratkovitsh DY (1982) Stochastic Models in an Engineering Hydrology. Moscow, Nauka. 184 p.

Ostrovsky LA and Stepanyants YA (1989) Do Internal Solitons Exist in the Ocean? Rev of Geophys 27(3), pp 293-310.

Palmén E and Newton CW (1969) Atmospheric Circulation Systems. Academic Press, NY.

Parsmar R and Stigebrandt A (1997) Observed Damping of Barotropic Seiches through Baroclinic Waves Drag in the Gullmar Fjord. JPO, 27(6), pp 849-857.

Pourahmadi M and Salehi H (1984) On subordination and Transformation of Harmonizable and Periodically Correlated Processes. Lect Notes Math, 1080, pp 195-213.

Privalsky VE (1985) Climatic Variability, Stochastic Models, Forecast, Spectra. Moskow, Nauka, 145 p.

Riehl H (1979) Climate and Weather in the Tropic. Acad Press London-NY-San Francisco.

Rodríguez AA (1994) Programa de Control de Malezas Acuáticas en el Lago de Chapala Jal. CNA-IMTA, Mexico, 241 p.

Rozhskov VA (1979) Metodi Veroyatnostnovo Analiza Okeanologicheskix Procesov. Leningrad, Gidrometeoizdat, 280 p.

Sandoval F (1994) Pasado y Futuro del Lago de Chapala. UNED, Guadalajara, México, 94 p.

Scorer RS (1978) Environmental Aerodynamics. NY-London-Sydney-Toronto.

Simons TJ (1984) Effect of Outflow Diversion on Calculation and Water quality of Lake Chapala. Report Project MKX CWS-01, 23 p.

Turner JS (1973) Buoyancy Effects in Fluid. Cambridge, at the University Press.

Volzinger NE, Klevanny KA, Pelinovsky EN (1989) Long-Wave Dynamics of the Coastal Zone. Leningrad, Hydrometeoizdat, 272.

Chapter 8

Nutrients and Eutrophication in Lake Chapala

JOSÉ DE ANDA[1] AND HARVEY SHEAR[2]

[1]*Centro de Investigación y Asistencia en Tecnología y Diseño del Estado de Jalisco, A. C., Normalistas 800 S. H., Guadalajara, C. P. 44270, Jalisco, Mexico. email: janda@ciatej.net.mx*
[2]*Environment Canada-Ontario Region, 4905 Dufferin Street, Toronto M3H 5T4, Canada. email: harvey.shear@ec.gc.ca*

Key words: Limnology, water quality, eutrophication, phosphorus, nitrogen

Abstract: The primary tributary to Lake Chapala is the Lerma river. Large quantities of domestic, agricultural, and industrial sewage from the entire Lerma-Chapala basin still flow untreated into the watershed and eventually into the lake, resulting in excessive inputs of phosphorus (P) and nitrogen (N), both known to cause eutrophication. Anthropogenic water demand from Lake Chapala surpasses the surface supply and groundwater recharge rate. This has resulted in a hydrologic imbalance in its basin. In addition, high nutrient concentrations in the lake have led to degraded water quality, resulting in growth of floating aquatic vegetation and blue green algae. Some studies establishing a P balance for the lake have been carried out, but none for an N balance. Spatial and temporal P and N distributions in the lake during the past 24 years are discussed in relation to hydrologic changes. A method based on the bathymetric features of the lake is proposed to estimate the amounts of total P and N. This method uses a digital elevation model (DEM) of the site and the water quality data monitored for 24 years in 28 lacustrine and 9 littoral monitoring stations. Many authors suggest that the most effective long-term measure for the control of eutrophication in a water body is a reduction in the input of external nutrients. Determining the nutrient balance of any lake or reservoir constitutes a very important tool to understand its chemical and biological processes. Similarly, the knowledge of existing relationships between hydrological variables, nutrient concentrations, and primary productivity could strongly help to determine the most appropriate restoration and clean-up policies wherever eutrophication problems are present. The different relationships are discussed in this chapter.

1. INTRODUCTION

The goals of the work that will be presented in this chapter are to describe the nutrient concentrations and loading trends for Lake Chapala during the period from 1974 to 1998.

Aparicio (2001) has described the hydrology of Lake Chapala. The volume, surface area, and depth have undergone significant variations over time, some of them caused by the climate, and others by the exploitation of the water resource as discussed below. Table 1 summarizes some of these variations.

Table 1 Relevant hydrological values for Lake Chapala

Values	Volume (Mm^3)	Surface area (km^2)	Mean depth (m)	Date
Minimum	940.37	658.93	1.43	July, 1955
Mean	5341.78	1,071.17	4.98	1934 to 1998
Maximum	9576.29	1,160.08	8.25	October, 1935

The importance of Lake Chapala is that it incorporates all possible beneficial uses such as recreation, agriculture, water supply, fishing, aquatic sports, etc. Unfortunately, large quantities of domestic, agricultural, and industrial sewage from the entire Lerma-Chapala basin still flow untreated into the watershed (Hansen *et al.*, 1995) and this may eventually reach the lake. In addition, erosion processes in the hills surrounding the lake due to deforestation and change in land uses constitute the main non-point nutrient source to the lake (de Anda *et al.*, 2000).

The existing in-basin nutrient concentrations promote the growth of floating aquatic vegetation, water hyacinth (*Eichhornia crassipes*) and bulrush (*Typha latifolia*) that have bloomed in the Lerma river, Lake Chapala, and Santiago river in recent years. Blooms of these weeds have, in the past, covered littoral areas, and have adversely impacted commercial and recreational fishing, waterfowl feeding, boating and swimming, and the aesthetic beauty of the lake (de Anda *et al.*, 1998). For example, in 1993 aquatic vegetation covered more than 135 km^2 representing about 13 % of the total surface of the lake (INEGI, 1995). Most of the water hyacinth present in the lake grows in the Lerma river, and is then carried to the lake during the rainfall season. Recent clean up measures have removed a significant part of the water hyacinths from the Lerma river, preventing most of the plants from entering the lake.

More recently, there have been blooms of blue green algae, identified as *Anabaena flos-aquae*, a cyano-bacterium species that could cause taste and odor problems in drinking water supply to the city of Guadalajara.

Mexican standards allow municipal sewage discharges of up to 5 mg L^{-1} for total P and 15 mg L^{-1} for total N, into lakes and reservoirs (SEMARNAP, 1996). In comparison, the Canada-US Great Lakes Water Quality Agreement stipulates a maximum discharge for P concentration of 0.5-1.0 mg L^{-1} for sewage treatment plants discharging > 4.5×10^6 L day^{-1} (Great Lakes Water Quality Agreement, 1987).

The United States Environmental Protection Agency (USEPA) water quality criteria states that to control algal growth, phosphates should not exceed 0.05 mg L^{-1} if streams discharge into lakes or reservoirs, 0.025 mg L^{-1} within a lake or reservoir, and 0.10 mg L^{-1} in streams or flowing waters which do not discharge into lakes or reservoirs (USEPA, 1999). Surface waters that are maintained at levels of 0.01 to 0.03 mg L^{-1} of total P tend to remain unpolluted by algal blooms.

Because of the elevated Mexican standards, waste water treatment plants along the Lerma river and the shoreline of the lake, are not required to have tertiary treatment to eliminate nutrient discharges. Furthermore, detergents, which contain phosphates, and several mixtures of NPK (nitrogen, phosphorus, potassium) fertilizers, are used in Mexico without regulation. The main uses of land along the shoreline of the lake and the part of the Lerma river close to the lake are agriculture and livestock. These activities frequently develop close to the lake, and they add high phosphorus and nitrogen loads, especially during the rain season.

In previous works, Limón *et al.* (1989) and Limón and Lind (1990) studied long and short-term variations of some physical and chemical parameters of Lake Chapala. Water quality parameters for the past three decades are reported in Table 2. These data show that the dissolved oxygen in the lake reaches values close to a 100% of saturation at the water surface level. The E-W and W-E wind velocities reach values above 9 km h^{-1} and, considering the shallow characteristics of the lake, provide enough circulation to maintain the waters well mixed (Simons, 1984). The lake water is moderately hard and persistently alkaline. Secchi disk and turbidity values show that the lake is highly turbid, limiting the primary productivity (Lind *et al.*, 1992a and 1992b). This turbidity is primarily caused by 1) sediment resuspension in the lake, 2) input of significant amounts of suspended solids from the Lerma river, and 3) erosion of the hills surrounding the lake. Chemical oxygen demand, total solids, total suspended solids, total dissolved solids, hardness and alkalinity have all increased significantly during the period of study. While average nitrate concentrations have decreased over the period of monitoring, average concentrations of organic nitrogen, ammonia, total phosphorus, and phosphates have shown an increasing tendency during the past decades.

2. METHODOLOGY

The common strategy to estimate amounts of nutrients in lakes normally relies on a simple equation (Ryding and Rast, 1989):

$$M(t) = V(t) \bullet C(t) \tag{1}$$

where:

$M(t)$ = total amount of the nutrient in the lake at time "t"
$V(t)$ = total volume of the lake at time "t"
$C(t)$ = average concentration of the nutrient at time "t"

Table 2 Lake Chapala water quality reported for three time periods*

Parameter	1974-79	1980-89	1990-98
Temperature (°C)	22.05	22.08	21.78
pH	8.56	8.66	8.54
Dissolved Oxygen (mg L^{-1})	6.84	6.87	7.13
Biological Oxygen Demand (mg L^{-1})	1.53	1.73	2.01
Chemical Oxygen Demand (mg L^{-1})	17.21	24.27	34.97
Total alkalinity (mg L^{-1})	186.85	247.44	293.25
Total hardness (mg L^{-1})	147.79	190.66	218.16
Total solids (mg L^{-1})	374.21	617.8	748.26
Total suspended solids (mg L^{-1})	22.44	34.57	47.09
Total dissolved solids (mg L^{-1})	351.77	583.23	698.97
Organic nitrogen (mg L^{-1})	0.96	-	1.35
Nitrate (mg L^{-1})	0.51	0.33	0.24
Ammonia (mg L^{-1})	0.14	0.12	0.43
Total phosphorus (mg L^{-1})	0.40	0.62	0.56
Phosphates (mg L^{-1})	0.17	0.40	0.37

* (CNA, 1998)

The average concentrations were estimated by calculating the average of all measured concentrations of nutrients in the lake at time "t". This strategy leads to uncertainty when applied to lakes having any of the following features:

- There are concentration gradients in the water column
- Lake waters are not well mixed
- Flow pathways in the lake produce a non-uniform distribution of the chemical components in the lake
- There are insufficient sampling points in the water body or they are not well distributed.

As explained above, Lake Chapala is wide and shallow and flow patterns change in relatively short time periods (Simons, 1984; Limón *et al.*, 1989). Previous studies (de Anda *et al.*, in prep.) demonstrate that the phosphorus concentrations in the lake in most cases follow a regular E-W distribution where concentrations are usually the highest at the eastern end of the lake because of the Lerma river discharges.

Considering these facts, it is not possible to apply a simple model to estimate the amount of every chemical species in the lake. A better way to make such estimations is to apply a grid model according to Equation 2.

$$M = A \sum_{k=1}^{K} [m \bullet h_k] \quad (2)$$

where:

M = amount of nutrient (P or N) in the lake
A = area of the kth interpolation cell
m = interpolated concentration of nutrient at the kth cell in lake
h = interpolated water column height at the kth cell in the lake
K = total number of cells in the grid covering the lake.

Equation 2 deals with the lake as a 2-D system, since this was considered the best approximation because Lake Chapala is very shallow. One of the most important considerations in the P and N distribution analysis is that vertical P and N gradients are not present in the water column of the lake. Data obtained in the past at various depths in the water column confirm this assumption (CNA, 1981).

Numerical calculations to solve Equation 2 were carried out using the ARC-INFO program supported in a UNIX system. Different numerical trials suggested a grid cell dimension of 500 m x 500 m. Results obtained with a smaller grid dimension were practically the same showing differences between 2 to 3%.

Data used to carry out the 2-D interpolation process were water quality data measured at 28 sampling stations distributed in the lake (Figure 1). These data were available on an at least four-monthly basis from 1974 to 1998. Additionally, concentrations measured in nine littoral sampling stations were also considered in the interpolation process. There is, therefore, sufficient information to make a first approximation of the amount of the different compounds of P and N in the lake.

Figure 2 shows the annual average volume and depth of Lake Chapala according to measurements by the CNA. According to this Figure, as well as in Figures 3 and 4, the shallowest lake depth recorded since 1934 occurred in the middle of the 1950s during one of the most severe droughts in Mexico in the 20th century. The 1950s drought put at risk the water supply for the city of Guadalajara, which at that time had a population of over half a million people.

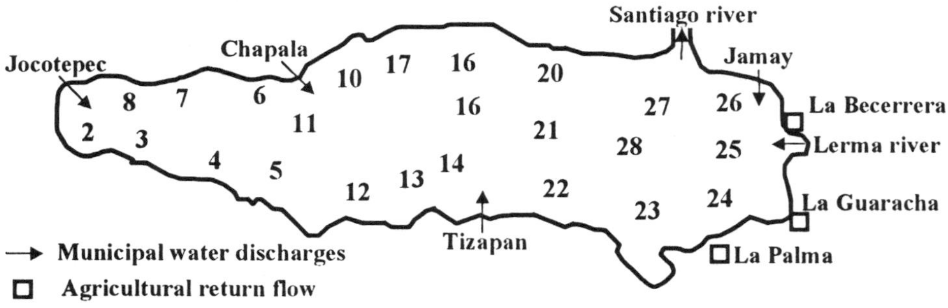

Figure 1 Sampling stations in Lake Chapala

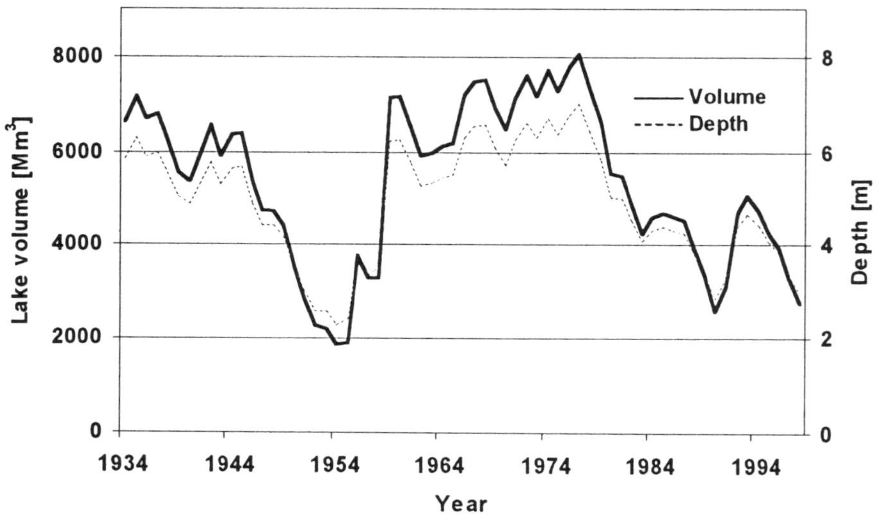

Figure 2 Annual average volume and depth of Lake Chapala

This Figure shows that hydrologic equilibrium of the system was also severely disturbed at the beginning of the 1990s. This second perturbation was caused by excessive water withdrawals in the Lerma river basin, as discussed by de Anda *et al.*, 1998.

Starting in the late 1970s the Lerma river inflow to Lake Chapala has been drastically reduced due to over exploitation upstream (Aparicio, 2001). In order to visualize the negative impact that reduction in flow from the Lerma river has had on the lake, we have divided the average lake volumes for each year by the annual average inflow from the Lerma river, to calculate a hydraulic residence time associated with the Lerma river inflow. These results are reported in Figure 3a.

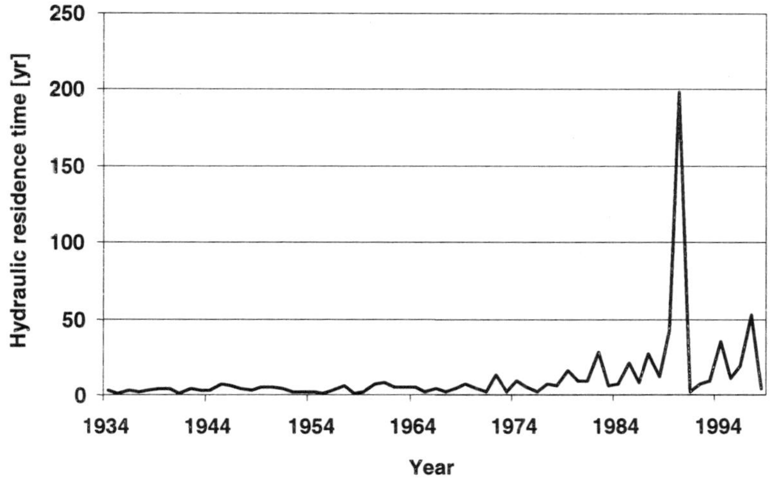

Figure 3a Estimated hydraulic residence time (Lerma river only)

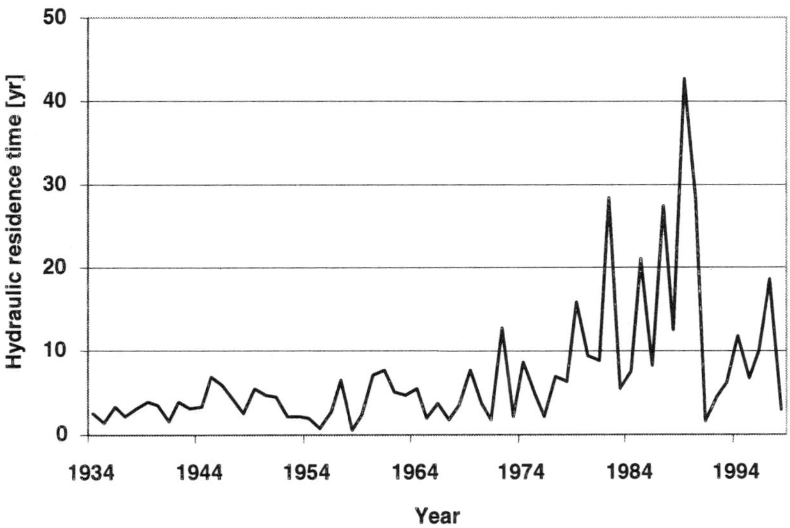

Figure 3b Estimated hydraulic residence time (Lerma and Duero rivers)

The results show an increase in the hydraulic residence time of inflow waters (lake volume/Lerma river inflow), confirming that the second perturbation of the water regime in the 1990s was caused by the use of water in the basin and not by a climatic effect, as suggested by Filonov (1998).

Taking into account the contribution from the Duero river, and calculating the inflow to the lake as the sum of the Lerma and Duero rivers, we have estimated a different scenario for hydraulic residence time (Figure 3b) where a maximum value of 40 years was reached in 1989.

3. RESULTS AND DISCUSSION

3.1 Nitrogen

In natural water and wastewater the forms of N of greater interest are, in order of decreasing oxidation state, nitrate, nitrite, ammonia, and organic nitrogen. All these forms of nitrogen, as well as nitrogen gas (N_2), are biochemically exchangeable and are components of the nitrogen cycle (Greenberg *et al.*, 1992).

To estimate the average amounts of N in Lake Chapala, we used Equation (2). Figure 4 shows the variations in the amounts of organic N (a), ammonia (b), and nitrate (c) in the lake during the 1974-1998 period. Concentrations were calculated by dividing the total amounts by the volume of the lake.

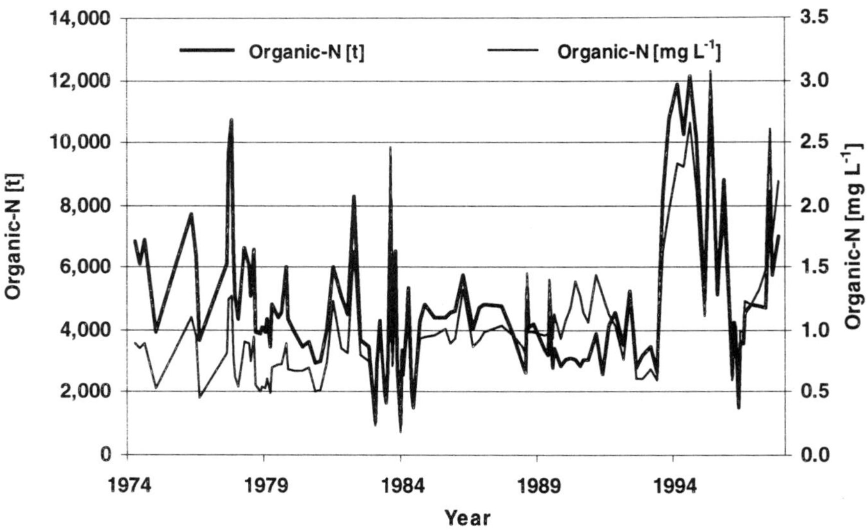

Figure 4a Amount and concentration of organic N in the lake

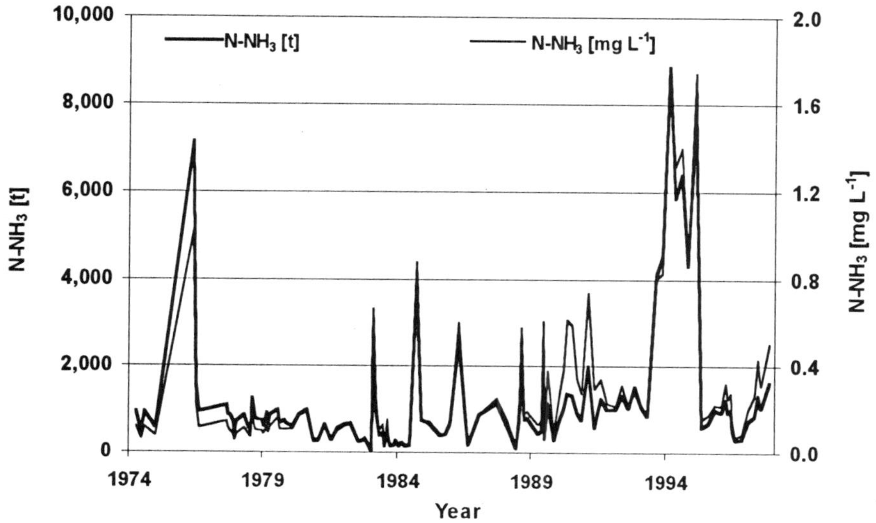

Figure 4b Amount and concentration of $N\text{-}NH_3$ in the lake

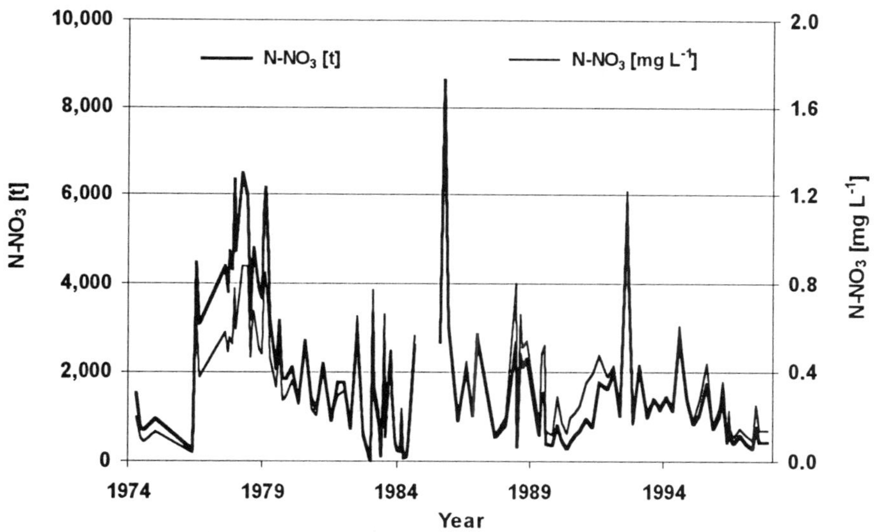

Figure 4c Amount and concentration of $N\text{-}NO_3$ in the lake

3.2 Phosphorus

Unlike N in lake systems, the only significant form of inorganic P is orthophosphate (HPO_4^{2-}). More than 90% of the P is bound as organic phosphates and cellular constituents in living particulate matter of seston, or associated with organic material (Wetzel, 1975).

To estimate the average amounts of total P and HPO_4^{2-} in Lake Chapala, we used Equation 2. Figure 5a shows the variation in these amounts in the lake during the 1974-1998 period. However, all these procedures, interpretation and uses must be done carefully (Thomann and Mueller, 1987).

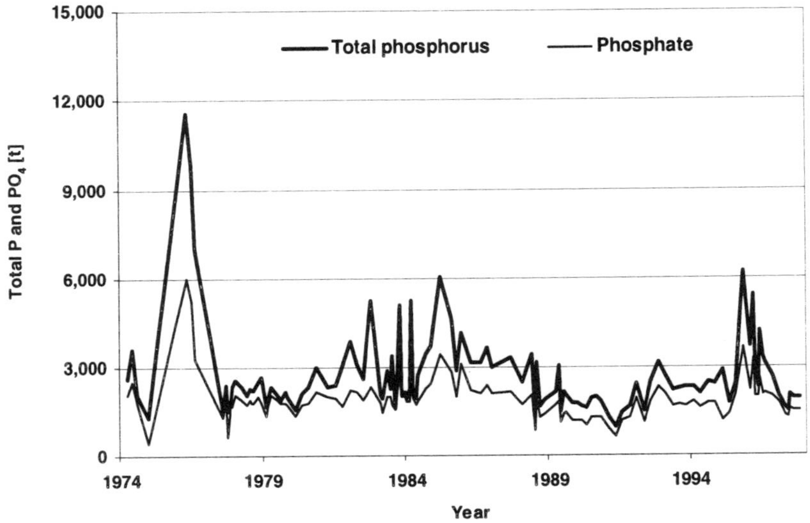

Figure 5a Amount of total P and PO_4^{3-} in the lake

Concentrations as shown in Figure 5b were estimated by dividing the total amounts by the volume of the lake. It is observed that there is no clear long-term trend for P concentrations in the lake, but there are some periods where P concentrations increase significantly (Figure 5b). The high levels of total P reached in 1976 were probably due to very high water levels in the lake, which in that year flooded highly fertile agricultural land around the lake. After this event, phosphorus probably attached to suspended clay particles and settled slowly to the lake bottom sediments. Also in 1976, the Mexican policy in favor of agriculture reduced considerably the prices of fertilizers. Fertilizer production and fertilizing programs increased substantially during this year.

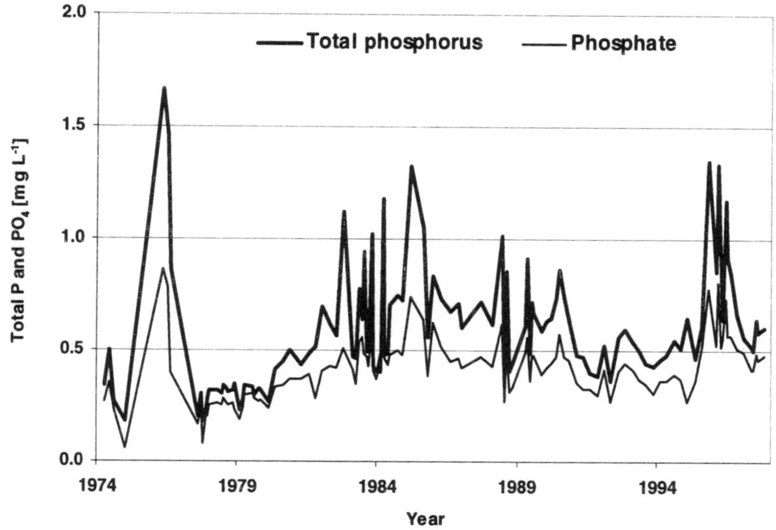

Figure 5b Concentration of total P and PO_4^{3-} in the lake

Solana-Espinosa *et al.* (1998) analyzed P in sediment samples, and found concentrations of about 200 mg kg^{-1} wet weight, with humidity values from 80 to 82%. This suggests that the sediments may be an important internal source/sink of P when pH and redox potentials change at the sediment/water interface (Boers, 1991) and due to resuspension and settling of sediments in the lake.

3.3 N/P ratio behavior in the lake

Plant biomass increases through the uptake of available P and N from the water. Such nutrients result from discharges of point and non-point sources. If nutrients discharged into water bodies are reduced, the pool of available nutrients (that can be used by plants) will be reduced and, as a consequence, so will the total plant biomass (Thomann and Mueller, 1987).

For lakes with non-point source nutrients inputs, P usually controls growth. Lakes that may have significant point source inputs may be limited in N concentrations, but can be made limited in P concentrations by removing P in sewage treatment plants. The N/P ratio is therefore a useful measure for understanding the relationship between N, P, and biomass. Using the amounts of N and P as estimated in the previous section, the N and P ratios were calculated. The tendency of this ratio is illustrated in Figure 6.

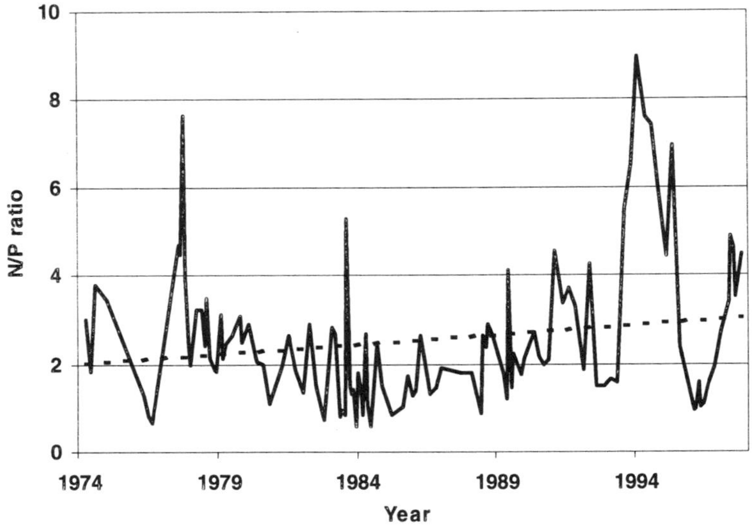

Figure 6 N/P ratio for Lake Chapala

An average N/P value of about 2.5 is observed during the entire monitoring period, with a slight tendency towards increase. N/P ratios below 4 may stimulate the growth of noxious blue-green algae in freshwater (Thomann and Mueller, 1987). These algae can obtain nitrogen for their growth by fixing atmospheric nitrogen. Such algae tend to form unsightly surface mats, which can result in serious impairment of water uses.

The N/P ratio is, however, not the whole story of eutrophication. In Lake Chapala the high amount of suspended sediments limits phytoplankton production processes and therefore there is not an agreement between the empirical models that relate biomass production or trophic status of nutrient concentrations. Such models assume that phytoplankton growth and biomass is controlled by phosphorus and nitrogen availability (Lind *et al.*, 1993).

However, in Lake Chapala phytoplankton production is limited in a higher degree by abiotic factors not related to nutrient availability, namely high inorganic turbidity, than by nutrient concentrations (Lind *et al.*, 1992a, 1992b; Dávalos-Lind, 1993).

4. CONCLUSIONS

The Lerma river basin is one of the most important agricultural areas in the country where the use fertilizers is elevated and consequently it is highly polluted by nutrients. From 1976 to 1977 ammonia and phosphates simultaneously reached high concentrations in the lake. In these years, fertilizers such as $(NH_4)_2HPO_4$ and urea NH_2CONH_2, were extensively used by farmers as a result of a government policy, which urged increase of agricultural production. The prices of fertilizers were reduced, and they were applied without restriction.

From 1994 to 1995 organic nitrogen and ammonia concentrations were high probably as a result of the degradation of organic matter. During this period, the lake experienced a maximum water level after the second period of low water volume. Water quality data in this period showed significant increases in Biological Oxygen Demand (BOD).

At the same time, the extensive water hyacinth growth rapidly vanished in 1995, probably because they sank out of the water column, resulting in high organic nitrogen (1.35 mg L^{-1} organic-N) in the water, and NH_4^+ (0.43 mg L^{-1} N-NH_4^+) as a result of a biological degradation of the organic matter and consumption of oxygen.

Table 2 shows a decreasing tendency in NO_3^- over the monitoring period. The biochemical mechanisms involved in this decline need to be further investigated.

Concentrations of total P and PO_4^{3-} are strongly related, having their main origin in wastewaters discharged into the Lerma river. Approximately 70% of total P occurs as PO_4^{3-}. Phosphorous sources are usually anthropogenic, although some may originate from the decomposition of organic matter.

Vollenweider (1975), Rast and Lee (1978), Salas and Limón (1985); Melack (1992), and Maniak (1997) have proposed some criteria to classify the trophic status of a lake based on the concentrations of P. Vollenweider (1975) proposed a graph that describes the trophic level of a lake or reservoir taking into account the relationship between flow, lake surface area, and annual P surface loading.

Data from Lake Chapala for three different time periods are shown in Figure 7. These values were estimated using monthly measured and estimated loading data from the Lerma river only (de Anda *et al.*, 2000). Subsequently, we obtained yearly surface phosphate loads by dividing the total estimated monthly loads by the surface area of the lake, and then summing over a 12-month period.

Each circle in Figure 7 was calculated by using the yearly average in each period. According to this Figure, Lake Chapala has maintained a consistent eutrophic status during the last decades.

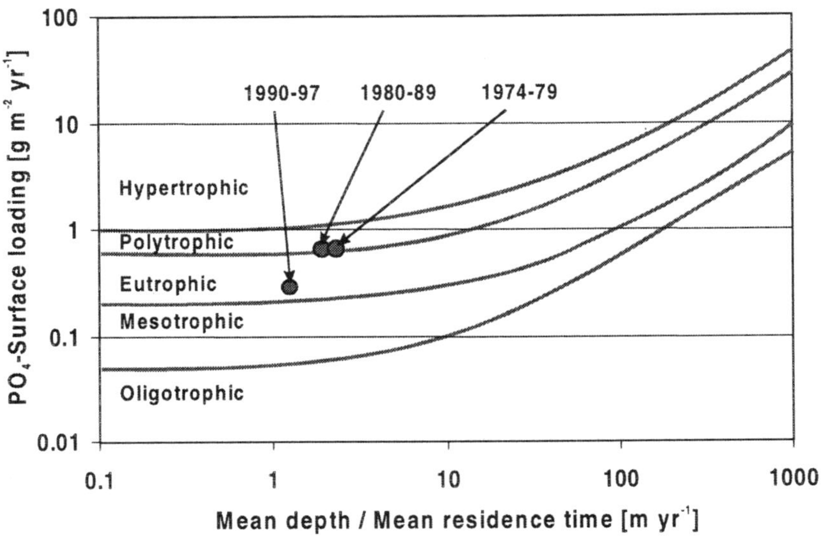

Figure 7 Trophic status of Lake Chapala

ACKNOWLEDGMENTS

This research was supported by following sponsors:

- Universidad Nacional Autónoma de México (UNAM). Programa de Apoyo a las Divisiones de Estudios de Posgrado 1997 (PADEP). Apoyo a proyectos de investigación. Tesis doctorales.
- Instituto Mexicano de Tecnología del Agua.
- Ing. J. Jesús Amezcua Cerda. Comisión Nacional del Agua (CNA). Centro de Estudios Limnológicos (CEL). Mexico.
- Environment Canada-Ontario Region, Toronto, Canada.
- Deutscher Akademischer Austauschdienst (DAAD). Germany.
- Prof. Dr.-Ing. Ulrich Maniak and co-workers. Technische Universität Braunschweig. Leichtweiss-Institut für Wasserbau, Abteilung Hydrologie und Wasserwirtschaft. Germany.

REFERENCES

Aparicio J (2001) Hydrology of the Lerma-Chapala Watershed *In: The Lerma-Chapala Watershed, Evaluation and Management* (Hansen AM and van Afferden M eds) Kluwer Academic/Plenum Publishers, London.

Boers PCM (1991) The Influence of pH on Phosphate Release from Lake Sediments. Wat Res, 25, 3, pp 309-311.

CNA (National Water Commission) (1981) Analytical Results and Vertical Distribution in Lake Chapala (Oct 1977-Oct 1981), Gerencia Regional Lerma-Santiago, Centro de Estudios Limnológicos, Guadalajara, Jal, Mexico. Unpublished.

CNA (National Water Commission) (1998) Water Quality Data in Lake Chapala. Gerencia Regional Lerma-Santiago. Centro de Estudios Limnológicos, Guadalajara, Jal, Mexico. Unpublished.

Dávalos-Lind LO and Lind OT (1993) The Changing State of Limnology in Mexico: Lake Chapala as an Example, Verh Internat Verein Limnol, 25 , pp 427-430

de Anda J, Quiñones-Cisneros S, French R and Guzmán M (1998) Hydrologic Balance of Lake Chapala (Mexico). Journal of the American Water Resources Association. 34(6), pp 1319-1331.

de Anda J, Shear H, Maniak U and Riedel G, (2000) Phosphorus Balance in Lake Chapala (Mexico). J Great Lakes Res, 26(2), pp 129-140.

de Anda J, Shear H, Maniak U and Riedel G (Submitted for publication) Phosphates in Lake Chapala (Mexico), Submitted to the Journal of Lakes and Reservoirs: Research and Management.

de Anda J, Shear H, Maniak U and Riedel G (In preparation) Sediment Transport to Lake Chapala (Mexico).

Filonov AE (1998) Will Lake Chapala Die? Can it be Saved?. Article in the Magazine "Teorema", September-November 1998, Mexico.

Great Lakes Water Quality Agreement (1987) Governments of Canada and the United States of America, Ottawa and Washington.

Greenberg AE, Clesceri LS, Eaton AD and Franson MA (eds.) (1992) Standard Methods for Examining Water and Waste Water. 18th edition, American Public Health Association, American Water Works Association, Water Environment Federation, Washington DC.

Hansen AM, León A and Bravo L (1995) Sources of Contamination and Enrichment of Metals in the Sediments of the Lerma-Chapala Basin. Ingeniería Hidráulica en México, 10(3), pp 55-69.

INEGI (National Institute of Statistics, Geography, and Data Management) (1995) Espaciomapa Guadalajara. Hoja F13-12, Esc 1:250,000, Image date: March-April, 1993.

Limón JG, Lind OT, Vodopich DS, Doyle R and Trotter BG (1989) Long- and Short-term Variation in the Physical and Chemical Limnology of a Large, Shallow, Turbid Tropical Lake (Lake Chapala, Mexico). Arch Hydrobiology, 83, pp 57-81.

Limón JG and Lind OT (1990) The Management of Lake Chapala (Mexico): Considerations After Significant Changes in the Water Regime. Lake and Reservoir Management, 6(1), pp 61-70.

Lind OT, Doyle R, Darrell S, Vodopich, Trotter BG, Glass J, Dávalos-Lind LO, and Limón JG (1992a) Lake Chapala: Factors Controlling Phytoplankton Production. Ingeniería Hidráulica en México. 7 (2/3), pp17-28.

Lind OT, Doyle R, Vodopich DS, Trotter BG, Limón G and Dávalos-Lind L (1992b) Clay Turbidity: Regulation of Phytoplankton Production in a Large, Nutrient-rich Tropical Lake. Limnol Oceanogr, 37(3), pp 549-565.

Lind OT, Terrell TT and Kimmel BL (1993) Problems in Reservoir Trophic-state Classification and Implications for Reservoir Management *In* Straskaba M, Tundisi JG & Duncan A (eds.) Comparative Reservoir Limnology and Water Quality Management, Kluwer Academic Publishers, the Netherlands, pp. 57-67.

Melack JM (1992) Eutrophication and Water Quality in Lakes in Tropical Alluvial Plains. Ingeniería Hidráulica en México, 7 (2/3): pp 142-147.

Maniak U (1997) Hydrology and Water Management. Introduction for Engineers, 4th Edition. Springer-Verlag, Berlin-Heidelberg-New York.

Rast W and Lee GF (1978) Summary Analysis of the North American (US portion) OECD Eutrophication Project: Nutrient Loading-lake Response Relationship and Trophic State Indices. Ecological Research Series, No EPA-600/3-78-008, US Environmental Protection Agency, Environmental Research Laboratory, Corvallis, Oregon, USA: 454 p

Ryding SO and Rast W (1989) The Control of Eutrophication of Lakes and Reservoirs. Man and the Biosphere Series (Jeffers JNR ed) Volume I, The Parthenon Publishing Group, Park Ridge, New Jersey.

Salas HJ, Limón G (1985) Report of the 3rd Meeting for the Regional Development Project of Simplified Methodologies for the Valuation of the Eutrophication in Temperate Tropical Lakes. Guadalajara, Jalisco, Mexico, World Health Organization, Panamerican Health Organization, Environmental Health Program. No report number.

SEMARNAP (Secretaría de Medio Ambiente, Recursos Naturales y Pesca) (1996) Diario Oficial de la Federación. Official Norm Establishing the Maximum Permissible Values of Contaminants in Wastewater Discharged into National Water Reservoirs. NOM-001-ECOL, Mexico City.

Simons TJ (1984) Effect of Outflow Diversion on Circulation and Water Quality of Lake Chapala. Centro de Estudios Limnológicos, Secretaría de Recursos Hidráulicos, Guadalajara, Jal, Mexico. Pan American Health Organization. Project MEX-CWS-010.

Solana-Espinosa G, Israde I, de Anda J and Zárate-del Valle PF (1998) Geochemistry Data at Lake Chapala, Jalisco, Western Mexico. (ECOS M95 and SIMORELOS 970306003) *In* 1ra Reunión Nal. de Ciencias de la Tierra, Soc Geológica Mex, pp 124.

Thomann RV and Mueller JA (1987) Principles of Surface Water Quality Modeling and Control, Ed Harper and Row, New York, pp 644.

USEPA (United States Environmental Protection Agency) (1999) Nutrient Criteria Technical Guidance Manual: Lakes and Reservoirs. EPA 822-D-99-001, April, 1999, Washington DC.

Wetzel RG (1975) Limnology. WB Saunders Co, Philadelphia.

Vollenweider RA (1975) Input-Output Models with Special Reference to Phosphorus Loading Concept in Limnology. Scweiz Zeit, Hydrol 37: 53-84.

Chapter 9

Phytoplankton and Bacterioplankton Production and Trophic Relations in Lake Chapala

A Review

LAURA DÁVALOS-LIND[1] AND OWEN T. LIND[2]

[1]*Chapala Ecology Station (Univ. Autónoma de Guadalajara and Baylor University), PO Box 97388, Waco, Texas 76798-7388, USA. email: laura_davalos-lind@baylor.edu*

[2]*Limnology Laboratory, Department of Biology, Baylor University. Waco, Texas, 76798-7388, USA. email: owen_lind@baylor.edu*

Key words: Phytoplankton, bacterioplankton, trophic relations, clay aggregates, primary production

Abstract: This chapter makes use of data from literature published during 15 years of research supported by NSF, USEPA, Baylor's University Research Committee and Univ. Autónoma de Guadalajara's Special Projects Office and Research Programs. A list of the literature can be found at the end of the chapter. This research has also been supported by the work of many graduate and undergraduate students and others interested in the conservation of Lake Chapala. The chapter integrates research related to production processes in Lake Chapala – primary and secondary production. In Lake Chapala average annual primary production has changed from 80 g C m^{-2} during 1983-84 to 100 g C m^{-2} during 1989-90. This change was due to a modification in the light climate in the eastern region of the lake. Production is governed by inorganic turbidity. Secchi transparency in the lake averaged 0.2 m in the most turbid region and 0.7 m in the least turbid region. Production is also affected by rainy and dry seasons, typical of the tropics. The lake becomes less turbid during the rainy season due to increased depth and lessened sediment resuspension. In laboratory bioassays and in-lake container assays using both artificial and native algae it was determined that the most limiting nutrient for the phytoplankton production is nitrogen, but the ultimate factor is the poor availability of light due to high clay turbidity. Phytoplankton chlorophyll a averaged 5.4 mg m^{-3} of the mixed water column for all regions in the period of 1983-84. Chlorophyll a at the shallowest and most turbid station averaged almost twice that of the other stations. Chlorophyll a increased through the rainy season (as transparency and inorganic N content increased). During 1989-90 chlorophyll values increased to an average of 13.9 mg m^{-3}. Again this change is due to a change in the light climate at the eastern region of the lake. As inorganic turbidity (clays) becomes a limiting factor for the light-dependent phytoplankton production it also can play other roles in the Lake Chapala ecosystem. Suspended clay particles provide surfaces for the adsorption of dissolved organic material. Heterotrophic bacteria can use this concentrated form of organic matter. The clay-organic matter aggregates are colonized by the bacteria and > 80% of the planktobacteria are attached to these aggregates. Bacterial numbers increment as turbidity increases and their biomass is significantly greater in more turbid regions of the lake. Mean annual bacterioplankton production was 67 mg C m^{-3} day^{-1} during 1989-90. The bacterioplankton production showed little seasonal variation in the lake. During the same period the area-weighted annual total phytoplankton production of the entire lake was $12.7x10^{10}$ g C yr^{-1} and the bacterioplankton production, $9.3x10^{10}$ g C yr^{-1}. Phytoplankton and bacterioplankton production are related. The suspended clay particles limit the phytoplankton production due to light reduction; this negatively impacts bacterioplankton production due to the reduction of organic-substrate supplied by phytoplankton. At the same time, clay facilitates bacterial production by concentrating dissolved organic matter that is used by the attached bacteria. The relationship between phytoplankton, bacterioplankton and suspended aggregates give Lake Chapala a unique food web that is in need of further investigation. The presence of the large clay-organic-bacteria aggregates and their relationship to the lake's food web can have interesting and serious implications.

1. INTRODUCTION

Phytoplankton production is the source of energy in the food web of most aquatic ecosystems. The knowledge of phytoplankton production is not only important for understanding the functioning of aquatic ecosystems but is also essential to understand the effects of a changing watershed. Perturbations directly in aquatic ecosystems or their watersheds can cause changes in the phytoplankton production of biomass and/or changes in the transfer of biomass through the food chain that makes the management of such ecosystems extremely complex. Those changes can bring shifts from oligo- to meso- to eu-trophic states. Too little or too much production, "undesirable" organisms causing taste and odor of water, unsafe water for drinking, or low quality and unsafe food source for humans and other species are some of the challenges managers can face (Wetzel, 1983; Lind, 1985; Horne and Goldman, 1994; Cole, 1994; Pourriot and Meybeck, 1995; Lampert and Sommer, 1997).

This review uses published data from 12 years of research in Lake Chapala. This is not a monitoring program but a series of answers to questions regarding the function of the Lake Chapala ecosystem. The first study was conducted from June 1983 to June 1984. It was initiated at the request of the Centro de Estudios Limnológicos (CEL) following concerns regarding possible eutrophication based on nutrient loading. This study sought to answer questions regarding the quantity of phytoplankton production, its variation in the lake and how it was regulated. The purpose was to determine the trophic state and to aid the effective planning of subsequent long-term studies related to the lake ecosystem's water quality and fishery. During this study samples were collected every two weeks from five stations almost equidistant from each other (Figure 1).

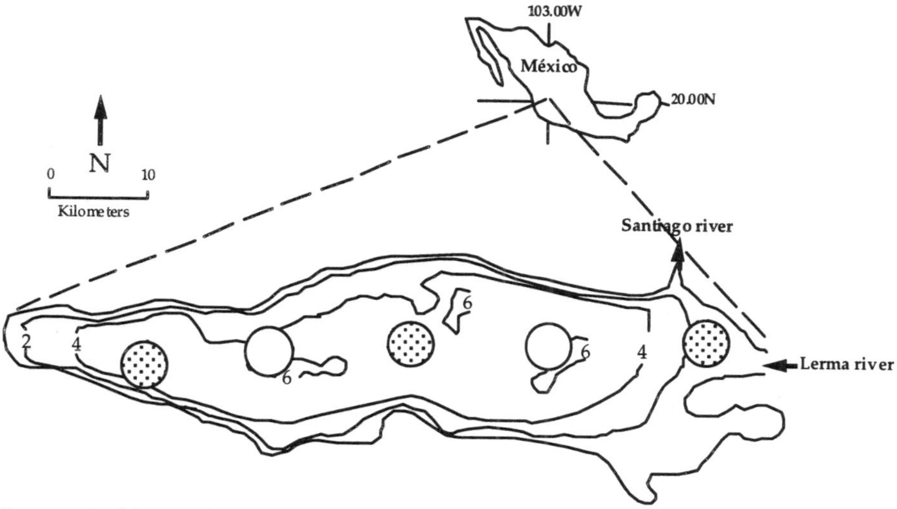

Figure 1 Map of Lake Chapala with isobats (m) showing location of sampling stations. Five stations (circles) were sampled during the 1983-84 study and three stations (shaded circles) were sampled additionally during the 1989-90 study

The discovery of the important role of clay in regulating primary production in Lake Chapala led to a second study (from June 1989 to July 1990). This study sought to answer the question of a compensatory role of the clay particles as an alternate trophic pathway when suspended sediments limit autotrophic production. During this study samples were collected every month from three stations representing the different lake environments (Figure 1).

The discovery of the significance of bacterial production-clay aggregate coupling led to a third study (from May 1995 to June 1997). This study was carried out to test the hypotheses regarding the movement of toxic metals into the lake's food web as facilitated by suspended clays in the water through the alternate trophic pathway. These published results (Lind *et al.*, 2000) do not pertain to the subject of this chapter and will therefore not be discussed here.

2. PHYTOPLANKTON PRODUCTION AND ITS VARIATION

Phytoplankton production in both studies was low and governed by inorganic turbidity. Seasonality in Lake Chapala is related to differences between the rainy (June-September) and dry seasons (October-May). The annual mean surface particulate primary production averaged 80 g C m^{-2} from June 1983 to June 1984 and compares to that of oligotrophic temperate lakes. Production increased to 100 g C m^{-2} from June 1989 to July 1990 placing the lake on the borderline of oligo-mesotrophic classification (Lind *et al.*, 1992, Lind *et al.*, 1997).

Phytoplankton production during 1983-84 was not homogeneous across the lake nor through time (Table 1). Integral water column production (P_{int}), the production measured in the water column, was lowest at the shallowest and most turbid East end of the lake with an average of 151 mg C m^{-2} day^{-1} and highest in the western station with an average of 249 mg C m^{-2} day^{-1}. The differences in production among stations were significant ($P < 0.05$) (Lind *et al.*, 1992, Lind *et al.*, 1997).

Table 1 Comparison of mean ± SD net daily phytoplankton particulate production and chlorophyll a at each region of Lake Chapala. Production notations are: P_{int} = integral production (mg C m^{-2} day^{-1}), P_{mix} = mean volumetric production within the mixed water column (mg C m^{-3} day^{-1}), Chloro = mean in the mixed water column (mg Chl *a* m^{-3}), P_{cap} = maximum Chl *a*-specific production (mg C (mg Chl *a*)$^{-1}$ h^{-1}). For 1983-84, n=23 except Central station where n=46. For 1989-90, n=16, na = data not available (modified from Lind *et al.*, 1997)

Region	Study	P_{int}	P_{mix}	Chloro	P_{cap}
West	1983-84	249±81	56±18	5.0±2.5	9.6±4.4
	1989-90	248±95	na	9.5±3.1	7.9±5.0
Center	1983-84	209±80	30±124	2.0±1.4	10.0±5.6
	1989-90	250±61	na	9.3±3.1	7.3±2.3
East	1983-84	151±93	86±73	9.3±3.4	10.2±6.0
	1989-90	262±108	na	23±15	6.8±2.6

Seasonal patterns of P_{int} during 1983-84 varied among stations (Figure 2). Each station showed a production maximum in the beginning of the rainy season but subsequent patterns varied for the different regions of the lake. The western region showed a general increase throughout the rainy season and reached a maximum at the beginning of the dry season. The central region production was generally higher and variable in the rainy season. The eastern region production was high at the beginning of the rainy season and then decreased sharply. In all cases production declined toward the end of the rainy season and remained low throughout the dry season. The maximum variation of P_{int} through the year was at the eastern region (CV = 62%) where the fluctuation of water level was severe.

In the 1989-90 study, P_{int} was more homogeneous across the lake primarily because of incremented production at the eastern region due to a more favorable light climate (Table 1). The central station had the least seasonal variation (CV = 24%).

P_{int} varied seasonally in the 1989-90 study. The patterns were similar to those of 1983-84. Each station had a maximum at the start of the rainy season, which then declined and remained low (Figure 2).

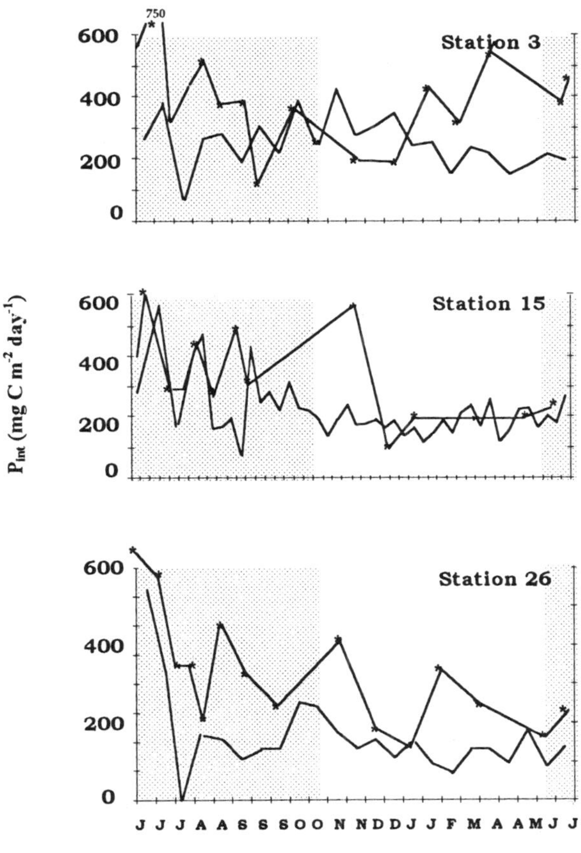

Figure 2 Seasonal patterns of integral water column phytoplankton production in Lake Chapala. Shaded areas represent rainy season. Station 3 represent the western region, Station 15, the central region, and station 26, the eastern region. Legends: Stars, the 1989-90 study; none, the 1983-84 study (modified from Lind *et al.*, 1992 and Lind *et al.*, 1997)

Phytoplankton biomass in 1983-84 was similar at the West and central stations (5.0 and 4.2 mg Chl *a* m^{-3} respectively) and much greater at the eastern station (9.3 mg Chl *a* m^{-3}). The biomass in 1989-90 showed the same pattern – although higher – similar at the West and central stations (9.5 and 9.3 mg Chl *a* m^{-3} respectively) and this time greatly increased at the eastern station to 23 mg Chl *a* m^{-3}. The seasonal variation in algal biomass was greatest at the West end of the lake (CV = 50%) and the seasonal pattern of Chl *a* was similar at each station. After an initial decline at the start of the rainy season (dilution), phytoplankton biomass increased through the rainy season and reached a maximum near the beginning of the dry season.

In both studies the phytoplankton community production per mean unit volume (P_{mix} , mg C m^{-3} day^{-1}) at the eastern region was much higher than at any other station because of the higher biomass and shallower water column, and the phytoplankton was more concentrated per unit volume of water. The central region had the lowest phytoplankton community production per unit volume (P_{mix}) with the deepest circulating water column (Table 1).

Even if there were large differences among regions of the lake in P_{int} and P_{mix} during 1983-84 (Table 1), the photosynthetic capacity of phytoplankton, P_{cap}, mg C (mg Chl *a*)$^{-1}$ h^{-1}, was similar among regions. P_{cap} was uniform and typically high. Mean P_{cap}, of 9.6-10.6 mg C (mg Chl *a*)$^{-1}$, equaled or exceeded that of other lakes. For example, in Lake Constance P_{cap} was 3.1 mg C (mg Chl *a*)$^{-1}$ (Tilzer 1983) and in a tropical African lake, this value was of 8.1 mg C (mg Chl *a*)$^{-1}$ (Robarts 1979). The intensity of illumination required to saturate and inhibit phytoplankton photosynthesis, raise as temperature increases, and photosynthetic capacity of phytoplankton adapts to the lake's light climate with a compensation occurring as the light climate becomes less favorable (Wetzel, 1983). During 1989-90 the photosynthetic capacity of phytoplankton diminished by an average of 26% in all stations (Table 1).

3. PHYTOPLANKTON PRODUCTION AND ITS REGULATION

Phytoplankton production in Lake Chapala was governed by light attenuation of inorganic turbidity and not from lack of essential elements such as nitrogen or phosphorus. Both elements are present in high concentrations (Limón *et al.*, 1989). The lake's high clay turbidity regulates the light climate that ultimately determines its trophic state and the food web associations.

Shallow lakes such as Lake Chapala have a potential for high production especially when located in tropical areas with long growing seasons and mild temperatures. Nevertheless, it is important to recognize the potential for sediment resuspension that results in light attenuation. The degree of resuspension is a function of fetch, wind velocity and the nature of the sediments. Lake Chapala has an elongate basin with ~ 75 km fetch that channels wind along its axis, and clays are small (0.5 μm mean diameter). In Lake Chapala the rainy season brings the greatest transparency over most of the lake – contrary to other lakes, but common in many other Mexican lakes and reservoirs (Lind *et al.*, 1992; Hernández-Avilés *et al.*, in press).

In both studies, Lake Chapala showed significant among-station differences in water transparency (Table 2). The lake has two optically different regions – a small eastern region and the remainder of the lake. In the eastern region, Secchi transparency averaged only 0.2 m, the vertical attenuation coefficient averaged 9.7 m^{-1}, and the photic depth was almost always < 1 m. The source of light attenuation could be either clay or phytoplankton. To distinguish the nature of light attenuation, we ran multiple regression analyses with turbidity as the dependent variable and inorganic suspended solids (clay), chlorophyll *a* and particulate organic carbon as independent variables. We found turbidity not as important to phytoplankton production as the ratio Z_{mix}:Z_{eu} (depth of mixing to depth of euphotic zone), which describes the time period a cell spends in the dark unable to photosynthesize. Z_{mix}:Z_{eu} was greatest at the mid-lake region. Phytoplankton in this region spends more time in the dark than at the other regions. For the eastern region in 1989-90 turbidity was caused mainly by phytoplankton. This was not the case in 1983-84. Turbidity increased due to algal blooms initiated by the supply of inorganic nitrogen and a favorable Z_{mix}:Z_{eu} caused by greater shallowness in this region compared with 1983-1984. For the other regions, turbidity was primarily due to clays and not plankton.

Table 2 Comparison of mean ± SD of selected physical and chemical characteristics of each sampled station in Lake Chapala during 1983-84 and 1989-90 studies. na= data not available (modified from Lind *et al.*, 1992 and 1997)

Characteristic	Study	Region		
		Western	Central	Eastern
Secchi depth (m)	1983-84	0.6 ± 0.20	0.6±0.1	0.2 ± 0.1
Turbidity (NTU)	1989-90	47±48	36±10	121±74
Total P (μg L^{-1})	1983-84	426±57	428±55	487±93
	1989-90	na	na	na
SRP (μg L^{-1})	1983-84	388±43	403±46	452±81
	1989-90	na	na	na
NO_3-N (μg L^{-1})	1983-84	53±56	190±185	255±356
	1989-90	235±136	188±138	235±187
NH_3-N (μg L^{-1})	1983-84	48±52	44±27	159±324
	1989-90	65±35	68±41	113±126

Algal bioassays made during 1983-84 and in studies continuing to date show that N is the growth-limiting nutrient, provided that adequate light is present (Dávalos *et al.*, 1989). During both productivity studies, inorganic nitrogen (ammonia and nitrate) was high and, although rather uniform across the lake, was typically greater near the region of river inflow. Although both P and N varied significantly throughout the year, the variability in N was much greater (Table 2). The laboratory assays indicate that the nutrients in the water can support much higher concentrations of algae than are actually present. The significance of these findings is important if the lake water level were to increase with a corresponding decrease in clay turbidity. The high concentration of nutrients such as nitrogen and phosphorus will be used by the phytoplankton in presence of more available light with a consequent increase in lake productivity.

In 1984 studies were conducted to evaluate the input of nitrogen by nitrogen-fixing cyanobacteria (Glass, 1987). Fixation of nitrogen was not found. This was attributed to wind-driven turbulence and its associated light deficiency. Nitrogen fixation is an expensive process that requires energy derived from photosynthesis. When sediments were removed from lake samples and cyanobacteria were artificially introduced, nitrogen fixation occurred. This could be due to the increased availability of light for photosynthesis or to the removal of inhibiting agents in the suspended clays.

Suspended clays also may influence production through adsorption-desorption reactions of nutrients and/or toxic substances. Laboratory bioassays were conducted on elutriates (water containing chemicals released by sediments) to study the effects of resuspended sediments on lake phytoplankton growth. Phytoplankton growth in elutriates of 100 and 200 mg L^{-1} of superficial lake sediments from the central region of the lake inhibited growth by 50% in either concentration. Elutriates of the same concentrations from the eastern region inhibited growth completely in both concentrations (Dávalos-Lind, 1996). The average concentration of inorganic suspended solids in the lake is approximately 100 mg L^{-1}.

The trophic relation between primary production and nutrient regeneration by grazers was studied in 1984. The regeneration of NH_3-N by Lake Chapala's zooplankton is an important contribution to the NH_3-N pool. It accounted for an average of 14.5% of ambient NH_3-N concentration. Regeneration by zooplankton provided an average of 33% of NH_3-N taken up by phytoplankton. Variability in regeneration by zooplankton, and uptake by phytoplankton was high both daily and monthly. Though regeneration, on average, can supply the nitrogen required to maintain phytoplankton production, it is insufficient by itself to meet nitrogen demands at all times of year (Trotter, 1988). The phytoplankton-zooplankton interactions are important to the nutrient dynamics and production processes in Lake Chapala and are in need of further investigation.

4. BACTERIOPLANKTON PRODUCTION AND ITS VARIATION

Bacterioplankton production in Lake Chapala was high. The production was favored by the high turbidity. Most of the bacteria live attached to clay-aggregates contrary to other aquatic systems where they are free floating (Lind and Dávalos-Lind, 1991).

Mean annual bacterioplankton production among the regions ranged from 59 to 75 mg C m^{-3} day^{-1} (Lind *et al.*, 1997). The annual coefficient of variation was less than 6% for the eastern and central regions and only 10% for the western region.

Shallow East-region bacterioplankton biomass was 302±107 μg C L^{-1}. In the central region it was 142±51 μg C L^{-1}, and at the western region, 180±100 μg C L^{-1}. The eastern region consistently had higher bacterial biomass compared with the other regions that were similar. The seasonal pattern in bacterial biomass closely paralleled that of phytoplankton chlorophyll *a* at the western region, moderately so at the central region, and least at the eastern region. Bacterioplankton abundance was high in the highly turbid eastern region of the lake (Table 3). The abundance was very similar in the central and western region $1.95\pm0.47\times10^{10}$ cells L^{-1} and $1.95\pm0.57\times10^{10}$ cells L^{-1}, respectively. The mean number of aggregate-associated bacteria ranged between four to seven times greater than the number of free bacteria at any station (Table 3). Up to 90% of the water column bacteria were associated with the lake's clay-organic matter-aggregates. Even at the least turbid condition, aggregate-associated bacteria were overwhelmingly important. The aggregate-associated bacteria cells were 56% larger than free-living bacteria. Because the clay particles in the lake adsorb dissolved organic matter and form large aggregates they are used efficiently as substrate by bacteria. The clay micelle serves as a nucleus for the aggregation of dissolved organic carbon (DOC) that can be used by heterotrophic bacteria. Arruda *et al.* (1983) suggested that in clay-rich water, up to 30 times more available organic matter is present on clay surfaces than is available via phytoplankton production. The relative importance of free-living vs. aggregate-associated bacteria to the lake ecosystem is poorly understood. Nevertheless it is clear that the clay-organic aggregate is important in regulating the success of Lake Chapala's planktonic bacterial community. No wet vs. dry season patterns in bacterial number, total bacterial biomass, or size of aggregate-associated bacteria at any region of the lake were found. As with phytoplankton biomass, bacterial biomass also increased from 1991 to 1997 – 120% and 433%, respectively (Lind and Dávalos-Lind, 1991).

Table 3 Annual mean (± SD) abundance (x 10^8 cells L^{-1}, * indicates total number of bacteria), size (μm^3) of free-living and aggregate-associated bacteria and turbidity (NTU) by station (modified from Lind and Dávalos-Lind, 1991 and Lind *et al.*, 1997)

Station	Abundance		Size		Turbidity
	Free	Attached	Free	Attached	
Western					
Year 1991	7.6±1.0	27.8±3.7	0.073±0.017	0.088±0.004	18.1±1.9
Year 1997*	195±0.57				
Central					
Year 1991	6.4±0.8	30.5±5.8	0.064±0.007	0.090±0.006	20.6±1.9
Year 1997*	195±0.47				
Eastern					
Year 1991	9.1±0.08	68.8±10.3	0.055±0.010	0.115±0.015	49.1±25.1
Year 1997*	405±1.18				

Mean specific bacterioplankton production; i. e., production per bacteria biomass, was much less at the eastern region, 0.009 μg C μg C^{-1} h^{-1}, than at the other two regions where it was similar, 0.019 and 0.020 μg C μg C^{-1} h^{-1} respectively. The turnover time of bacteria for the eastern region was 4.6 days and at the central and western regions was 2.2 days. The specific bacterioplankton production was greater when bacterioplankton biomass was low, suggesting food-resource limitation. This is not unexpected in a lake with low primary production where P_{cap} (photosynthetic specific production, mg C (mg Chl $a)^{-1}$ h^{-1}) for the eastern region has decreased by 33% from 1983-84 to 1989-90. Specific bacterial production allows evaluation of environmental conditions as they affect the individual organism and thus reflect the "health" of the producers.

Clay aggregates may not only adsorb organic matter but also pollutants such as metals and pesticides. Lind *et al.* (2000) addressed this effect.

5. BACTERIOPLANKTON PRODUCTION AND ITS REGULATION

In Lake Chapala, bacterioplankton production was regulated both by the clay turbidity and the supply of organic substrate. Suspended clay micelles adsorb dissolved organic matter and form aggregates rich in nutrients. The bacteria, which attach to the aggregates and use the adsorbed nutrients, are larger in volume and number than the free-living bacteria (Table 3).

Dissolved organic carbon (DOC) was negatively, positively, or not correlated with phytoplankton production or biomass depending upon the region (Lind *et al.*, 1997). The large amounts of DOC (eastern region = 28±10, central region = 25±7 and western region = 23±6 mg L^{-1}) and the fact that DOC was inconsistently correlated with phytoplankton production or chlorophyll suggests another source of organic carbon exists, which is important to bacterial production. Total organic carbon (TOC) rather than DOC was the variable that explained changes in cell size and biomass of the bacterial community since most organic matter in the lake was associated with clays. For the western and central regions particulate organic carbon (POC) was a good estimator of the organic matter on the clays available for use by the associated bacteria.

The importance of clay-adsorbed organic substrate was demonstrated by elutriation studies. The metabolic activity of bacteria was greatly increased when grown in elutriated lake water. Elutriates of sediments from the eastern and central region caused 146% and 197% increase in metabolic activity, respectively (Dávalos-Lind, 1996). As discussed before elutriates from the same regions inhibited phytoplankton growth. The western region was not included in this study.

Multiple-regression analysis indicated that variation in the number of aggregates explains variation in the number of bacteria; he higher the number of aggregates the higher the number of attached bacteria. Whereas variation in the organic matter component of aggregates (TOC) explains variation in cell size (Table 4).

Table 4 Percent variation (r^2) in the listed dependent variable explained by turbidity and turbidity + TOC (from Lind and Dávalos-Lind, 1991)

Dependent variable	Turbidity	Turbidity + TOC	F-ratio (2,12, df)
Aggregate associated			
Cell size	3.2	46.3	5.2
Cell number	44.6	50.1	6.0
Biomass	64.6	83.3	29.8
Free living			
Cell size	0	0	---
Cell number	9.1	9.1	1.3
Biomass	7.6	19.9	1.5

Specific bacterial production was determined primarily by POC at the western and central regions. Phytoplankton production at optimal optical depth (P_{opt}) was also important but secondary at these regions. For the eastern region phytoplankton production was the only factor that could explain variation in bacterial specific production, in this region both high phytoplankton production and biomass occurred (Lind *et al.*, 1997)

6. TROPHIC RELATIONS IN LAKE CHAPALA

The trophic relations in Lake Chapala were presented in a conceptual model where the double role of suspended clay is introduced (Lind *et al.*, 1997). Clay modifies the autotrophic food web pathway by limiting the light penetration in the water and at the same time favors the heterotrophic food web by concentrating organic matter (Figure 3).

The autotrophic pathway is in the open water via zooplankton and planktivorous fishes. The observed low rates of primary production due to light limitation in Lake Chapala (< 100 g C m^{-2} yr^{-1}) were insufficient to account for the lake's reported fishery of almost $9x10^6$ kg yr^{-1}. Several methods of relating phytoplankton production to fish production exist in the literature. By applying several of these the inadequacy of phytoplankton production (1983-84) to support Lake Chapala's food web was evident. Table 5 shows the predicted annual fishery (kg) using six different models. These models calculate a fish production smaller that the reported fish harvest (Lind *et al.*, 1994). A substantial quantity of organic carbon available to the upper trophic levels of Lake Chapala must come from sources other than direct grazer harvest of autochthonous primary particulate production. Some is likely in the form of particulate and dissolved organic carbon entering from the Lerma river as pulsed flow at the onset of the rainy season. Much of this comes from beds of water hyacinth (*Eichhornia crassipes*) although this amount remains to be quantified.

Table 5 Application of various fishery models to Lake Chapala. The 1980-87 average annual fishery for Lake Chapala = $8.7x10^6$ kg (from Lind *et al.*, 1994)

Model used	Predicted annual fishery (kg)
OGLESBY (1977) Primary Productivity Commercial Fishery	$1.1x10^5$
Trophic Dynamic (10% efficiency, 35% harvest efficiency)	$1.3x10^5$
OGLESBY (1977) Algal Biomass Commercial Fishery	$6.4x10^5$
JONES AND HOYER (1982) Algal Biomass Sport Fishery	$1.3x10^6$
MELACK (1976) Primary Production African Lake	$5.4x10^6$
OGLESBY (1977) Morphoedaphic Index Large Lake	$2.0x10^7$

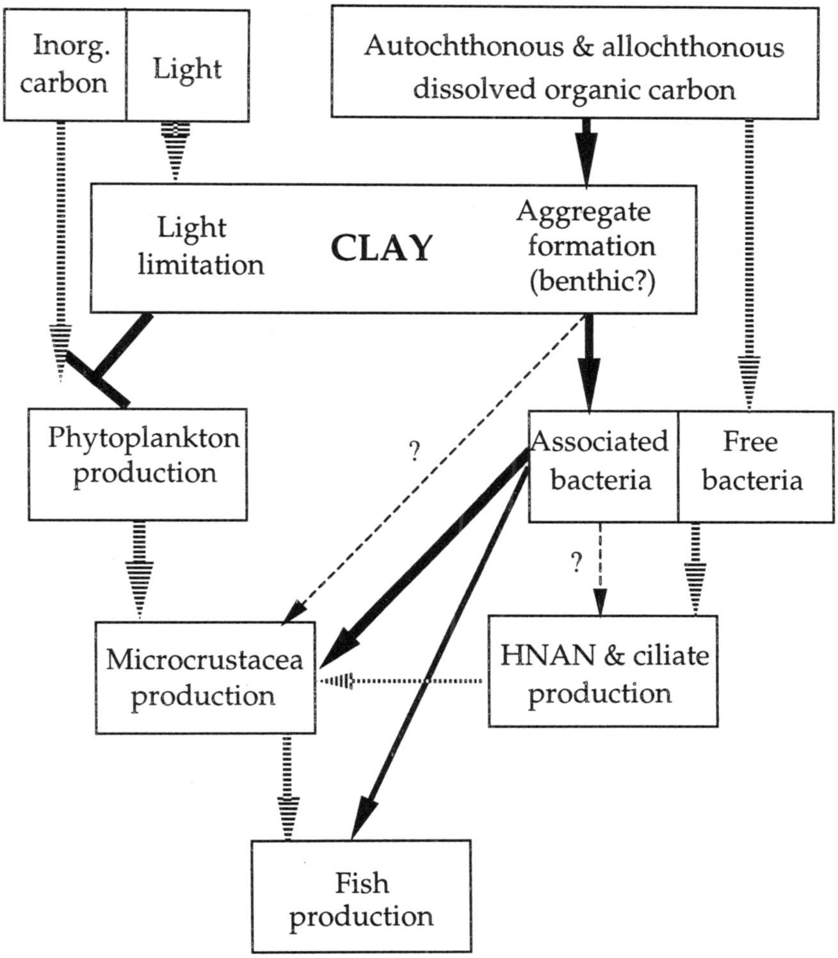

Figure 3 Conceptual model of the role of suspended clay in modifying aquatic food-web processes. Inhibition of phytoplankton photosynthesis is partially offset by increased bacterioplankton production due to concentration of DOC onto clay aggregates whose surfaces and bacteria colonize interiors. The clay-organic-bacteria aggregate (COBA) reaches a size capable of being grazed by larger zooplankton. From Lind *et al.*, 1997

Water-column bacterioplankton production exceeded phytoplankton production for nine central and western region dates. With only one exception, these occurred during the dry season. The area-weighted annual total phytoplankton production of the entire lake was 12.7×10^{10} g C yr^{-1}. When compared to the lake-wide area-weighted annual bacterioplankton production of 9.3×10^{10} g C yr^{-1} (Table 6), it is observed that the latter represent 42% of the total small-particle production. The phytoplankton production was low due to the very shallow photic depth, and bacterioplankton production represent a significant portion of Lake Chapala's primary production.

Table 6 Comparison of annual total (particulate and soluble) phytoplankton autotrophic production. From Lind *et al.*, 1997

Sampling Region	Production ($g\ C\ m^{-2}\ yr^{-1}$)	
	Phytoplankton	Bacterioplankton
East	125	24
Central	139	114
West	138	113
Lake (area-weighted*) $g\ C\ yr^{-1}$	12.7×10^{10}	9.3×10^{10}

*used 30%, 55%, and 15% for West, Central and East Regions respectively

In laboratory experiments the significance to higher trophic levels of the clay-bacteria-organic matter-aggregates (COBA's) was measured. The size of the COBA's is equal of that of some phytoplankton cells. Fish such as *Oreochromis niloticus* and *Goodea atripinnis* were tested and were not able to consume the COBA's, and they always showed a preference for zooplankters. But the endemic *Chirostoma* spp. was capable of consuming the COBA's. *Chirostoma* spp. had the ability to switch food sources from zooplankton to COBA's when given the choice and when their first choice (zooplankton) was depleted (Dávalos-Lind *et al.*, 1992). Further studies are needed but it is important to recognize the COBA's potential as a compensatory food source in the light limited and low productive Lake Chapala.

The presence of COBA's in lakes is a significant factor in limnetic processes. We still have much to learn about how they modify Lake Chapala's limnology including their role in concentrating pollutants (metal, pesticides, etc.) and facilitating their movement into food webs.

7. CONCLUSIONS

High clay turbidity is the main characteristic governing phyto- and bacterioplankton production of Lake Chapala. Clays have two roles in the lake's ecosystem. One is their interference with sunlight penetration. At the nitrogen and phosphorus concentrations in the lake, models predict that the lake is eutrophic or hyper-eutrophic. The high clay concentration in the lake is the key regulating factor of primary production – it is what keeps the lake from being hyper-eutrophic. The other role is the ability to form COBA's, which provide a compensatory food source for the biota. Besides adsorbing organic matter, clays can also concentrate ionic or polar pollutants. If the COBA's constitute an alternate food source for the Lake Chapala food web they also potentially constitute a toxic food source for the lake food web, including man. Management practices, both for water quality and for fisheries, which fail to recognize the importance of the COBA's may be inappropriate and fail.

REFERENCES

Arruda JA, Marzolf GR and Faulk RT (1983) The Role of Suspended Sediments in the Nutrition of Zooplankton in Turbid Reservoirs. Ecology. 64:1225-1235.

Cole GA (1994) Textbook of Limnology. Waveland Press Inc, USA, 412 pp.

Dávalos L, Lind OT and Doyle R (1989) Evaluation of Phytoplankton-limiting Factors in Lake Chapala, Mexico: Turbidity and the Spatial and Temporal Variation in Algal Assay Response. Lake and Reservoir Management, 5(2):99-104.

Dávalos-Lind L (1996) Phytoplankton and Bacterioplankton Stress by Sediment-borne Pollutants. Journal of Aquatic Ecosystem Health, 5: 99-105.

Dávalos-Lind L, Sada R, Guerra A, Velarde G, L, Orozco J, Chrzanowski T and Lind OT (1992) La Producción Bacteriana y su Importancia en la Cadena Trófica en el Lago de Chapala. Ingeniería Hidráulica en Mexico, May/December.

Glass JA (1987) Biological Nitrogen Fixation in a Nitrogen Limited Tropical Lake, Lake Chapala, Mexico. MS Thesis. Baylor University.

Hernández-Avilés S, Bernal Brooks F, Velarde G, Ortiz D, Lind OT and Dávalos-Lind L (in press) The Algal Growth Potential and Limiting Nutrient of 30 Lakes and Reservoirs of the Mesa Central of Mexico. Verh Internat Verein Limnol, p. 27.

Horne AJ and Goldman CR (1994) Limnology. 2nd Edition. McGraw-Hill Inc, USA, 576 p.

Lampert W and Sommer U (1997) Limnoecology: The Ecology of Lakes and Streams. Translated by Haney JF. Oxford University Press, USA, 382 pp.

Limón JG, Lind OT, Vodopich DS, Doyle RD and Trotter BG (1989) Long- and Short- term Variation in the Physical and Chemical Limnology of a Large, Shallow, Turbid Tropical Lake (Lake Chapala, Mexico). Arch Hydrobiol Suppl, 83 (1): 57-83.

Lind OT, Chrzanowski TH and Dávalos-Lind L (1997) Clay Turbidity and the Relative Production of Bacterioplankton and Phytoplankton. Hydrobiologia 353:1-18.

Lind OT (1985) Handbook of Common Methods in Limnology. 2nd Edition, Kendall/Hunt Publishing Company, USA, 199 pp.

Lind OT and Dávalos-Lind L (1991) Association of Turbidity and Organic Carbon with Bacterial Abundance and Cell Size in a Large, Turbid, Tropical Lake. Limnol Oceanogr, 36(6):1200-1208.

Lind OT, Dávalos-Lind L, Chrzanowski T and Limón J (1994) Inorganic Turbidity and the Failure of Fishery Models. Int Revue ges, Hydrobiol, 79:7-16.

Lind OT, Doyle R, Vodopich DS, Trotter BG, Limón JG and Dávalos-Lind L (1992) Clay Turbidity: Regulation of Phytoplankton Production in a Large, Nutrient-rich Tropical Lake. Limnol Oceanogr, 37(3): 549-565.

Lind O, Dávalos-Lind L and Ford T (2000) Clay and the Movement of Metals into Food Fishes. J Environ Sci Health, Part A35(7): 1171-1182.

Pourriot R and Meybeck M (1995) Limnologie Générale. Masson SA, France, 956 pp.

Robarts R (1979) Underwater Light Penetration, Chlorophyll a and Primary Production in a Tropical African Lake (Lake McIlwaine, Rhodesia). Arch Hydrobiol, 86:423-444.

Tilzer MM (1983) The Importance of Fractional Light Absorption by Photosynthetic Pigments for Phytoplankton Productivity in Lake Constance. Limnol Oceanogr, 28:833-846.

Trotter BG (1988) The Role of Zooplankton-phytoplankton Community Interactions in the Nutrient Dynamics of Lake Chapala, Mexico. MS Thesis, Baylor University.

Wetzel RG (1983) Limnology. 2nd Edition, CBS College Publishing, USA, 767 pp.

Chapter 10

Fish Fauna of Lake Chapala

Past and Present

RODRIGO MONCAYO-ESTRADA[1] AND HÉCTOR RENÉ BUELNA-OSBEN[2]

Programa de Ecología Acuática y Pesquera, Centro Interdisciplinario de Investigación para el Desarrollo Integral Regional, Instituto Politécnico Nacional Unidad Michoacán (CIIDIR IPN MICHOACÁN), Justo Sierra 28, Jiquilpan, 59510 Michoacán, Mexico.
email: [1]rmoncayo@hotmail.com; [2]rbuelna@mail.ciidirmich.ipn.mx

Key words: Ichthyofauna, fish community structure, fishery

Abstract: Lake Chapala is one of the most important reservoirs, which reflects the origin, evolution, and biogeography of freshwater fish fauna in central Mexico. It is the area where more diversity of the endemic subfamily of *Goodeinae* ("pintillas" and "tiros"), with nine species, is found. This zone outstands for the evolution and sympatric specialization of "charales" and "pescados blancos", (*Atherinidae*) which include eight species. It is also the habitat of endemic "bagres" (*Ictaluridae*), different native carps (*Cyprinidae*), and historically, it served as the habitat of lampreys (*Petromyzontidae*). The ichthyofauna of the lake is also very important, at least seventeen of the species provide on-the-spot work of artisan fishery for 1,699 fishermen belonging to 86 fisheries' co-operatives, and is the local and regional food source. Unfortunately, Lake Chapala presents a high level of environmental degradation, due to a profound modification of the ecosystem. As a result of this fact we can confirm the actual presence of 27 out of 39 registered species, including both native and exotic. Likewise, before 1990 there were high-volume captures (13,710.78 tons per year), however the present production has decreased by a factor of ten. The aforementioned problem cannot be attributed to a single cause, but to a synergetic effect that involves different factors combined with the lack of regulation and pursuit of plans for sustainable development. A contemplated conservation initiative for the restoration of the lake is not the best way to solve the problem. Recommendations are formulated regarding future studies and activities on the situation of the ichthyofauna in the lake.

1. INTRODUCTION

Mexico boasts an extensive and interesting ichthyofauna due to its highly varied physical geography, large latitudinal extension geological history. Particularly in the southern third part of the Mexican Plateau of the westward flowing Lerma-Santiago river system, endemism has reached taxonomical levels as subfamily and genus (Barbour, 1973a). Said diversity is described in ichthyo-geographical terms as the Lerma province (Álvarez, 1972; Miller, 1986), which include the Lerma river basin, lakes Patzcuaro, Zirahuen, Cuitzeo, Chapala and some others with varying degrees of ecological significance. The abundance and variety of the Lerma fish fauna is not only due to the great diversity of water bodies, ranging from warm to cold, from stagnant to swift and turbulent, from ponds to large rivers and from shallow to deep lakes, but also to the diverse origin of fishes.

In particular, Lake Chapala (the largest natural lake in Mexico) provides an excellent example for describing ichthyofauna in the Lerma river and its characteristics. Lake Chapala holds the most distinctive element of fish fauna in central Mexico, the endemic subfamily *Goodeinae*. Some species of marine origin are permanent freshwater residents in the lake (*Atherinidae*; *Chirostoma*). A group of native carps arrived from the western mountain track and established itself in the lake (*Algansea, Notropis* and *Yuriria*). A few components are characteristic genera of the North American ichthyofauna (*Scartomyzon* and *Ictalurus*), manifested in related groups in the Bravo river basin. There is one group (*Poeciliopsis*), which represents the rich and varied ichthyofauna of Central America. It includes species related to the anadromous fishes of the Pacific coast of North America, and which are relicts in lacustrine environments (the lamprey genera *Lampetra*). The evolution of its ichthyofauna has been documented; there also exists an important fossil record in various locations in and around the actual Lake Chapala basin, from which the presence of an ancient lake of larger dimensions can be inferred (Miller and Smith, 1986; Table 1).

The presence of fish is important for the socio-economical development of the area, since it supports traditional fishery activities for commercial and/or subsistence purposes. This is the main activity of small-scale fishermen, and the captures are the main source of protein in their diet. At the same time, fishermen's families and other members of the community take part in the various activities ranging from capture to the post-harvest process, creating jobs for many people, thereby resulting in regional development. Generally speaking, the total investment in fishing equipment is low, and the capture, which is achieved with the consumption of a ton of fuel, is far greater than that registered by the industrialized fishing sector (Sainsbury *et al.,* 1988).

Table 1 Composition of ichthyofauna in Lake Chapala

Known fossil records related to the Chapala Formation[1]	Native fish of Lake Chapala	Exotic fish of Lake Chapala
Cyprinidae	***Petromyzontidae***	***Cyprinidae***
Algansea sp.	*Lampetra spadiceus* [2]**	*Algansea lacustris*[2, 4]
Notropis sallei	***Cyprinidae***	*Carassius auratus*[4]
Yuriria alta	*Algansea popoche* [3,4]*	*Ctenopharyngodon idellus* [4]
Catostomidae	*Algansea tincella* [3]	*Cyprinus carpio* [4]
Scartomyzon sp.	*Notropis calientis* [2]	***Poecilidae***
Ictaluridae	*Notropis sallei*[2]	*Poecilia sphenops*
Ictalurus dugesi	*Yuriria alta* [4]	*Poecilia reticulata*
Ictalurus spodius	***Catostomidae***	*Xiphophorus helleri*[2]
Atherinidae	*Scartomyzon austrinus* [2,4]	***Ictaluridae***
Chirostoma sphyraena	***Ictaluridae***	*Ictalurus punctatus* [3, 4]
Chirostoma spp.	*Ictalurus dugesi* [3, 4]	***Atherinidae***
Salmonidae	*Ictalurus ochoterenai* [3, 4]	*Chirostoma estor* [2, 4]
Oncorhynchus sp.	***Atherinidae***	***Salmonidae***
Rhabdofario sp.	*Chirostoma arge* [2, 4]	*Oncorhynchus mykiss*[2, 4]
Salmo australis	*Chirostoma consocium* [4]	***Centrarchidae***
Centrarchidae	*Chirostoma chapalae* [4]	*Lepomis macrochirus* [2, 4]
Micropterus relictus	*Chirostoma jordani* [4]	*Micropterus salmoides* [2, 4]
Goodeinae	*Chirostoma labarcae* [4]	***Cichlidae***
Alloophorus robustus	*Chirostoma lucius* [4]	*Oreochromis aureus*[4]
Chapalichthys encaustus	*Chirostoma promelas* [3, 4,] *	*Tilapia rendalli*[4]
Goodea atripinnis	*Chirostoma sphyraena* [4]	
Tapatia occidentalis	***Poecilidae***	
	Poeciliopsis infans	
	Goodeinae	
	Allotoca dugesi [2]*	
	Alloophorus robustus	
	Chapalichthys encaustus	
	Goodea atripinnis	
	Skiffia bilineata [3]*	
	Skiffia lermae [2]*	
	Skiffia multipunctata [2]	
	Xenotoca variata	
	Zoogoneticus quitzeoensis	

[1] Data from Smith *et al.* (1975); Miller (1986), and Miller and Smith (1986)
[2] None collected or observed in recent years
[3] rare
[4] fishing resource

Both water quality and ichthyofauna of Lake Chapala have been modified by human activities (Guzmán-Arroyo, 1995). Many areas have suffered serious deterioration – some have even disappeared – due to the steady advance of civilization, as a consequence of the unbalanced extraction and diversion of water (Aparicio, 2001). Several pollutants, including inorganic and organic substances as well as industrial waste have a negative impact on the ecosystem (Hansen *et al.,* 1995). Likewise, at the present time, over-exploitation of fishing resources gives rise to serious problems, which are intensified by the low productivity of the system. With no doubt, the conservation and restoration of the natural environment presents a challenge that must be faced both by the population in the basin and nationwide. The main objective of this chapter is not only to evaluate the past and present situation of the ichthyofauna, but also to analyze the community's structure and the fisheries. At the end, the chapter will propose alternative measures regarding future actions and studies.

2. MATERIALS AND METHODS

Parallel to an intensive review of literature on the subject, the following four museums were surveyed in order to obtain historic records of fishes in Lake Chapala: (1) Escuela Nacional de Ciencias Biológicas, Instituto Politécnico Nacional, Mexico City (ENCB); (2) Colección de Peces, Universidad Michoacana de San Nicolás de Hidalgo, Morelia (CPUM; our acronym); (3) University of Michigan Museum of Zoology, Ann Arbor (UMMZ); (4) Florida State Museum (UF).

Fish population clumps are not easy to study because they represent populations whose presence usually cannot be seen from the surface, and boundaries cannot be effectively located. They also change their distribution with time – even during sampling (Whittaker, 1975). To overcome these inconveniences, sampling locations were selected according to a systematic design proposed by Eberhardt and Thomas (1991). The current status of fish fauna was determined through field surveys using a standard boat mid water trawl in 15 localities in the lake during one year (Figure 1). Sampling was repeated every month from June 1997 to June 1998, with the purpose of constructing a database for future work on the structure of the fish community (confronting hypothesis) and fishing resources (gathering alternate information, e.g. biomass).

The identification of attributes and variations that define the characteristics and fishing impacts in Lake Chapala were provided by the literature review, official fishing statistics, technical reports and secondary non-official information. Fieldwork was also performed on a monthly basis at several fishing communities from June 1997 to June 1998 in order to obtain additional information. A database was created to support the general discussions on exotic species. For this purpose, frequency distributions of "mojarras" samples (*Oreochromis aureus*) from the island of Petatán were registered and analyzed every month from July 1991 to July 1997 (Figure 1).

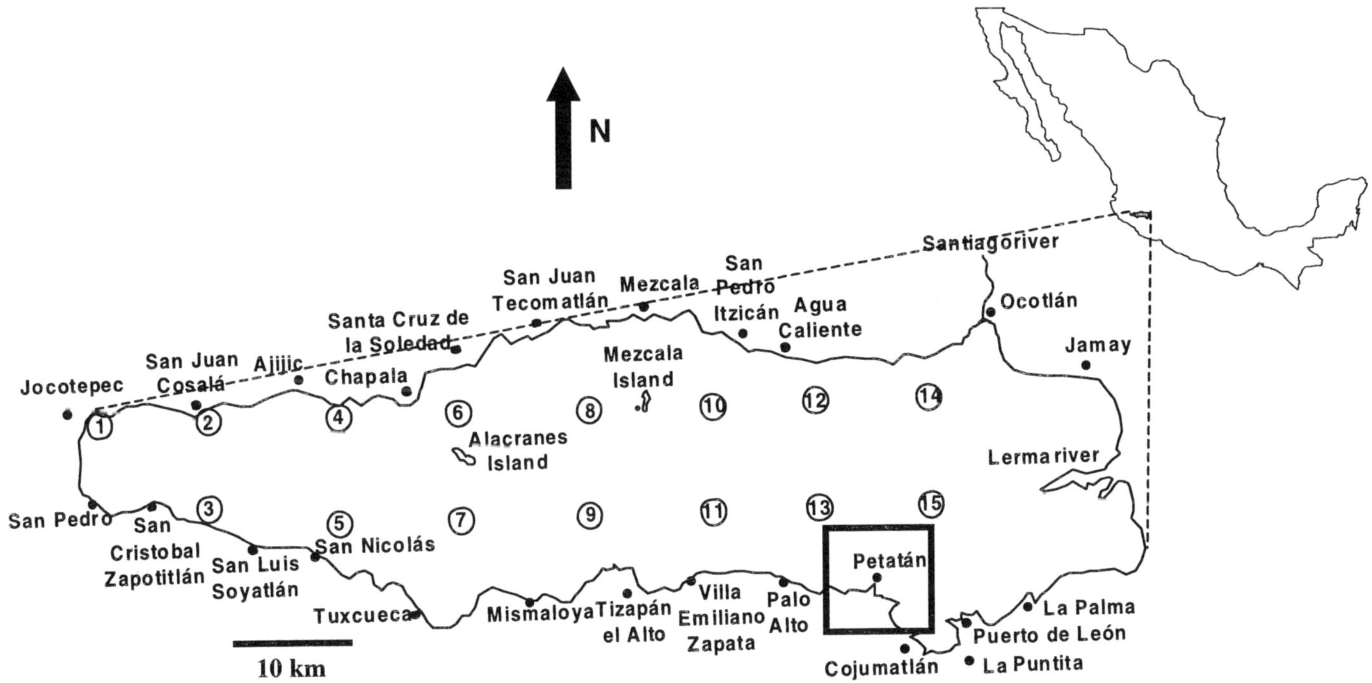

Figure 1 Lake Chapala with survey locations and main towns. Petatán island is enclosed in the square

3. COMPOSITION OF THE FISH COMMUNITY

The specific composition of the ichthyofauna in Lake Chapala includes 39 species – 27 native and 12 exotic (Table 1). Nevertheless, the tendency and present situation of the fish community indicate a decrease in diversity, considering that some species have neither been collected nor seen in recent years. For this reason we can only confirm the presence of 27 species: 19 native and 8 exotic (Table 1). Despite this fact, the lake still represents the most diverse fish fauna ecosystem in the Lerma-Chapala basin.

Analyzing some of the fish families in the lake, we found interesting evolutionary aspects. In general terms, the case of the *Goodeinae*, represent an outstanding example of adaptive radiation on a small scale. In this sense the most obvious characteristic includes a large variety of trophic adaptations, which have produced a diversity of nutritional types. We can classify the species in Lake Chapala as filter feeders (*Goodea*), carnivores zooplankton feeders (*Allotoca*), periphiton feeders (*Zoogoneticus*) fish feeders (*Alloophorus*), preferential omnivores-carnivorous (*Xenotoca*), preferential herbivorous (*Skiffia*), or exclusively herbivores (*Chapalichthys*). Moreover, the degree of differentiation can reach the intra-lacustrine level within the population of a single species (as in the case of *Goodea atripinnis*). Here, the lack of any apparent differentiation suggests that stochastic forces are responsible for the structure of local populations (White and Turner, 1985). Other isolating mechanisms of the goodein species, which do not involve feeding aspects but rather reproductive behavior, include exhibition-performing rituals during courtship. These taxonomic interpretations of morphological information are also confirmed by ethological data (Fitzsimons, 1976).

Miller and Smith (1986) suggested that the origin of goodeins in this area is due to the following: (1) presence, in large quantities, of the Miocene fossil (*Tapatia occidentalis*) in the Santa Rosa deposit, located near Lake Chapala (this fossil is the oldest ever documented in the central Mexican ichthyofauna, Figure 2); (2) evidence of other fossil-bearing strata around the lake (Jocotepec, Ajijic), which include different species of the subfamily; and (3), the importance of the lake and its surroundings in the current distribution of this subfamily, which comprises the area of greatest concentration of genera (11/17, Figure 2; Uyeno *et al.,* 1983).

When analyzing the present state of goodeins in the lake and using it to assess the ecosystem's health, an important level of environmental degradation is revealed. Plentiful taxa (*Goodea atripinnis* and *Chapalichthys encaustus*), considered to be tolerant to polluted waters (Soto-Galera *et al.,* 1998), and some sensitive species, such as the *Allotoca dugesi* and *Skiffia multipunctata*, have not been captured recently (Table 1).

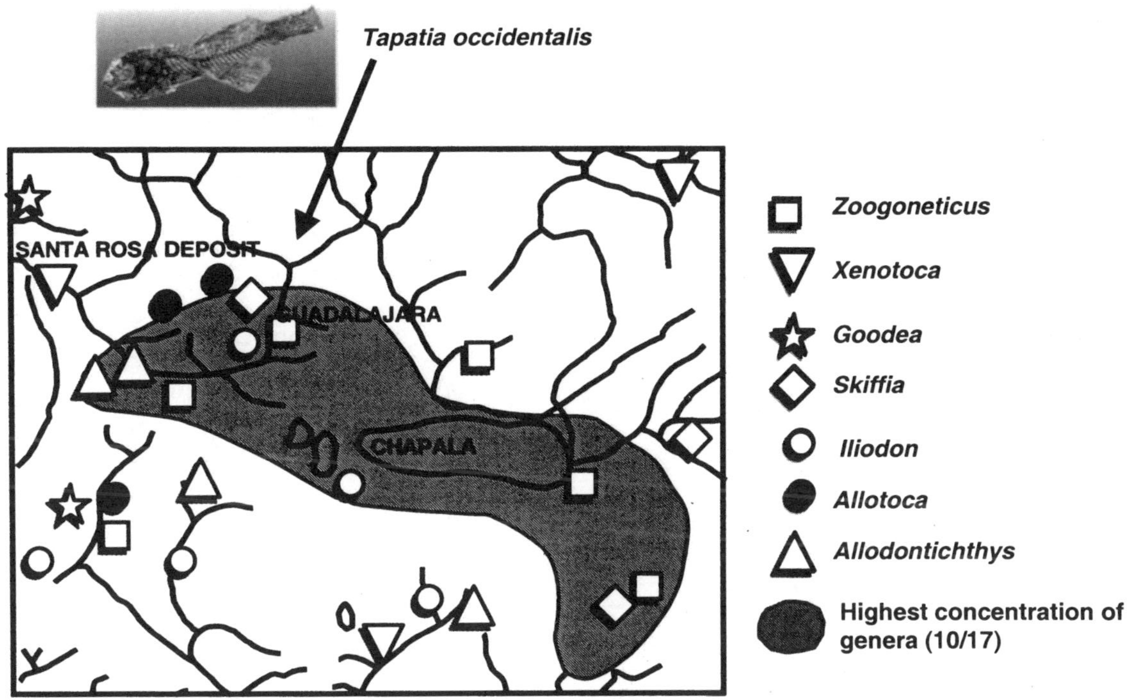

Figure 2 Location of the mentioned goodein fossil *Tapatia occidentalis*, and distribution of the genera of the Goodeidae in the area surrounding Lake Chapala. Genera in the area of highest concentration: *Allodontichthys, Alloophorus, Allotoca, Ameca, Chapalichthys, Goodea, Ilyodon, Skiffia, Xenotoca* and *Zoogoneticus*. *Modified from Webb (1998)*

With the eight species of the genus *Chirostoma* ("charales" and "pescados blancos") in Lake Chapala, we find the highest degree of diversity of the *Atherinidae* family in freshwater, including the most rigorously documented example of the three species of "pescados blancos": *Chirostoma sphyraena, C. lucius* and *C. promelas,* which are considered part of a monophyletic group (Barbour and Chernoff, 1984). The difference in body length is probably the most important factor, which has allowed the co-existence of the different species (Barbour, 1973a); they are all carnivores and live in an area, which lacks a major diversity of habitat. This has allowed crossing of ecological boundaries due to feeding on different prey sizes. Trophic specialization has evolved in the genus *Chirostoma* as a result of competition between individuals of the same size, being congeners or not (Barbour, 1973b). It would be interesting to observe the competitive effect, which may have occurred (in the sense of broadening or narrowing of food niches), in the Lake Chapala community owing to changes in the ecosystem.

The actual status of the atherinids in the lake reflects the presence of all but one species (*Chirostoma* arge), and a decrease in abundance of others. Unfortunately, three species (*C. promelas, C. lucius,* and *C. chapalae*) are only sporadically found in the captures. Likewise, when identifying atherinids, a possible hybridization is reported between *C. consocium* and *C. chapalae*, as well as between *C. lucius* and *C. sphyraena* (3; UMMZ 209281, 209282). It is important to mention this fact, since the relative abundance of hybrids has been used as an indicator of environmental degradation (Karr, 1981). Loss of diversity of habitat caused by the modification of the lake margin, re-suspension of sediment and erosion, have all inhibited the reproductive isolating mechanisms among fishes that use specific substrata for spawning (Fausch *et al.,* 1990).

The present situation of both species is critical; they rarely appear in recent collects or commercial fishing nets along the lake margin. Two aspects of their natural history versus two events contributing to the degradation of the environment have had a lot to do with this: first, the preferential distribution of *Cyprinidae* in the mouth of rivers and spring areas where aquatic vegetation was found. From 1909 to 1912, a dike was constructed near the Lerma river delta, isolating at least 50,000 hectares of swamp.

The above-mentioned project destroyed a large part of the standing vegetation area of the lake (Goldman, 1951). Second, an important migratory movement towards the tributaries of the lake for the spawning phase (Barbour and Miller, 1978), in opposition to the reduction in water flow and the increase in sediment and contaminants brought by the Lerma and Tizapán rivers.

The present situation of the cyprinids *Notropis calientis* and *Yuriria alta* in Lake Chapala is not well known. They are still captured eventually by commercial fishing nets. Both species are defined as having an intermediate and tolerant "tolerance level", respectively, to environmental degradation (Soto-Galera *et al.,* 1998). This gives them a better chance of survival as compared to more sensitive species. With regard to *Notropis lermae* – a synonym of *Notropis sallei* (Chernoff and Miller, 1981) – although there are no current collections they should be positive (De Buen, 1946), as the species has been known in the area from the fossil records of the Chapala Formation (Pliocene) (Smith *et al.,* 1975; Miller and Smith, 1986), and has been caught in the Duero river (Ledesma, 1987). Finally, the report of the cyprinid *Falcularius chapalae* by De Buen (1946) is in fact a synonym for *Yuriria alta* (Gilbert, 1998).

The *Petromyzontidae* family includes the species *Lampetra spadicie* that appears to be a southern geographic relict derived from a common anadromous ancestor with a wider primary range of distribution (Miller and Smith, 1986). This lamprey has been reported in Lake Chapala and in the lower region of the Lerma river including the Duero river (Lyons *et al.,* 1994). Actually, this species has not been caught in the lake. The scarcity of *L. spadicea* – and all the fish in the lake – is related to major environmental changes that have taken place in the Lerma-Chapala basin. Due to the increase in water extraction from the Lerma river, Lake Chapala has experienced severe reductions in water levels during the past 15 years. This has resulted in increasing amounts of dissolved and suspended solids (Limón-Macías *et al.,* 1989; Limón-Macías and Lind, 1990; Guzmán-Arroyo, 1995). Together with these changes in the ecosystem, there has been a reduction in various host species of the lamprey. At the present time, it is uncommon to capture a "pescado blanco" (*Chirostoma lucius*) or a "bagre" in the lake weighing more than 0.5 kg, and the large cyprinid, the popocha (*Algansea popoche*), is practically extinct (Guzmán-Arroyo, 1995).

This antagonistic ecological background has motivated some federal institutions such as the "Comisión Nacional para el Conocimiento y Uso de la Biodiversidad" (CONABIO – from the initials in Spanish) to include Lake Chapala in its catalogue of Prioritary Hydrologic Regions (the mission of the CONABIO is to co-ordinate actions of conservation and research, to preserve biological resources). More specifically, the region is defined as Chapala-Cajititlán-Sayula (key number 58), which covers an area of 6,150.42 km^2. Moreover, Lake Chapala is divided into three zones: AAB, a region of high biodiversity; AV, a region of different uses for the productive sector; and AA, a threatened region.

4. ICHTHYOFAUNA STRUCTURE

The study of the community structure is important to understand how the ichthyofauna, defined as a living system of interacting species populations, is organized (Whittaker, 1975). The most important part of this and many other ecological and fisheries studies is the evaluation of the abundance of a particular species' population (Krebs, 1989). The spatial structure must be considered, since it directly affects the response of organisms to prevailing uses and impacts.

To make a more concrete analysis of the community structure, it is useful to rely on an example – the atherinids, one of the most typical ichthyofauna elements of Lake Chapala. The most dominant species throughout the year is *Chirostoma consocium* ("charal de rancho") followed by *C. jordani* ("charal cambray"), which has a wider distribution in the Lerma river basin. The density (number of individual elements of a species per area unit) of the above mentioned species is of approximately 308 "charales" of more than 6.5 cm in standard length, per hectare. This size is specified because it represents the size of the onset of maturity of the "charal" (Guzmán-Arroyo *et al.,* 1995). The number of "charales" grouped by length classes' intervals of 0.6 cm is presented in Figure 3. It should be noticed that the seasonal arrangement of data shows a sequential change in a statistic mode. This tendency shows in detail the reproductive behavior and recruitment of the population, in accordance to vulnerability to the fishing art employed. At the beginning of spring and summer, it is common to capture large-sized organisms, due to a strong tendency to group in shoals. Perhaps the main reason for this behavior is that the higher peak of reproduction occurs in March (Guzmán-Arroyo *et al.,* 1995). During summer, water movement is induced by the inflow of the Lerma river to the lake. During Autumn, we observed recruiting of individuals from lakeside areas to the fishing population as a result of the above-mentioned reproductive peak (March), together with the winter capture of medium-sized organisms, which are recruits to the reproductive population. The low number of large-sized specimens in the winter can be explained by the fact that it is the most intense period of fishery of *C. consocium.*

The nature and development of spatial and temporal heterogeneity is affected by physical gradients, particularly by bathymetrical differences. While species *C. consocium* is dominant throughout the lake, there is an increase in relative abundance of *C. labarcae* in deeper waters (Figure 4). In general terms there is an antagonistic behavior between depth and temperature with respect to more or less warming, which depends on the depth of the water column. This tends to be related to the distribution of the above-mentioned fishes (Figure 4).

An increase in relative abundance of *C. jordani* in localities one and four (among the shallowest in the lake) was observed (Figure 4). A possible explanation for this phenomenon may be the structure of the lake bottom. These locations are situated close to the towns of Jocotepec and Ajijic respectively, and are characterized by the presence of sandbanks.

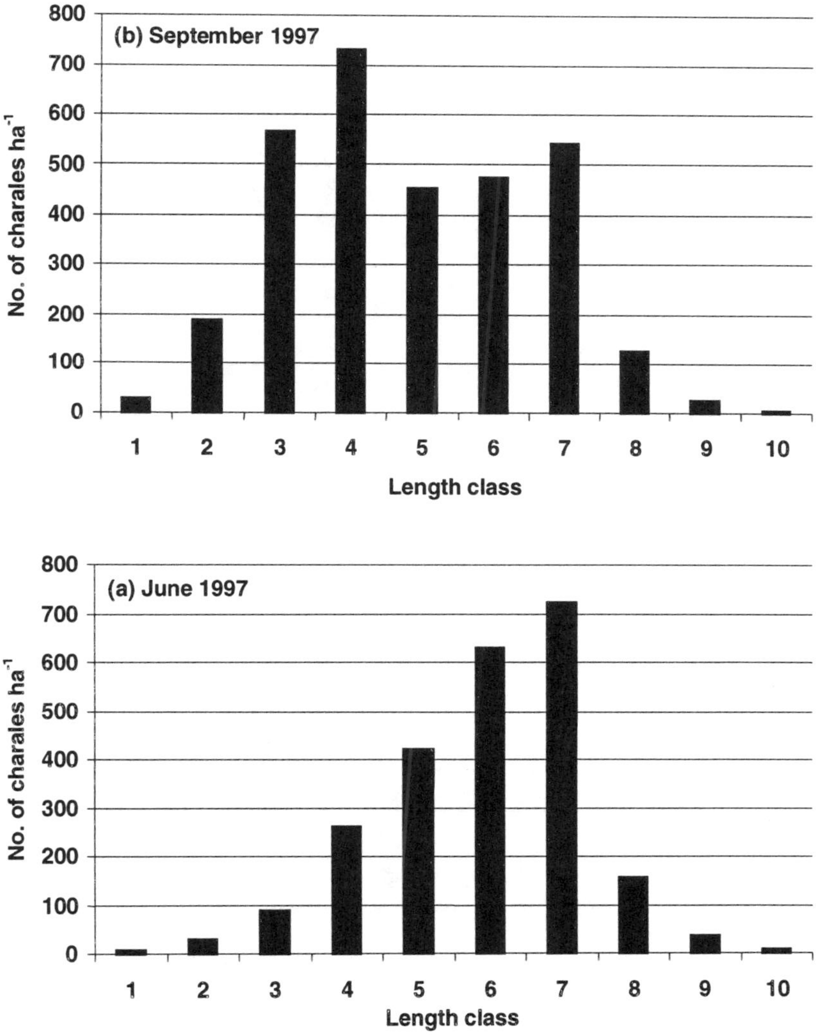

Figure 3 Length composition of the "charal de rancho" (Chirostoma consocium) in Lake Chapala. The graphs show numbers of fishes in 0.6-cm length classes (a) June (b) September, 1997

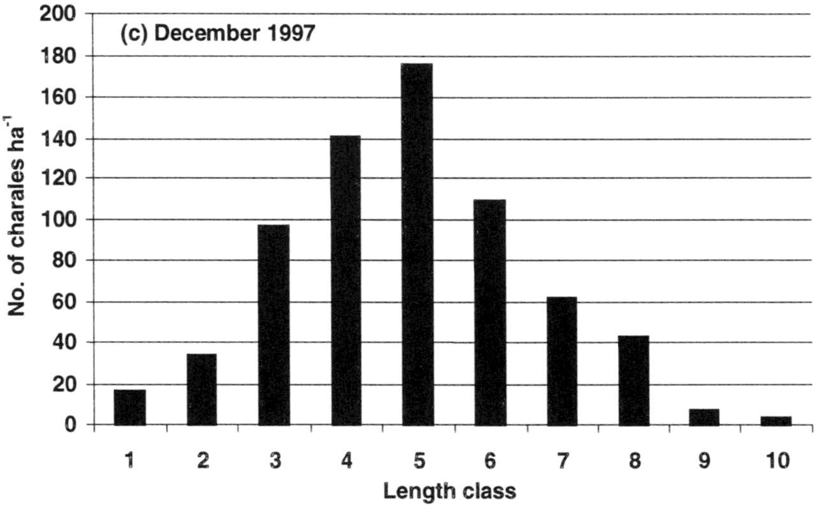

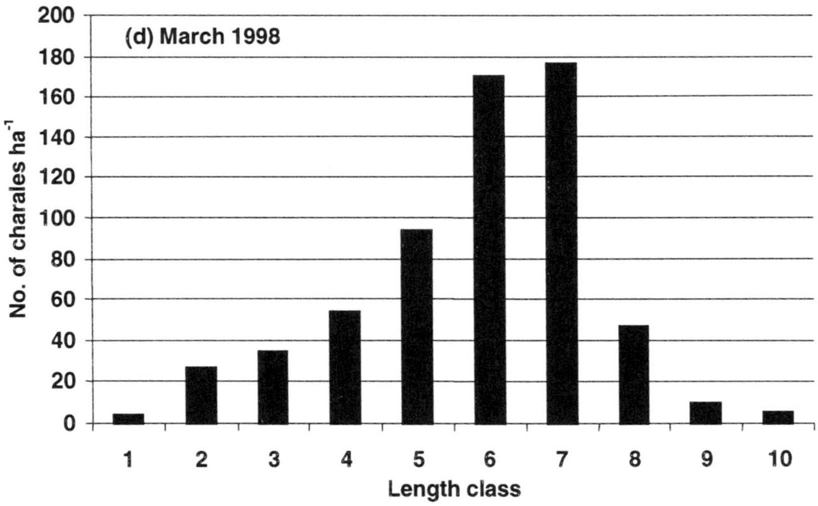

Figure 3, cont. Length compositions of "charal de rancho" (Chirostoma consocium) in Lake Chapala. The graphs show numbers of fishes in 0.6-cm length classes (c) December, 1997 (d) March, 1998

Such distribution is also common for fish populations in other lakes, such as Cuitzeo, where areas preferred by *C. jordani* are described as shallow-lenthic, lacking submerged vegetation and with a sandy bed (Chacón-Torres, 1980).

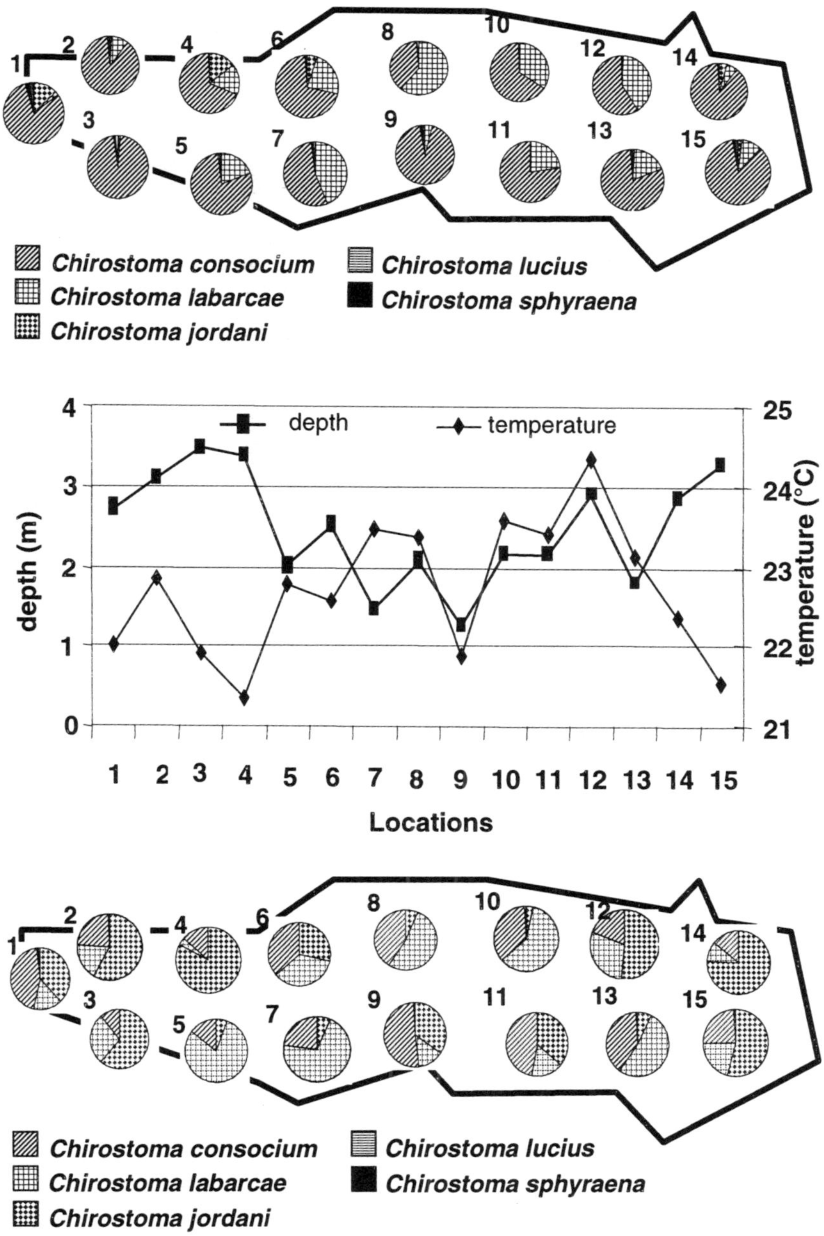

Figure 4 Top graph: Lake Chapala relative abundance of atherinids species in June 1997 Lake Chapala. Middle: comparison between depth and temperature in 15 localities in the lake. Bottom graph: relative abundance values in December 1997

As the seasons pass, spatial distribution changes in a process of substitution of some species by others, as part of the behavioral aspects of their life cycles. Thus, the species *C. consocium* and *C. labarcae* are replaced in relative abundance by *C. jordani* (Figure 4), according to preferences of habitat. This is a consequence of reproductive migrations of the above-mentioned species, from feeding grounds to those of reproduction around the islands and diverse locations along the lake margin. By analyzing the time-related structure of the community (Figure 5), we can appreciate the fluctuation of different incidental peaks among those species, which alternate with time. This is observed in two species: *C. consocium* and *C. labarcae* with two maximum peaks, one during the summer and the other in autumn. However, they do not coincide on same months or intensity, suggesting inter-posed behavioral patterns. The *C. jordani* presents an evident increase in winter, dominating the afore-mentioned groups, which can be directly related to the reproductive phase already discussed. As a consequence, part of the adaptation process can be confirmed to ensure the co-existence of the species, according to their situation in time and space, as well as the search for different substrates for spawning.

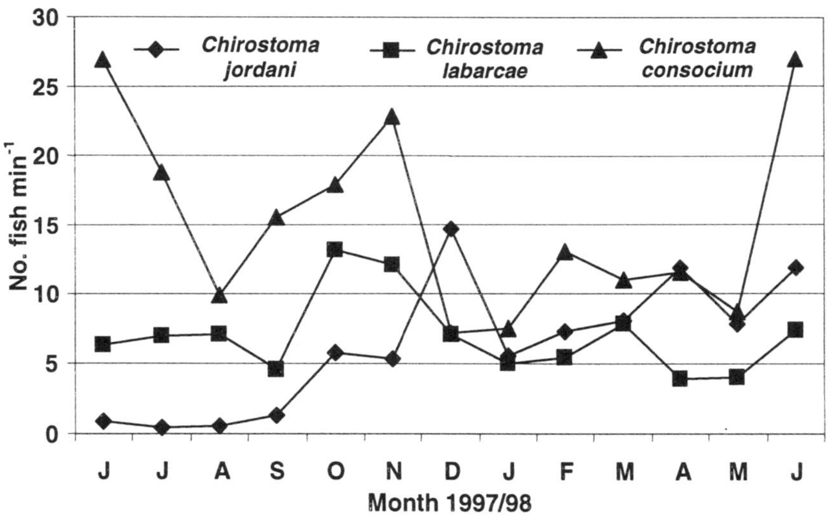

Figure 5 Capture per unit of effort (CPUE) of the most abundant species of charales

5. ICHTHYOFAUNA FISHERIES

The ichthyofauna of Lake Chapala is an important resource for the rural economy and food supply. In large parts of this region, fish represent an essential and almost irreplaceable source of high quality and low-cost animal protein, important for the dietary balance in these communities. Fishing resources from the lake, which at the present time boast 17 species (Table 1), are usually exploited from small boats using unsophisticated fishing arts and poor infrastructure for storage and protection of the capture. Traditional boats are constructed with pine planks and are waterproofed by means of coal tar and tow. However, the modernization of more than 80 percent of the traditional fishing fleet includes purchase of fiberglass boats, which are stronger and last longer. Means of propulsion include oars for short distances and outboard gasoline engines.

Nowadays, fishing methods include the use of transparent single and multi-filament nylon nets (Figure 6). The shape and size of the mesh of the gill nets vary from 3" to 6" according to the target species and their size. Nevertheless, some fishermen use nets of 2" mesh in the low-level production seasons. Long lines are used fixed or near the bed so that species' size can be selected by hook size and type of bait. Fishermen use "camaroncitos" (*Cambarelus*) as bait when fishing for "carpas" (*Cyprinus*) and sometimes for "bagres" (*Ictalurus*). Traps are highly selective, they are designed according to the type and size of species, and their positioning requires extensive knowledge of the local conditions.

Traps used for fishing "charales" (*Chirostoma*) in Lake Chapala can be summarized into four different types, according to the materials and shape used (Figure 6). The "mangueadora" net is an active fishing art with meshes small enough to prevent the desired fish from gilling (filtering nets). This type of net is set and managed from a boat by a team of two people (Figure 6). They are used as transverse barriers across the surface currents' flow. The "rancho charalero" is a modification of the coastline, frequented by the *Chirostoma* species, creating areas separated by rocks (Figure 6). The size of these areas is proportional to the diameter of an "atarraya" net to capture the "charales" following spawning (Figure 6). Knowledge of the migratory behavior of species is important when using "rancho charaleros" and "mangueadora" nets. A more detailed description of the fishing methods used in the lake has been published by Guzmán-Arroyo and Ortiz-Martínez (1995). Much of the production is consumed or commercialized on a local basis, sales being made of the raw, fresh fish or after some processing, such as filleting, salting and/or drying.

According to Official Fishery Statistics (SEMARNAP, 1999) there are 1,699 fishermen in Lake Chapala, belonging to 86 fishery organizations that use 51,400 fishing arts. These numbers, compared to those of 1987 (Guzmán-Arroyo and Ortiz-Martínez, 1995), show a decrease in the number of people involved in this activity (2,041), an increase in the number of fishery organizations (69) and the development of more than twice the number of existing methods of capture (25,000). Considering the above, the decrease in the number of fishermen (17 percent) does not imply a considerable reduction in the captures, because those who leave the fishing activity are not those who have sustained a high fishing mortality level ("F").

A greater number of fishery co-operatives and unions do not necessarily imply better organization but rather a head-on collision of interests of the members and the formation of new groups. Finally, although there have been fundamental changes in fishing methods, an increase in the fishing effort is inconsistent with the reality of the lakes' fishing resources: reported production has suffered a fall from 13,710.78 tons in 1990 to 3,484.018 tons in 1998 (SEMARNAP, 1999). Unfortunately, a diagnosis of the present situation estimates that the resources are being over-exploited. Furthermore, the evaluation of the state of fishery sustainability in continental waters in Mexico during 1997-1998 shows an alarming 100% deterioration rate with demands of an urgent implementation of a recovering strategy (INP, 1998).

Figure 6 Different fishing methods used in Lake Chapala: a) agallera (gill net); b) nazas (traps with different construction materials); b') cuna (a recently introduced type of trap); c) mangueadora (type of seine net); d) atarraya; e) rancho charalero

Analyzing the captures of "charales" (*Chirostoma*) from the lake, we observe a common pattern in the development of fishing vis-à-vis the native species. It is possible to identify three stages (Figure 7): (I) a sort of steady state, in which the stock shows a more or less steady production; (II) rise and fall followed by a period of larger captures; (III) the stock shows a tendency towards total collapse without any clear indication of recovery. Regarding the first stage, there is a slight decrease in production in the first half of the 1950's, perhaps due to an important change in the lake's volume (Figure 7), since there were no substantial changes in the fishing effort. The increase in the second stage (1960 to 1980) was due to the incorporation of more fishermen (= greater effort), an improvement in new technological equipment (synthetic material, propulsion with outboard gasoline motor and the introduction of fishing methods – such as mangueadora) and better all-round fishing efficiency (Lyons *et al.*, 1998). The fall in the production in the middle 1980's could have been a consequence of the eradication of the "chinchorro" net (seine net), and its corresponding capture-quote in fishery statistics. Volume, however, was recovered in a short time, as captures of exotic species gradually replaced those of native varieties. Finally, the sudden fall in the third stage can be a result of a synergetic effect involving (1) variation in water levels in the lake due to natural fluctuations and excessive water extraction in different areas of the Lerma-Chapala basin. This is due to the increase in water demand in the agricultural, industrial, and urban sectors and has been discussed by Aparicio (2001). (2) Increased levels of pollution and environmental degradation as can be seen by the unusual massive water hyacinth growths; and (3), the absence of a real and integrated fishery management, evidenced by the lack of control of the fishing effort.

Another indicator of the abovementioned absence of fishery management is the over-exploitation of the native "bagre" (*Ictalurus dugesi*). This was attributed by Menchaca-Hernández *et al.* (1986) to the combination of low-level survival-rate of the population and high-level exploitation-rate plus the capture of small individual specimens. With the absence of any concrete action, we are left with the fact that this species is rarely caught in commercial fishery. Above all, the periods of the official closed season (SEPESCA, 1989) do not include the mainly caught species (*Chirostoma consocium*), but rather *C. chapalae* (Catalán-Romero and Menchaca-Hernández, 1993). As a consequence, those of lower value have replaced the most valued and higher-priced species. There is a change from a minimum contribution of exotic species to represent the greater percentage of the captures (Figure 8). With reference to the benefit derived from these exotic species, there is also a lack of regulation and planning for its sustainable use and development.

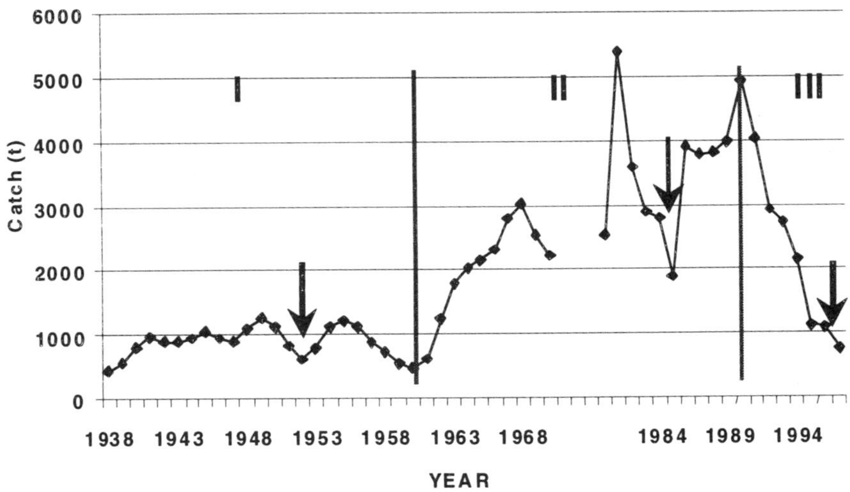

Figure 7a Trend of charales' captures in Lake Chapala from 1938 to 1997 (no data available for 1971-1979) (Guzmán-Arroyo and Ortiz-Martínez, 1995, and SEMARNAP, 1999). Roman numbers represent the three stages and arrows indicate declining periods (see text for more information)

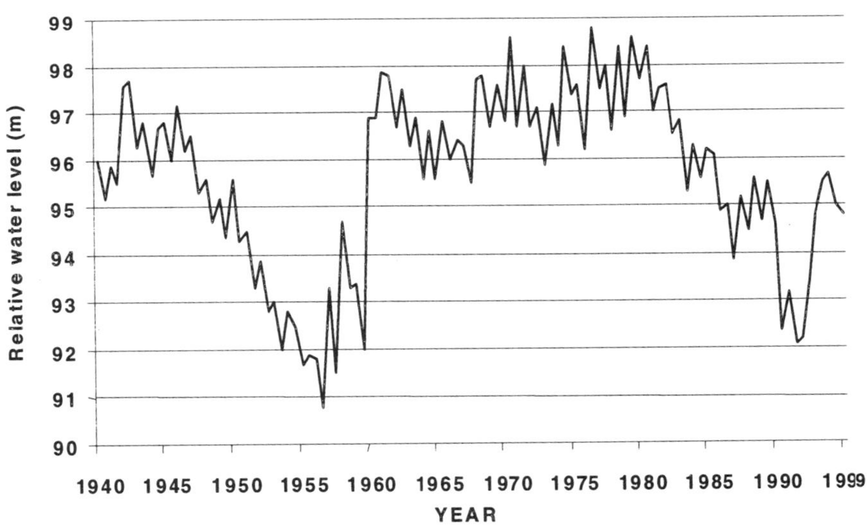

Figure 7b Variation in relative water level in Lake Chapala from 1940 to 1994 (CNA, 1994)

The size of gill net mesh permitted for the fishing of "mojarras" (*Oreochromis*) is of 3½", the smallest size of the capture being 20 cm

(SEPESCA, 1989). However, fishermen use smaller scale mesh (mainly 3"). Certainly, the capture increases in short time periods, but the fish are smaller (about 19 cm) (Figure 9). Consequently, there is a decrease in the resource and a potential problem of dwarfism among the population. On the other hand, if fishermen had to use a wider mesh, larger fish would be caught (21 to 23 cm) (Figure 9) which would sell at better prices.

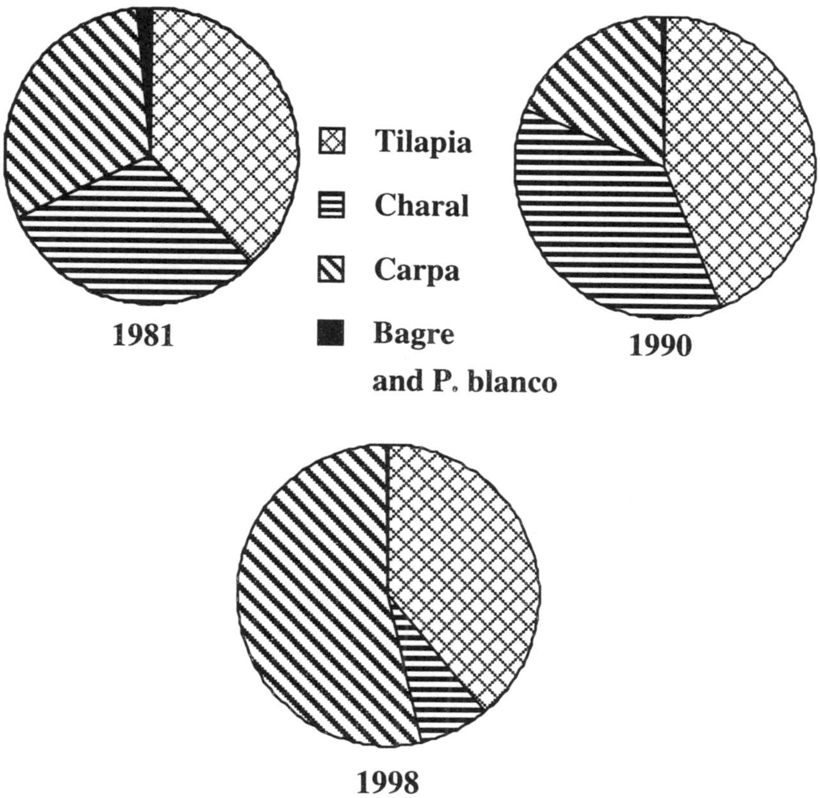

Figure 8 Relative abundance of commercial species in Lake Chapala during three different years (SEMARNAP, 1999)

Fishing of exotic species ("mojarras", "carpas" and "bagres") has developed in a special way in Mexico. The primary intention was to optimize fish production and to provide people with high-quality protein (here, the erroneous concept of "extensive aqua-culture" was used). Another justification was the need of filling empty niches of the lacustrine habitat.

Two points should be considered: first, the primary intention was indeed praiseworthy, but no conscientious analysis was made regarding the effect on the ecosystem. As already explained, the impact of the exotic species resulted in the displacement of those of high natural and economic value, rebounding this somehow on the economy of the fishery communities. Moreover, this type of policy somewhat looses validity if we consider the success in the culture of "mojarras", "carpas" and "bagres" in ponds, pools and floating cages. Perhaps a realistic aquaculture program may provide a better alternative without damaging the ecosystem or losing sight of the main objective: food and nutrition. Second, the concept of the niche is intrinsic to the species. In its simplest definition, it relates to the activity and integration of an organism in a particular space and time. It is therefore mistaken to talk about of "an empty niche". The standpoint regarding the "pescados blancos" (*Chirostoma*) was that the transfaunation of the species was justified due to its high price (Rosas, 1976). However, the "pescado blanco" as in the case of the "lobina negra" (*Microterus salmoides*) are top predators creating possible after-effects like depredation and competition.

The artisan type of fishery is facing difficulties, which are contributing to a margination of their living standard. Government policies commonly tend to concentrate resources in industrialized sectors with modern, large-scale, commercial and export fisheries. Thus, small-scale fishermen are facing great difficulties in or find it impossible to obtain credit, extension services, marketing assistance, or any other similar help from development programs (Sainsbury *et al.,* 1988). While artisan-fishing practices may not be as productive on a per-unit-basis as those used in more advanced temperate zone fisheries, substantial fishing effort is generated by the magnitude of operating units. In addition to this, there are cases where the resources available for fishery (e. g. water) are used in other more competitive areas. It is expected that fish production will be affected by a reduction in the availability of water quality and quantity. The major challenge is to guarantee an open and fair competition for resources, in a way such that fisheries can gain access to resources and maintain their production (FAO, 1999). Another problem is that there exist divisions among many of the official branch offices involved in fisheries. Those who collect data do usually not have the technical infrastructure to analyze the information, while those who carry out the analysis have poor field experience. The general recommendation would be that, when dealing with any analysis requiring data, efforts should be made to familiarize the personnel with the data-collection procedures (Hilborn and Walters, 1992). This would help to improve the analysis of field data and to provide realistic information for decision makers.

6. FUTURE STUDIES AND ACTIONS

Perhaps the most important issue is to improve water quality and to increase the water level in Lake Chapala. Also, we should work to establish a realistic national policy with the objective of maintaining the integrity and rational use of this ecosystem. A re-structuring of the Advisory Board for the Lerma-Chapala-Santiago river basin together with new proposals for strategic planning (taking into account past errors) is required, with the purpose of higher quality management to protect the lake.

Integration of productive activities – especially with better coordination of strategic plans and management of resources shared by the fishermen and others – is urgently needed, with the aim of facilitating a sustainable fishing production (FAO, 1999). In addition to the management plans derived from this, we should implement education, extensive diffusion of information and participation of the population directly or indirectly related to the resource at every level of civil and governmental organization.

The functional understanding and management of ecosystems is not beyond existing capabilities (Blair, 1964). This kind of research, however, is more expensive and therefore it is usually not feasible in developing countries. Moreover, in order to optimize and improve resources to attain a sustained ecosystem, we should seriously consider the establishment of integrated research programs.

To fulfill the proposed plans, the first critical step should be inter-institutional co-ordination, since the active participation of the public and private sectors traditionally presents a low level of communication. Communication is either restricted or does not exist, and projects are repeated. To avoid this, and with an integrated research program, an intensive study of selected segments of the ecosystem could be attained. Thereby critical information gaps will be filled with the purpose of achieving real solutions to problems.

The general idea and perspective, in the research of the ichthyofauna of Lake Chapala, is that the species should be studied as living systems more than as preserved specimens. This concept would open an extensive and integrated action field, in which the composition, evolution mechanisms, structure and function of the community, and dynamics of exploited populations, would be taken into account as contributors to the functioning of the natural system of species populations. In this respect, we are still at the stage of enumerating present species, albeit with some knowledge of individual distribution. However, it is assumed that knowledge of interactions at different organizational levels, are essentially the present and future well-being of humanity (Blair, 1964).

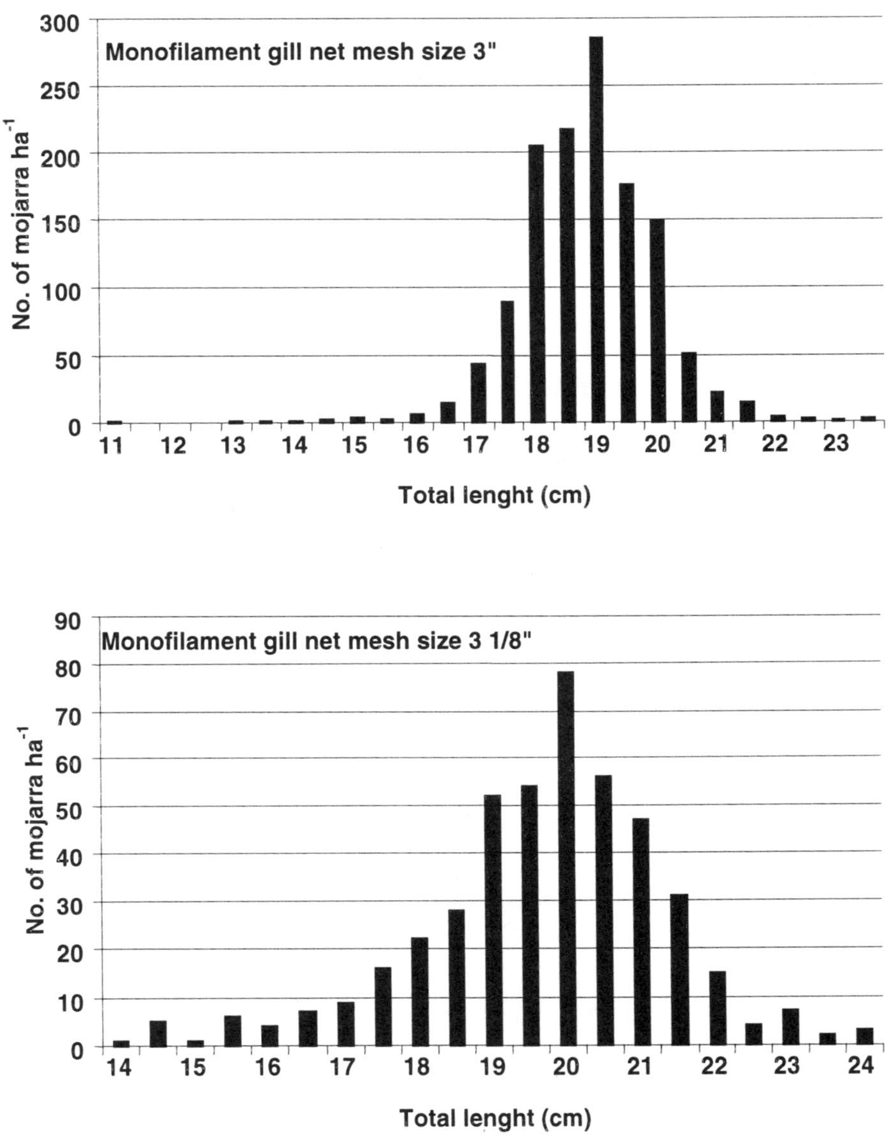

Figure 9 Length compositions of mojarra (Oreochromis) in different gill net mesh sizes. The graphs show the frequency distributions of fishes in 0.5-cm length classes. (a) 3" gill mesh size (b) 3 ⅛" gill mesh size

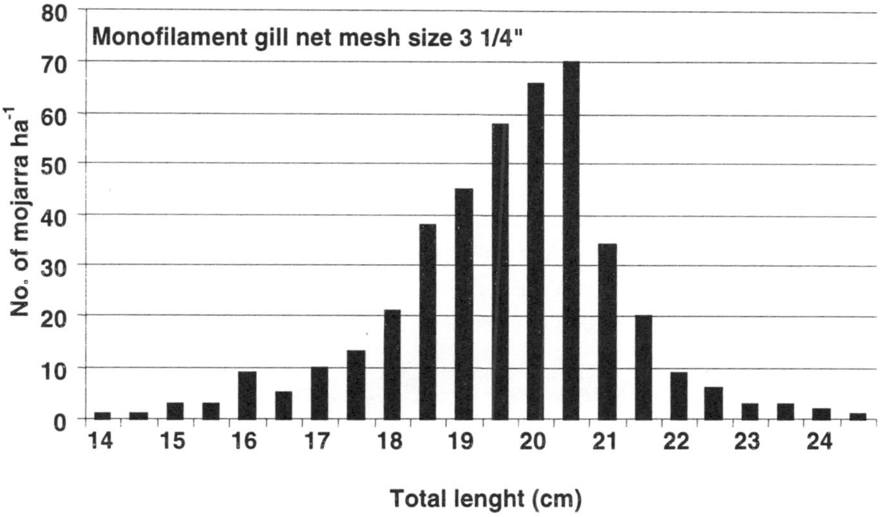

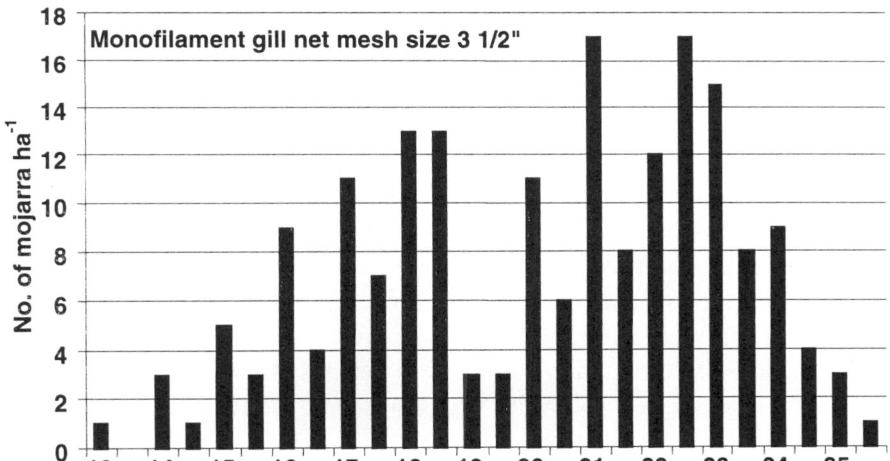

Figure 9, cont. Length compositions of mojarra (Oreochromis) in different gill net mesh sizes. The graphs show the frequency distributions of fishes in 0.5-cm length classes. (c) 3 ¼" gill mesh size (b) 3 ½ gill mesh size

Because of this, it is necessary to complete the research in the areas of auto ecology, natural history, and synecology. Since sensitive and conscientious management begins with a solid-based, fundamental understanding of the species and communities' ecological relations and natural history (Brussard, 1991). The key factors in this program should include the habitat diversity, distribution patterns of ecological assemblages and differences in productivity.

It is important to test hypotheses concerning the persistence of populations. Here, one of the major principles of conservation biology dictates that stochastic processes, rather than deterministic ones, govern the dynamics of small populations (Brussard, 1991). At the same time, two vital topics must be addressed: first, it is necessary to discover to what extent biological diversity patterns are important to determine the behavior of ecological systems (i. e. response to climatic changes, rates of nutrient flow, or responses to contaminants). Second, it is necessary to understand how ecological processes interact with physical and chemical factors and parameters and how they are modified on both ecological and evolutionary time scales (Lubchenco *et al.,* 1991).

A successful plan for the conservation of biodiversity and development, as well as fishery management, would require continuous monitoring of the status of available resources (population, communities, etc.). It is important to enhance the study of the dynamic behavior of the community in the long term, as there is strong evidence that environmental factors and natural changes are as important as fishing mortality (Hilborn and Walters, 1992). The present study was conducted during a year affected by "El Niño", and different studies have shown that this phenomenon influences the thermal regime of Lake Chapala (Tereshchenko *et al.,* 1999).

Reliable information is required for adequate fishery administration (FAO, 1999): (1) status, trends and biodiversity of fishery resources; (2) current production and potential increases of fisheries capture; (3) socio-economic situation of fish producers; (4) resource utilization techniques; and (5) distribution/consumption of products. The biological basis for sustainability is the analysis of the reproductive surplus, which is calculated from the balance between births, deaths, and somatic growth (Hilborn *et al.,* 1996).

Generally, exploited resources are not friendly and homogeneous unit stocks, as conveniently considered by theorists. Furthermore, analyzed stocks are embedded in ecosystems, which are certainly not constant over time (Walters, 1984). The present tendency of fisheries' management is to consider greater holism, multiple species interactions and broad-scale physical factors (Bostford *et al.*, 1997).

Proposals for a solution vis-à-vis the lack of sustainability of fisheries must be changed in two aspects, which are the root of the problem of over-fishing: (1) reduction of uncertainty in predicting the effects of management and (2) reduction of pressure on managers for increased harvest (Bostford *et al.*, 1997). Uncertainty in this context is defined as "the incompleteness of knowledge about the state or processes (past, present, and future) of nature" (FAO, 1995). The evaluation of this uncertainty is an important step, because it is a real feature of the status of fish stocks and the scientific proposal regarding management (Rosenberg and Restrepo, 1994). Another related concept should be the decrease of "risk" (Francis and Shotton, 1997). With reference to fisheries management, this has two distinct stages: the first is related to the formulation of advice for fisheries managers, and the second deals with the ways in which managers use advice to make decisions.

Recently, considerable interest has been shown in the concept of "adaptive management", where the decisions taken (in the face of uncertainty) are visualized as a sequence of more or less informative "experiments", regarding the system's potential (Walters, 1992). This is the kind of action that shows us that we should start in reverse order – contrary to traditional thinking – bearing in mind the following context: (1) policy options – (2) response models (of resources to possible policy changes) – (3) seek help from data (emphasizing the spatial sense).

Another promising protocol is the development of ecosystem models together with an estimate of the network characteristics. Certainly, analyses of the ecosystem have generated a great number of excellent scientific papers. However, there have been few studies that have resulted in the presentation of balanced models of the whole system. Actually, some models can provide such approximation (Christensen and Pauly, 1992). This vision of a greater holism needs to expand to fisheries' management, and include economical aspects and political sciences (Bostford *et al.*, 1997).

ACKNOWLEDGEMENTS

We thank the many people who helped us to collect and summarize the data reported, especially Salvador Becerra-Muñoz, Juan Manuel Catalán-Romero, Adán Mejia-Muratalla, Jesús Morales-Castellanos, Elizabeth del Río-Núñez. We are particularly grateful to the fishermen of Lake Chapala for sharing their experience. This project was supported partly by the IPN-CEGEPI projects number 970468 and 990008 and by the CONACyT-SIMORELOS project 19980306009. We wish to thank P. Esteinou for her help during translation of the manuscript.

REFERENCES

Álvarez J (1972) Ictiología Michoacana V. Origen y distribución de la Ictiofauna Dulceacuícola de Michoacán. An Esc Nac Cienc Biol México. 19:155-161. 1 lámina.

Aparicio FJ (2001) Hydrology of the Lerma-Chapala Watershed. *In: The Lerma-Chapala Watershed: Evaluation and Management* (Hansen AM and van Afferden M eds) Kluwer Academic/Plenum Publishers, London.

Barbour CD (1973a) The Systematics and Evolution of the Genus Chirostoma Swainson (Pisces, Atherinidae). Tulane Studies in Zoology and Botany. 18(3): 96-141.

Barbour CD (1973b) A Biogeographical History of Chirostoma (Pisces, Atherinidae): A Species Flocks from the Mexican Plateau. Copeia (3): 533-556.

Barbour CD and Chernoff B (1984) Comparative Morphology and Morphometrics of the Pescados Blancos (Genus: Chirostoma) from Lake Chapala, Mexico *In* Echelle AA, and Kornfield I (eds.) Evolution of Fish Species Flocks. University of Maine at Orono Press.

Barbour CD and Miller RR (1978) A Revision of the Mexican Cyprinid Fish Genus Algansea. Misc Pub Museum of Zoology, University of Michigan 155:1-72.

Blair WF (1964) The Case for Ecology. BioScience, 14(7): 17-19.

Bostford LW, Castilla JC and Peterson CH (1997) The Management of Fisheries and Marine Ecosystems. Science, 227:509-515.

Brussard PF (1991) The Role of Ecology in Biological Conservation. Ecological Applications, 1(1): 6-12.

Catalán-Romero JM and Menchaca-Hernández JA (1993) Fecundidad, Período de Reproducción, Parasitismo y Crecimiento del Charal Chirostoma Chapalae Jordan & Snyder, en el Lago de Chapala. Technical Report CIIDIR-IPN-MICH 3799/006. 59 pp.

Chacón-Torres A (1980) Contribución al Conocimiento de la Ecología y Composición de la Ictiofauna del Lago de Cuitzeo, Mich. B Sc Thesis, Universidad Michoacán San Nicolás de Hidalgo University, Michoacán, México.

Chernoff B and Miller RR (1981) Systematics and Variation of the Aztec Shiner, Notropis sallei, a Cyprinid Fish from Central Mexico. Proc Biol Soc Wash, 94(1): 18-36.

Christensen V and Pauly D (1992) ECOPATH II – A Software for Balancing Steady-State Ecosystem Models and Calculating Network Characteristics. Ecological Modelling, 61:169-185.

De Buen F (1946) Investigaciones sobre Ictiología Mexicana, III. La Ictiofauna del Lago de Chapala con Descripción de una Nueva Especie (Haustor ochoterenai De Buen). Anales del Instituto de Biología, Universidad Nacional Autónoma de México, 17(1-2): 261-281.

Eberhardt LL and Thomas JM (1991) Designing Environmental Field Studies. Ecological Monographs 61(1): 53-73.

FAO (Food and Agriculture Organization) (1995) Precautionary Approach to Fisheries. Part 1: Guidelines on the Recautionary Approach to Capture Fisheries and Species Introduction. FAO Fish. Tech. Pap. No 350/1. 51 pp.

FAO (Food and Agriculture Organization) (1999) Integrated Resource Management for Sustainable Inland Fish Production. Fisheries Department Committee on Fisheries (23rd Session), COFI/99/2. 8 pp.

Fausch KD, Lyons J, Karr JR and Angermeier PL (1990) Fish Communities as Indicators of Environmental Degradation. pp. 123-144. In: S. M. Adams (ed.). Biological Indicators of Stress in Fish. Symposium 8, American Fisheries Society.

Fitzsimons JM (1976) Ethological Isolating Mechanisms in Goodeid Fishes of the Genus *Xenotoca* (Cyprinodontiformes, Osteichthyes). Bull of South Cal Acad Sc 75(2): 84-99.

Francis RICC and Shotton R (1997) "Risk" in Fisheries Management: A Review. Can J Fish Aquat Sci, 54:1699-1715.

Gilbert CR (1998) Type Catalog of Recent and Fossil North American Freshwater Fishes: Families *Cyprinidae*, *Catostomidae*, *Ictaluridae*, *Centrarchidae* and *Elassomatidae*. Florida Museum of Natural History, Special Publication No 1. i-11 + 1-284.

Goldman EA (1951) Biological Investigations in Mexico. Smith. Misc Collect, 115:I-xiii + 1-476.

Guzmán-Arroyo M (1995) La Pesca en el Lago de Chapala: Hacia su Ordenamiento y Explotación Racional. Universidad de Guadalajara - Comisión Nacional del Agua Press, 302 pp.

Guzmán-Arroyo M and Ortiz-Martínez JM (1995) Producción Pesquera pp. 211-229 *In* Guzmán-Arroyo (ed.) La Pesca en el Lago de Chapala: Hacia su Ordenamiento y Explotación Racional. Universidad de Guadalajara-Comisión Nacional del Agua Press.

Guzmán-Arroyo M, Galaviz-Solís S and Morelos-López MG (1995) Ciclos reproductivos. pp. 93-117. In Guzmán-Arroyo (ed.). La pesca en el lago de Chapala: hacia su ordenamiento y explotación racional. Universidad de Guadalajara-Comisión Nacional del Agua Press.

Hansen AM, León-Zavala A and Bravo-Inclán L (1995) Pollution Sources and Metal Enrichment in Sediment Lerma-Chapala Watershed. Ingeniería Hidráulica en México. 10(3)55-69.

Hilborn R and Walters CJ (1992) Quantitative Stock Assessment. Chapman & Hall, London, 570pp.

Hilborn R, Walters CJ and Ludwig D (1996) Sustainable Exploitation of Renewable Resources. Annu Rev Ecol Syst, 20:45-67

INP (Instituto Nacional de la Pesca) (1998) Sustentabilidad y Pesca Responsable en México Evaluación y Manejo. Executive Summary, INP-SEMARNAP, 26pp.

Karr JR (1981) Assessment of Biotic Integrity Using Fish Communities. Fisheries (Bethesda), 6(6):21-27.

Krebs CJ (1989) Ecological Methodology. Harpes Collins Publishers, 654 p.

Ledesma-Ayala CL (1987) Estudio Ictiológico del Río Duero, Michoacán. B Sc Thesis, Nuevo León Authonomous University, Nuevo León, México.

Limón-Macías JG, Lind OT, Vodopich DS, Doyle R and Trotter BG (1989) Long- and Short-Term Variation in the Physical and Chemical Limnology of a Large, Shallow, Turbid Tropical Lake (Lake Chapala, Mexico). Arch Hydrobiol./Suppl. 83 (Monographische Beitrage), 1:57-81.

Limón-Macías JG and Lind OT (1990) The Management of Lake Chapala (Mexico): Consideration After Significant Changes in the Water Regime. Lake Reservoir Management, 6:61-70.

Lubchenco J, Olson AM, Brubaker LB, Carpenter SR, Holland MM, Hubbell SP, Levin SA, MacMahon JA, Matson PA, Melillo JM, Mooney HA, Peterson CH, Pulliam HR, Real LA, Regal PJ and Risser PG (1991) The Sustainable Biosphere Initiative: An Ecological Research Agenda. Ecology, 72(2):371-412.

Lyons J, Cochran PA, Polaco OJ and Merino-Nambo E (1994) Distribution and Abundance of the Mexican Lampreys (*Petromyzontidae*: *Lampetra:* Subgenus *Tetrapleurodon*). The Southwestern Naturalist. 39(2):105-113.

Lyons J, González-Hernández G, Soto-Galera E and Guzmán-Arroyo M (1998) Decline of Freshwater Fishes and Fisheries in Selected Drainages of West-central Mexico. Fisheries, 23(4)10-18.

Menchaca-Hernández JA, Arzate-Maldonado O, Mejía-Muratalla A and Diego-Bautista A (1986) Estimación de Algunos Parámetros Poblacionales del Bagre (*Ictalurus dugesi*) en el Lago de Chapala en los Años 1978 a 1985. Technical Report CIIDIR-IPN-MICH 3818/025. 25 pp.

Miller RR (1986) Composition and Derivation of the Freshwater Fish Fauna of México. An Esc Nac Cienc Biol, Mexico, 30:121-153.

Miller RR. and Smith JM (1986) Origin and Geography of the Fishes of Central Mexico. pp. 487-517 *In* Hocutt CH y Wiley EO (eds.) The Zoogeography of North American Freshwater Fishes. John Wiley & Sons, New York.

Norma Oficial Mexicana (1994) Peces. Mexican Federal Government, México, DF, Rule NOM-ECOL-O59-94. Published in the Federation Official Diary, date May 16, 1994.

Rosas MM (1976) Peces Dulceacuícolas que se Explotan en México y Datos Sobre su Cultivo. Editorial Tercer Mundo, Instituto Nacional de la Pesca, 135 pp.

Rosenberg AA and Restrepo VR (1994) Uncertainly and Risk Evaluation in Stock Assessment Advice for US Marine Fisheries. Can J Fish Aquat Sci, 51:2715-2720.

Sainsbury JC, Walter HA, Alevizon W, Ben-Yami M, Bohnsack J, Drew SC, Hansen JE, Harper E, Kubaryk JM, Lipuma E, Lyons AJ, McCreight, McVey JP, Meltzoff SK, Polovina JJ and Shortall III JW (1988) Fisheries Technologies for Developing Countries. National Academy Press, Washington DC, 168 pp.

SEPESCA (Secretaría de Pesca) (1989) Análisis del Programa de Acción Inmediata para el Desarrollo Acuícola-Pesquero del Lago de Chapala. Sría Pesca, Del Fed Pesca Jalisco, 27 pp.

SEMARNAP (Secretaría de Medio Ambiente, Recursos Naturales y Pesca) (1999) Anuario Estadístico de Pesca 1998, Mexico City, 241 pp.

Smith ML, Cavender TM and Miller RR (1975) Climatic and Biogeographic Significance of a Fish Fauna from the Late Pliocene-early Pleistocene of the Lake Chapala Basin (Jalisco, Mexico). Univ Mich Mus Paleon, Papers on Paleontology, 12:29-38.

Soto-Galera E, Díaz-Pardo E, López-López E and Lyons J (1998) Fish as Indicators of Environmental Quality in Río Lerma Basin, Mexico. Aquatic Ecosystem Health and Management (Elsevier), 1: 267-276.

Tereshchenko IE, Filonov AE, Monzón CO, Gallegos AG and Rodríguez RS (1999) Influencia del Niño 1997-98 en el Régimen Térmico del Lago de Chapala, México. Boletín Informativo (summary and program). GEOS Unión Geofísica Mexicana AC, Época II, 19(4):227.

Uyeno T, Miller RR and Fitzsimons JM (1983) Karyology of the Cyprinodontoid Fishes of the Mexican Family *Goodeidae*. Copeia (2): 497-510.

Walters CJ (1984) Managing Fisheries Under Biological Uncertainty, pp. 263-274 *In* May RM (ed.) Explotation of Marine Communities. Dahlem Konferenzen, Berlin.

Walters CJ (1992) Perspectives on Adaptative Policy Design in Fisheries Management pp. 249-262 In Jain SK and Botsford LW (eds.) Applied Population Biology. Kluwer Academic Publishers.

Webb SA (1998) A phylogenetic analysis of the Goodeidae (Teleostei: Cyprinodontiformes). Ph. D. Thesis, University of Michigan, USA, USA.

White MM and Turner BJ (1985) Intralacustrine Differentiation in Two Species of Goodeid Fishes. Copeia, (1): 112-118.

Whittaker RH (1975) Communities and Ecosystems. 2nd Edition. MacMillan Publishing Co, Inc, New York, 440 pp.

SECTION IV

SOCIAL AND ECONOMICAL ASPECTS

Chapter 11

On the Shore: Social Awareness Regarding Lake Chapala and its Environment

Basic Elements for the Outlining of Communication and Social Participation Strategies

PABLO CHÁVEZ-HERNÁNDEZ
Instituto Mexicano de Tecnología del Agua, Paseo Cuauhnáhuac 8532, Jiutepec, 62550 Mor., Mexico. email: pchavez@cenca.imta.mx

Key words: Social organization, participation, communication, systemic communication approach, dialectic communication models

Abstract: General Background. This chapter offers an empirical approximation to the communications processes observed and analyzed in regions of three shore municipalities: Jocotepec, Tizapán el Alto and Chapala. The observations, analysis and the resulting reflections took place during the exploratory phase of a communications project. The document does not analyze nor evaluate the project; it focuses on the observable events and the study objects that appeared during the project. Peripheral References. The project mentioned here was named *Water and Social Life on the Chapala Shore*. Outlined in 1990, it was carried out in 1993 by the Lake Communications Office under the methodological and instrumental supervision of the Mexican Water Technology Institute. The observations, analysis and reflections correspond to that period. The project did not prosper: in 1994 with budget cuts and restructuring of the projects run by the agencies involved in the operations, including the National Water Commission, the project was cancelled. However, the exploratory phase provided a characterization of the social actors, the productive, economic and organizational scenarios, and a conceptualization of increasingly complex notion of communication networks. The audiovisual monograph, a key result of the project, use testimonials with a self-diagnostic focus, describe the history, conflictive aspects and expectation for change among the shore communities. The monographic content provides, on the one hand, random samples of the potential for participation and conflict that center around access to resources and their social administration; and on the other, it illustrates the reflexive tasks based on a complexity paradigm. A detailed discussion is provided. The document characterizes, in communications terms, shore individuals, groups and communities and their interactions with natural resources. For this purpose, individual and group perceptions are described and analyzed, explanations proposed of routes that result in social meanings; an analysis made of the convergence and divergence phenomena, recreation of awareness, communications and social and institutional participation. Agricultural, fishing and commercial activities from an environmentalist viewpoint will be covered; consideration is given to the differences and similarities between the official local or sub-regional presentations and the nonofficial discussions on a local scale. Jocotepec frames references with a primarily anthropological and cultural view point; Tizapán el Alto presents an approximation to the socioeconomic and, in part, of the organizational aspects; the municipality of Chapala provides the background to a review of the Los Alacranes island that brings out the political and environmental aspects. The images obtained during the exploration of the areas are conjugated in a dialogue; from here interpretation conjectures are formed to conciliate the diversity characteristic of the shore into a possible communications analysis unit. Based on this unit, suggestions are made for a communications strategy that corresponds to one objective: dimensioning and restarting the social, adjustment and transformational capacities around an "eco-organizational" project based on cultural identities, understanding and a self-reflexive comprehension of society as a component of the environmental order.

1. INTRODUCTION

1.1 Some general aspects

Any proposal for solving the water quantity and quality problems in water basin must consider a controlled and well-informed intervention by socially defined groups that live in these areas. Frequently the failure to recognize or to harmonize the diversity of social actors, interests, attitudes and abilities complicates, hinders or delays efficient planning or implementation of integrated actions.

It is, of course, necessary to have ample knowledge of the many differences that define the social groups involved in a proper water management. Along that line it would be possible to design an adequate communication strategy that shows consideration for the distinctive features of all the social actors involved in the sanitation of the Lake Chapala basin and the rational use of its resources. Within these differences it is necessary to identify and understand the different ways in which each part of the community not only perceives and values the water resources, but also with what other natural resources each of these groups is related.

Furthermore, in order to apprehend and understand the social meaning of the vital liquid we must understand the different forms in which each social group accounts for, recognizes and values the relationship between their natural environment and the productive, economic, technical, technological and political aspects of the community

And furthermore: the historical relationship and how the various social groups, and their diverse surroundings conceive it have proved to be essential to achieve an understanding of the present awareness and appraisal of the water resources. Indeed, the history of adjustments and transformations between human groups and their surroundings, their ensuing interdependencies and mutual conditioning, has greatly contributed in the explanation of the constantly adapting meanings ascribed to water.

Any communicative and participatory proposal regarding the adequate use of the water basin's resources cannot deny these considerations. Trying to avoid or to ignore them would impede the implementation of a proposal in spite of its technical or technological feasibility; it would diminish or hamper a project's realization. And, in turn, the basic premises for sustainable development must offer some guidance concerning the kind of action the situation calls for.

1.2 The specific situation

In 1990, the need for a solid, socially constituted basis from where to start working on the recovery, maintenance and appropriate use of the Lerma-Chapala basin resources, in compliance with established sustainability criteria, gave rise to a diagnostic study on the social awareness concerning the natural resources' state of the art, mainly water and soil. This survey also produced descriptive information about the local socioeconomic, technically productive, political, cultural and organizational environment and, whenever possible, these descriptions also included data on the local historical background. This article sums up the research experience in the Chapala basin, reduced to only part of three municipalities: Jocotepec, Tizapán el Alto (which includes Mismaloya, a fishermen community in the municipality's neighborhood) and the Los Alacranes island in the Chapala municipality.

The diagnosis was considered decisive for the susceptibility of joining civilian and governmental initiatives within an acceptable framework of an appropriate task distribution. This susceptibility would be determined by the interrelationship between the participatory potential and the conflict potential that dominate the areas and the groups involved. It was also believed that the degree of participation would have its immediate expression in the willingness of the community to intervene systematically in the assigned tasks.

Figure 1 Traditional fishing with *tumbo* in Lake Chapala South shore at San Luis Soyatlán county. The *tumbo* is an extended stationary fishing art for in-water captures

Said research of possibilities was realized entirely on experimental grounds. Regarding the scope of the social disposition to be discovered, no specific *a priori* assumptions were made. Far from being a demonstration of expected social behavior, this exploration was meant to yield a chart of perceptions and values related with the lake, its shore and its resources. Thus its aim was a description of what was being observed and not meant to present evidence for any hypothesis.

Regarding the communication, organizational and participatory processes some premises were put forward in order to direct the search for information and it's understanding. These premises related to the systemic communication approach, according to dialectic communication models (Serrano, 1986, 1994) and the autopoyetic organization processes (Maturana and Varela, 1984). Once the first part of the research was concluded, it was possible to make interpretations and considerations regarding phenomena of self-awareness and self-perception, and regarding some collective learning (Etkin and Schvarstein, 1989) and adaptation processes (Maturana and Varela, 1984).

The research was accomplished without consulting any books or writings about the shore; it was of an entirely empirical nature. Furthermore, the observational and analytical approach was based on qualitative, not quantitative parameters. Social life on the shore was reconstructed from direct testimonies and from an interdisciplinary perspective including methods taken from anthropology, sociology, oral history and communication sciences. Once the fieldwork and the corresponding reports were concluded, communication, organization and participation aspects were analyzed within the complexity paradigm (Morin, 1980).

The research was done and recorded with the help of audiovisual means and edited according to formats typical of the documentary genre.

2. METHODOLOGY

2.1 The research

The criteria used to choose the sites of research were based on the cultural diversity and geo-political differentiation. Along these guidelines the communities were selected randomly in two diametrically separate areas: on the northern and on the southern shore. Jocotepec and the Los Alacranes island are on the northern shore while Tizapán el Alto and Mismaloya are located on the southern shore of the lake.

The research was based on participatory observation processes combined with in depth interview techniques. Information sources were established through a direct cohabitation process with several community members, popular organization representatives and some authorities from the local municipality. No videotaped interview was improvised. To create a real communicative level, all the interviews on tape are the result of a previously well established relationship beyond the limits of a mere interview: we shared in the field labor, knew the families on a daily life basis, so we could unify action with conversation, informative propositions with communication facts. Established relationships turned the formal interview sessions into efficient and satisfactory communication acts.

The initial questionnaires had to be expanded; detailed and adapted according to whatever relevant information we received during the preparation, realization and follow-up of each interview. It happened that when interviewing one person, his answers and the topics he would bring up would lead us to other informants we hadn't thought of before and so we discovered different and new topics to elaborate on. This proved to be an adequate procedure to detect and contact very dynamic local communication networks.

Farmers, fishermen, shopkeepers, opinion leaders and some of the government officials were interviewed according to questionnaires based on two goals: a portrayal of important social actors and those that have significant communicative roles, and the description of environmental, productive, economic and political surroundings. In some cases the interviews reveal personal judgments, in others they were more like descriptions of the actual situation of the Mexican farmers after the announcement of macro-structural reforms (for instance, the signing of the North American Free Trade Agreement; the reforms in the Constitution's 27th Article). Also stories of past events were recorded, related with population growth and migratory processes, with the community's infrastructure, job opportunities and their water use, for irrigation as well as for human consumption.

One of the strategies followed to organize and build up all the testimonial material was the search for complementary and contradictory narratives: totally opposed points of view of the same events and also coinciding opinions were put together in the final arrangement of the audiovisual presentation. The documentaries allowed to generate a kind of diagnosis by the participants themselves: the inhabitants stating their own perspectives, debating their own expectations, questioning their own organization forms, expressing their own solutions to productive, technological and ecological problems in the area.

2.2 Documentation

The work done at an audiovisual level helped in that it united all the information that was obtained during the research. The accumulation method differed from conventional procedures used to videotape commercial spots or news bulletins. This method was based on the communication experiences related with integrated rural development. It is related with what is called didactic video recording and was worked out during the second half of the seventies, in the Mexican tropics.

The method can be distinguished in at least two features: this approach yields an insight based on community knowledge and its own cultural beliefs; in this way there is no need for persuasion nor for imposing ideas through opinions that may be instructive, prescriptive or compulsory; this kind of insight requires an in depth relationship between the audiovisual production team and the research object, generating dialogue, discussion and analysis with and among groups who often defend deviant interests. The method not only allows for the accumulation of opinions, it also stimulates reasoning and reflection. It is meant to recreate certain perceptions, to establish old and new concepts with a critical eye. Its implementation for this study allows for an applied research. The methodology consists of four major steps: research; the establishment of audiovisual scripts; the video recording and its edition. The research methodology has been described in section 2.1.

Interviews, scripts and video images were constructed within anthropological, sociological and historical dimensions, their visual contents primarily being focused on characters, places, techniques and technologies. Due to this construction, scripts and image sequences are intended to reveal problematic situations in the filmed scenarios in order to take away the veil of triviality from daily life, showing relations of interdependency and situations of mutual conditioning between, for instance, economic and ecological, productive and political or administrative aspects. The scripts, together with the video images and the editions, present oral history as a collective learning experience, as a means to achieve reflection and knowledge.

3. RESULTS

3.1 Jocotepec

3.1.1 Myths and rituals

The lake is part of a mythical vision that explains the community's origin, dating from before the Spaniards' arrival, as is the case of other shore settlements. It is considered a founding place that attracts ancient civilizations and, according to some of the present day inhabitants, its refill capacity seemed to be everlasting. This mythical conception of the lake was proved to be still very much alive, even at the end of the twentieth century.

This modern mythical conception incorporates two aspects: the regeneration power of nature itself, without human interference, and the inhabitants' willingness to participate disinterestedly in any rehabilitation works in case this would be required. So it happened that, on the one hand, some inhabitants expressed their trust in the potential of the atmospheric conditions and rain periods, or the aquifers natural refill to restore the basin's water reserves, while on the other hand it was stated that no particular political interests at all would affect the willingness to organize actions to improve or maintain the water volumes, or to work together for the lake's sanitation.

So this mythification gives rise to two positions. One that idealizes the lake's hydrologic potential in an almost complete negation, or at least neglect, of its progressive eutrophication. And the other one that recognizes the cause-effect dynamics between the basin's water resources and the human activities involving their use. But even this position idealized the social participation, in case it would be required, to solve the problems related to the lake by supposing this to be devoid of any political connotation. Both positions sometimes were heard within the same group or even expressed by one and the same person. The myth of everlasting water abundance would be conjugated with an unconditional social solidarity myth, a politics proof environmental awareness.

The indigenous people from the Jocotepec area were said to realize sacrificial ceremonies before the Spanish conquest. According to oral tradition, human offers were made to Lake Chapala. As may be expected, this tradition disappeared shortly after the Spanish arrived. But even in 1991, the social interest in the lake seemed to reveal some ritual tendencies. These tendencies of course had shifted from a religious environment to other society spheres: those of the civil organizations and environmental politics. Indeed, one of the "Sociedad Pro-defensa del Lago de Chapala" (Lake Chapala Defense Association) representatives was from Jocotepec. From there, this group promoted regular on shore cleaning and reforestation campaigns and also mediated with other civil and governmental organizations with similar interests in other areas of the lakeshore.

At the local schools for secondary education both the participation in these campaigns and the realization of meetings and conferences on the present evolution and state of the lake were promoted. Some social leaders, however, would criticize the effectiveness of these actions; they mentioned the meetings' lack of continuity and the cleaning and reforestation campaigns' contingent character. According to these critical observations, the trees planted in the reforestation actions were not suited for the basin's climate. Nevertheless, every now and then some activities were organized. Their ritual nature could be observed in two aspects: repeated collective actions though these didn't solve the problems they were supposed to and allegedly were intended to solve; and repeated speeches that reinforced the shore inhabitants' involvement in the recovery of lake and shore, and at the same time hardly ever would agree in their contents with the work really done.

Finally, myths and rituals, old and new, operated as intermediary forces between the lake and the human settlements; they also originated social practices and thus had a cohesive value in that they conjugated the willingness of several groups guiding them towards one reproductive goal: groups that maintain themselves united for a common project.

As far as can been seen, modern myths and rituals were creating contradictions that, although acknowledged, could not be overcome. The myths and rituals subsisted because of their organizational force but they were not enhanced at a prepositional level, These interpretations later became more significant in other coastal areas: the myths and rituals potential enabling individual decision making while simultaneously limiting awareness or social action was increasingly noticeable in different aspects in several of the villages surrounding the lake.

3.1.2 Vision and reason

Whatever their socio-economic status, the Jocotepec inhabitants fairly much agreed in their hydrographic and geopolitical vision of the lake and its shores. This vision's more remarkable sides were: the idea of a regulatory functioning of the water body and its environment concerning the climatic or atmospheric regional, not just local, balance. The comparatively disproportionate water quantities required for different uses by the coastal villages and demanded for the city of Guadalajara; the lake's varying water level immediate harmful effects for the coastal villages in comparison with the indirect impact these have in the State's capital; the region's productive and tourist potential considered as an economic option at a State or regional level.

Figure 2 Jocotepec central plaza. Lake Chapala North shore of Lake Chapala. "Jocotepec" means *sour fruit hill*

Thus the lake is considered a global patrimony. The more so because the shore inhabitants recognized the many connections their region sustains with other parts of the country, specially with those that are part of the Lerma-

Chapala basin. This global perception implied another feature, remarkable though for the outsider, obvious to the inhabitants: from their point of view, although the shore inhabitants never were the first ones to interfere heavily with water quantity and quality, they were the first ones in suffering the consequences of the lowering water level and the increasing water pollution. Consequently, the most interested party in rehabilitating the lake was the inhabitants themselves. This awareness seemed to guarantee a quick and positive reaction among the shore inhabitants regarding whatever civilian or governmental calls for social participation in any conservationist project.

Thus, a socially coordinated action seemed to be assured in advance. Nevertheless this apparent assurance was not entirely without doubts; the shore inhabitants were aware of some discrepancy between their vision of the problem and its solutions, and the vision of a federal government agency. Towards the end of the nineties, when the water deposits were getting low, the National Water Commission (CNA) decreed an interdiction on water extraction from the Lake Chapala basin. In compliance with this decree, tens of small water pumps used by the local low scale vegetable growers were taken away and quite a few households saw themselves severely affected in their daily economies. The Jocotepec inhabitants considered the taking away of their pumps an unjustified act in spite of the fact that it was an official and previously announced decree. They saw no justice in it because, comparatively much more water was lost in the realization of the water pumping from Lake Chapala basin to Guadalajara, while the government hadn't taken any efficient measures to repair the considerable water losses caused by this deficient water transport. Thus, some critical Jocotepec inhabitants considered the taking away of the private pumps from the farmers as a disproportionately radical cover-up action.

Thus, official attempts and measures to solve the water shortage problems within the urbanized areas were minimized, even by the people who were interviewed, as mere promotional events: the Guadalajara water crisis was faced by informative or inductive campaigns that promoted water saving. On the other hand, according to the Jocotepec inhabitants the industrial section was not required to pay the proper fines in proportion to the polluting impact of their wastewater discharged to the Lerma river, though legally these activities ought to be taxed. In Jocotepec the people insisted on three topics: the government showed a lack of interest in solving visible inefficiency problems regarding the water distribution for human consumption, and at the same time would impose high fines on the small level water extraction by the shore-dwellers; the urban area was clearly favored, their excessive water use was not punished but merely observed and discouraged; governmental rules were not equally applied because the industries would not be punished for their improper handling of polluting elements. For all these reasons the Jocotepec inhabitants called for social justice.

Though these points could be discussed, in this paper we do not refer to the justice or injustice of governmental decisions; neither will we comment on their legal or legislative implications. This sort of revision isn't considered to be part of the essay. First, we want to emphasize that, generally speaking, *a priori*, the prohibition decree regarding the extraction of lake water seemed reasonable to all. Secondly, in particular, *a posteriori*, there are reasons to agree both with the shore inhabitants when they complained about the injustice of the taking away of their equipment and the subsequent crop loss, and with the governmental action according to the established rules. In the third place, agreement with one or the other position depended from what was given priority in times of a critically low water deposit level. In the fourth place, the governmental and civilian views were not the same though all parties agreed on the necessity of an official decree. In the fifth place, in view of these different visions and argumentations what was needed was a shared comprehension and understanding of these differences, diverse principles and distinct ways of constructing visions and argumentations.

From the above mentioned five points another three followed: 1. Within the global order around the lake, the most fragile structures where those of the horticulturists, not the industries nor the state capital's urban population; nevertheless, the horticultural activity was the most affected one by resolutions on the macro level. Indeed, there should have been a mediating structure between the more fragile and the more solid regional structures, generating decisions not in an informative and reactive way, but more communicative and active. 2. Visions, reasons, decisions and actions are elements modeled and adjusted through negotiation and management; in turn, mediation, negotiation and management may be creatively developed in the field of comprehension and understanding. Concerning the prohibition, the established precepts, the adopted resolutions, the decisions and the value judgments about the presence or lack of justice caused a kind of communication gap in Jocotepec between the decision making government apparatus and the population. 3. While these structures, both the more solid and the more fragile ones, are being integrated in the increasingly global machinery, the complexity of the coupling requires dynamic, flexible communicational networks that can join actors with heterogeneous visions. That is where common projects are realized and administrated while allowing the differences to interact.

3.2 Tizapán el Alto

3.2.1 Historical trends and present day discussions

Between 1992 and 1993, social relationships in Tizapán el Alto were dominated by conflict situations and critical organizational politics. The way agriculture developed in the area may partially explain this situation. Rather intensive irrigational farming is done in Tizapán from ancient times. The La Pasión river comes from San José de Gracia, Michoacán and before it flows into Lake Chapala it runs through the municipality's main village. In the nineteenth century, still as part of hacienda grounds, the river was channeled to be used for irrigation of crops. During many years the San Francisco Hacienda administered the water. Certain quarrels between the villagers and some of the landowners permeated the legislative structure of the Porfirio Díaz epoch; society claimed equal shares of the costly liquid.

Since 1910, after the first agrarian redistribution, water control for agricultural purposes caused conflict among some families and agricultural producers. These disagreements had fatal victims and incited rivalries, mixed with simultaneous struggle for land. During several decades, gaining access to water and land was a constant cause of dissension. Some of these quarrels would subsist, perhaps in a less violent way, for at least two generations. Thus, the Tizapán inhabitants trained their ability of criticism through constant confrontation, as protagonists or audience: in 1993, this inclination towards controversy had become a distinctive feature of group representatives and opinion leader. Indeed, confrontation and a controversy could be considered part of their nature, built upon the historical conflicts around water and land.

In counterbalance to this tendency to confrontation, the Tizapán inhabitants and their organizations showed a well-developed capacity to negotiate and form cohesive, though closed, groups. Their negotiation capacity was distinguished by very narrow allowances. This indicates a mentality fit to achieve efficient commercial transactions, a tendency to a scrupulous administration of strategic information considered as a privileged tool to develop competitive relations. The inhabitants' economic and productive past played an important role in this character building. The main municipality center was part of the commercial corridor that ran from the upland areas to Guadalajara in the thirties and forties of the twentieth century. Lake Chapala was used to transport merchandise from the upland areas to Guadalajara, before the construction of the first roads; Tizapán was an obligatory transit point and the center of intense commercial exchange.

On the other hand, farming was prosperous in the seventies, at the expense of irrigation and local horticultural activities. In due time the resources and market operations got under firm control of hermetic groups, some of them part of the traditional local groups in power, considered authoritarian. The inhabitants' economies and living conditions grew increasingly apart. In the nineties quite a few of the interviewed persons took this situation to be part of vertical control relations that dated from the former century. According to some inhabitants, the North American Free Trade Agreement, the changes applied to Article 27 of the Mexican Constitution, and the Law on National Waters supported the social as well as the economic disparity. Whatever the reasons, the access to the resources and their management never stopped being a cause for controversies.

Figure 3 Onion harvest at Tizapán el Alto. The church is named after the local holy saint. The word "Tizapán" means on white land, over tizate stone

In this context, the population perceived from governmental institutions in charge of water management a privileged treatment towards their counterpart civilian organizations; they questioned the legitimacy and equality of the irrigator's representatives and even the impartiality of local and federal governments. A mechanism seemed to have grown that stimulated asymmetrical productive relations and unfair organizational competence. The inhabitants mistrusted any community project with governmental interaction.

In 1993, the disagreement with the ways in which the municipality recently manifested itself found an expression, not only in the town's political and party life but also in spheres that were not directly involved with political institutions. The explanation for this dissatisfaction and protest against the performance of productive, commercial, and even religious groups is to be found in the historically grown skepticism regarding any group representative authority. Thus it was not surprising to find the population fragmented in many aspects of its social life; this fragmentation among productive or commercial groups implied the disapproval of projects for changes conceived among them.

The Tizapán inhabitants sought to transform the vertical decision making structures rooted in the past; an attempt to democratize municipal life marked the community's activities: the inhabitants wanted to establish different communication networks, but functionally they were unable to do so. Yet, the present was a very much-discussed topic. The unilateral control of the change processes would have to remain in the past; and the population felt a historical commitment with this high-priority aspiration.

3.2.2 The lake: out of sight?

In Tizapán, Lake Chapala, being the usufruct source, only entered the discussion of the present at a secondary level, due to two reasons: the quantity, variety and weight of the conflicts affected the community relationships within the municipality; and, secondly, contrary to the northwestern shore, on this side the lake attracted no tourism at all.

The information received about the lake indicated the lack of continuity in government projects and erroneous decision-making regarding residual water treatment. Indeed, the former Ministry of Fisheries (Secretaría de Pesca) suspended the building of a big piscicultural center for the breeding and rearing of endangered fish species. The people interviewed on the subject denounced that productive areas were idly occupied by unfinished constructions, which was all that remained from projects abandoned without any clear explanation. The location of the municipal center's wastewater treatment plant's location was considered illogical and it's functioning inefficient: the main urban density was located downhill from the plant, out of its catchment area. These critics emphasized the federal and state government's inability to deal with the lake's problems. But they also disqualified the municipal government. All the official declarations seemed to be discordant or lacked practical consistency.

For the fishermen the Lake Chapala, as a disputed resource, was also relayed to a secondary level. Certainly they experimented the consequences of pollution; coveted species such as catfish and white fish, had drastically diminished. Eventual expansion of water lily made navigation and fishery more difficult or even prevented it. But fishing possibilities were not completely obliterated by either of these phenomena. Anyhow, the lake's natural hydrodynamics confined the bigger water lily masses towards the opposite northwestern shore. Finally, according to some fishermen, it was possible to find technological devises and marketing strategies that allowed to deal with the changes of aquatic fauna and habitats.

The more important aspects for the fishermen were of a organizational and economic nature. The three local cooperatives had not been able to unite, in a sustainable and profitable way, their productivity with market flows. In a continuous process of capital depreciation the fishermen were unable to improve their household incomes and were looking for jobs in other economic sectors. Actually, resources, technologies, and methods or attempts to organize the fishing activities were having a very hard time. The fishermen's life could be typified as one of subsistence and infra-subsistence.

Furthermore, as a result from the social conflict situation and disqualification of the municipality's organizational proposals, the fishing cooperatives impeded their own progress because of dissension that affected their negotiating capabilities, group cohesion and vision to undertake joint projects. And even worse: the fishermen considered themselves a forgotten group regarding municipal politics favoring economic and social improvement.

In 1993, all this mounted to one single question in Tizapán: Did Lake Chapala have a place in the future projections and previsions? Apparently not. Yet this does not mean that the inhabitants would refuse to participate when asked to attend problems related with the lake. Actually, the fishermen collaborated with the northern coast inhabitants removing the water lilies during 1993; and other inhabitants participated in periodic cleaning campaigns. But even so their interventions were merely to solve immediate problems. The information and communicative activities did not involve the lake because that would suggest the previous overcoming of accumulated conflicts or at least attending them. The perceptual patrons and communication processes were generally conditioned to confront adverse socio-political and organizational tendencies.

The Tizapán el Alto communication networks were oriented towards dissipation, not convergence, of the social forces; they tended to be more exclusive than inclusive. The information was restricted, only given or interpreted with defensive or offensive purposes; this strategy was efficient in preserving the groups, but not in uniting them. They went through an identity crisis, their structures were being rearranged process, their negotiation and agreement interests were regulated without reciprocities among groups and thus any joint action planning or coordination was thwarted. This hampered their convocation capacity as well as the consolidation of global projects. Also, the improper fulfillment of their goals, or the display of uncoordinated activities, affected their credibility and caused mistrust. Compared with the past, the inhabitants had achieved more tolerance concerning differences in thought and performance, as well as a growing capacity to criticize, but their communicative ways produced dissension or, at best, generated a very fragile consensus. Rumors and public opinion were the most common means to give or withhold credibility to these actions.

In order to put an end to this vicious circle it was necessary to orient the people's reflective capacities towards a social pact, based on a self-critical ethical observation of the whole town. Reflection, ethical self-observation and the social pact should take place within a creative and new perception of what the town had been, had become, and could be in the future. This was not to be achieved in the short term because of the weight of social conflicts that would make the participatory and communicative processes more trying and would paralyze the debate. This prevented the groups to reach a new order of reflection, criticism and action. If this situation could be overcome, the lake and other resources might be thought of as provisional and thereby involve a more prepositional dynamics in the community; if things would work out, a new relationship could grow between the villagers and their natural environment.

3.3 Los Alacranes Island

3.3.1 Three acts and one scenario

On the island a triple exploitation occurred: commercial activities, tourism and fishery. There were three restaurants and some small charales farms, areas destined to the breeding and rearing of this species. A boatman cooperative was in charge of the transportation of tourists between Chapala and the island. Nor restaurant owners, nor boatmen inhabited the isle; the fishermen lived there only temporarily, during the fishing period and the charales reproductive phase. The tourism was said to damage the charales breeding; the boats carrying the tourists were said to affect the fishery.

The island offered a contrasting appearance. The tourist and commercial organizations properly maintained the restaurants section. In this area a small zoo was built with a double purpose: as an additional tourist attraction and as the start of a small size ecological project, supported by the municipality. The island's fishing section was notably different in organization. Poorly improvised houses, irregular sanitary provisions and a careless arrangement of catching devices gear made the place look quite uncared-for. This was opposed to what the tourist activities required, thus the fishing activities were detrimental for the tourist section.

The boatmen, restaurant owners and fishermen would argue over mutual claims. The more persistent group in these mutual arguments was the tourist transportation cooperative. Their leaders claimed serious damage was done to the rock formation around the island where the fish were bred and reared. The fisherman had dug out part of the rock formation on the shore so the charales fish breeding would be protected and it provided a place where they could keep their catching devices. Additionally, the boatmen had other complaints about the fishermen: the lack of waste disposal and the presence of excrement outdoors, for instance. The boatman used a discourse in defense of the environment to achieve governmental action to correct the fishermen's ways of being.

Figure 4 Placing of *nazas* in a charales farm. Los Alacranes island, Chapala county. North shore. *Nazas* are fishing arts originally designed by local fishermen; they serve a double function: guaranteeing the capture of adult specimen and favoring the species' reproduction

The charales fishermen, on the other hand, accused the boatmen of gasoline leakage in the water around the isle, damage to the charales farms due to the presence of tourists, alleging that the tourist activities were affecting the fishing activities directed towards the preservation of an endangered species. The fishermen's discourse emphasized the importance of sustainable fisheries in the lake in the debates with the tourist transportation sector, and in defense of their own activities.

The restaurant owners were not really involved in the constant accusations between boatmen and fishermen. Actually, the restaurants on the island encouraged the tourist flow and obviously meant an increase in the boatmen's income; they, in turn, provided the restaurants with constant clientele. The restaurant owners, however, were accused of improper waste management; they passed this responsibility to the municipality because these had promised a regular waste transportation service from the island, which never existed. The boatmen tried to get the restaurant owners on their side against the fishermen, something they could not achieve because of the interdependency between the two groups: the food supply, either for direct consumption or for sale, anticipated possible conflicts.

What in the beginning seemed no more than a little disagreement among users of the island grew into a bigger conflict. Municipal, state and federal governments tried to regularize the uses of the island with the participation of the parties in dissension. Yet these regulations where not respected; there really did not seem to exist a real solution for the island's simultaneous use by these three kinds of users. Apparently, the commercial, tourist and fishing activities could not fit in one and the same scenery.

In spite of its very small scale, the island's situation bore an important meaning: it concentrated organizational, economic, political and social pressures developed during the functioning of heterogeneous structures and systems. At this small scale, the relations on the island, the mutual effects of different uses in the same space, the interactions among the users, all reflected phenomena discovered at a major scale in the shore towns we had visited. Perhaps the nature of the behavior and processes observed on the island was similar to a larger scale behavior and homologous processes in the whole Lerma-Chapala basin.

3.3.2 Communication or in-communication between two shores: order and complexity

The island's environmental features offer the best possible habitat for the charal, being the most appropriate place in the whole area to breed this species. It is also a privileged fishing spot for the fishermen since no other place in the whole lake allows for fishing in such a compact area. Furthermore, the charal is a much wanted species on the market and this particular capturing – rearing method offers a better productivity guarantee than other methods used with other species. The charales fishermen have built artisan-catching devices that allow catching and reproducing the species. The charales fishing techniques, catching devices and methods are traditionally famous along the whole shore. This recognition emphasizes the role they play in the species' protection. Around 1993-1994, the former Ministry of Fisheries, through its Chapala office, carried out an evaluation and research project concerning the fishermen's methods and tools efficiency and effectiveness. So the fishermen could reckon with an institutional basis that allowed them to acquire more importance regarding their participation in the growing conflict with the other island users.

The island also was a privileged place for the boatmen. Compared to excursions further away from the main piers, the ones between Chapala and the island were profitable because they paid less for fuel. The boatmen and the restaurant owners got institutional support from the Ministry for Tourism who has an office in the town center. In view of the profits Chapala receives from the tourist business, the boatmen emphasized this fact to their own benefits.

At a less visible conflict level there was a territorial dispute of the island. This had other than just physical dimensions. Indeed, the island was thought of as a strategic element for the economy based on the charal breeding and fishing research. At the same time, the organization of the island's exploitation involved political decisions, environmental norms, municipal regulation, inter-sectoral agreements and inter-institutional coordination. Civilian and governmental structures, different in size, symmetry and dependency, ought to achieve equilibria among its participants, and also between the components of different structures. To achieve this coupling, sufficient flexibility and firmness are required to solve an apparent incompatibility in the island's uses. In the conflict's more profound structures the civilian and governmental organizations' identities, apparent in the heat of the debates, could be questioned. The values by which their actuation, the degree of their freedom or tolerance and the scope of their actions were guided always took part, sometimes hardly perceptible, in the debates.

The situation of conflict on the island put into evidence the alliances between the users and the governmental structures. When claiming the institutional relationship to minimize the differences, the users would confound personal interests with institutional ones. There were no clear divisions between civilian and governmental spheres. The conflict situation had a curious dimension because all parties would appeal to the preservation of the island's natural resources while in the end each debate was provoked by the economic hegemony among the commercial activities, fishery and tourism related to municipal administration and planning, with the way priorities were established at a municipal level and with the municipal politics regarding each sector.

During these conflicts, while seeking this hegemony, different power structures would be related in a complex association between social and productive forces, between the technical and ethical authority of the organizations. In the debate's critical moments the institutions felt obliged to pronounce judgment or give technical advise to help decision-making. The parties often perceived partiality in the technical decisions, complaining that the responsible agencies would pronounce a favorable judgment for the groups belonging to their own sector, whichever it was, the fishery, commercial or tourist sector. On the other hand the parties in conflict would notice a kind of competence among the official institutions in displaying a higher technical or political authority. This had some socially important consequences: in the first place it affected the idea of the institutions' moral authority, discrediting their public image and technical capacities. And secondly, because the technical judgments could result detrimental or beneficial for any of the parties involved: if the boatmen canceled the island route they would harm the families whose incomes depended on the restaurants and tourism; if the fishermen left the island this also would have severe economic consequences.

The island's case became analytically important for several reasons. We will mention three of them: In the first place it shows how, when an exclusively environmental discourse imply political aspirations, without openly saying so, its apparent presentation hides the true interests or the conflicts real cause, such as political leadership, economic sectoral hegemony, environmental preservation, or other interests. In this way it becomes difficult to spot the cause of the problem, making it more difficult to design an efficient strategy that attends the causes and not the symptoms. In the second place, it shows how there is a growing participation of social and institutional actors in disagreements that make the actions ambiguous or deviates them from their own conjuncture, so paralyzing or postponing the realization of balanced and self-sustaining actions destined to improve economic and social conditions or environmental preservation. In the third place, it shows the municipal government's vulnerability in being the place where the conflicts that state, federal or local agencies couldn't solve converge, obstructing its management mechanisms and administrative structures. Usually the population's critics and pressure are oriented towards the regulatory force closest to themselves: the municipality.

To proceed with an efficient organization of the diverse island's uses it would be necessary to: 1. know the logical parameters according to which each of the civilian or governmental actors understood the given situation and how they would wish it to be; 2. understand the cultural framework in which each actor's interests and actions were fitted; 3. understand the chaotic succession of events regarding each actor's transformation and permanency strategies; 4. re-organize the variety of knowledge, discourse and acts in a union of dual values, obvious on the island: the strength of being delicate and the delicacy of being strong that allow living beings and their habitat to form self similar structures, self-reproductive, through one another, each one producing the other one.

As long as this knowledge and understanding did not evolve, the actors would go on generating lots of varied information without communication; they would have a presence but no effective participation. What would be lacking was a multi-rational reflection and action scheme by the means of which they might be able to adapt to a complex order.

4. CONCLUSIONS

The following proposals are not of a strictly conclusive nature. On the one hand, the project that gave access to the Chapala shore was suspended in its exploratory and documentary phases; we did not feedback the diagnosis with the shore inhabitants, neither was any consecutive communicative action undertaken. On the other hand, it must be taken into account that the research took place from 1991 to 1994, so some of the following considerations might have become obsolete.

The social conflict situation that does not refer directly to the water quality and quantity could be obstructing projects that are centered on these factors. The shifts of controversial moods between one and the other view of social life are very dynamic, due to the various interests and structures in contact.

With this in mind, the communication projects and those to impulse social participation could lose effectiveness if they are elaborated from a reductionism viewpoint. In order to understand social or communicative phenomena it is recommended to put aside any kind of reductionism, be it at a history-graphical, organizational, communicational, sociological, and technological or any other level. This research shows both, that it is pertinent to work with interdisciplinary schemes in inter-sectoral spheres, and also that this is very difficult to achieve.

Communicational possibilities and the shore inhabitants' participation potential show a social willingness to tackle the lake problems with an integrated approach, through inter-sectoral actions. Yet it must be recognized that this mere willingness is not enough to undertake a collectively sustainable action. The social conflict and lack of efficient communication could defeat the group's willingness to act. Both possibilities may be reproduced within the complex cause-effect relationship created between the civilian and governmental structures.

The step towards a disposition for sustainable action asks for transformations in the functioning of civilian and governmental organisms. Administrative, informative and management networks must be changed. But the more profound changes affect the organizations' ethics and the ways in which concerted forces guide the organisms. The shore communication networks must be revitalized with that purpose.

The social willingness and the ability to comply with interdisciplinary tasks around the lake depend on at least four factors: 1. The degree of coherence perceived by the users' sectors in both the governmental discourse and their own groups. 2. The degree of congruence, observed by the inhabitants, between the official proposals concerning a rational use of the resources and the actions derived from these proposals. 3. The propriety of actions that seek a sustainable use of the resources in an economic framework that harms people's reproductive strategies, their adaptation capacities and possibilities for evolution. 4. The projects' appropriateness and continuity.

Any project that promotes communication and social participation must contribute to the creation of a legitimate, credible and trusty relationship within and among the social organizations of Lake Chapala, within and among the governmental structures involved, and between these two public spheres. Ignoring this would mean the dispersion of the social forces, putting a heavier mediating burden on public opinion and impeding a growing awareness that will allow civilians and government organisms to develop a new view of the basin.

5. SUMMARY

5.1 Goals

- The shore inhabitant's perception of Lake Chapala and its natural environment, between 1990 and 1994, were identified and the social and communicative actors, productive, economic, organizational and political contexts featured.
- The local networks were recognized, taking into account both the cultural diversity and unity of the shore.
- The social feasibility to undertake or promote actions to achieve the recovery, conservation and appropriate use of the natural resources were carefully considered.

5.2 Methods

Three municipal centers were chosen as sample cases. Though it was an arbitrary selection, it was tried to chart perceptions and behaviors on geographically opposed sites: the northern and southern lakeshores. Field research was conducted from a participatory observation approach. In order to collect field testimonies in-depth interviews were used. The methodological use of audiovisual means allowed unifying historical, political, anthropological, sociological and communicative dimensions.

5.3 Main results

The main results are a preliminary diagnosis and an audiovisual documentary of the three municipalities' social life. The diagnosis includes: mapping of different perceptions concerning the ways the natural resources are approached and managed; historical views on the population and the water and soil uses; critical descriptions concerning the fishing and farming organizations' operations; reflections on how the area is inserted in the national economy and politics.

5.4 Conclusions

Between 1990 and 1994, the social willingness to solve the problems around the lake in a concerted way depended on at least three factors: The degree of coherence perceived by the users' sectors in the governmental discourse. The degree of congruence between the official proposals concerning a rational use of the resources and the actions derived from these proposals. Any communication and social participation project must generate legitimacy, credibility and confidence in and among its social and governmental actors.

ACKNOWLEDGMENTS

To Mercedes Escamilla, because, while in charge of the Oficina de Comunicación del Lago and promoting scientific research, her actions and negotiations on the shore gave rise to the Water and Social Life on Lake Chapala Shore project, on whose information this report is based. Her understanding of interdisciplinary approaches was an important support for the dynamic interactions with several organisms and a variety of cultural habits.

To Jorge Martínez Ruiz, coordinator of Communication, Participation and Information Technology of the Mexican Institute of Water Technology. The reflections and conceptual horizons shared with him helped a lot in the continuous reframing of theoretical principles, epistemologically adequate to systematize research processes of an empirical nature. His critical attitude enriched many of our observations and interpretations.

To Roberto Menéndez Frione, adjunct coordinator of Communication Technology of the Mexican Institute of Water Technology, because of the support he has lend to achieve an efficient organization of the Water and Social Life on Lake Chapala Shore project's research and documentation; he offered us the necessary logistics without which the project would have lacked efficiency; his understanding of the conditions was essential for its realization.

To the Lake Chapala shore families and organizations that collaborated with us in this research and documentation project; by giving us their words and their time they showed us their confidence in the possibility to improve their people's conditions.

REFERENCES

Books

Etkin J and Schvarstein L (1989) Identidad y Cambio en las Organizaciones. Editorial Paidós. Buenos Aires, Barcelona, México.

Serrano MM (1994) La Producción Social de la Comunicación. Alianza Editorial, Madrid, 1986. México.

Maturana H and Varela F (1984) El Árbol del Conocimiento. Las Bases Biológicas del Entendimiento Humano. Editorial Universitaria, Santiago de Chile.

Morín E (1980) El Método. Tomo II, La Vida de la Vida, Ediciones Cátedra, Madrid.

Videos

"Agua y Vida Social en la Ribera de Chapala. Notes for the Elaboration of a Diagnosis" Elaboration: Instituto Mexicano de Tecnología del Agua. Production: Oficina de Comunicación del Lago. Jocotepec, Cerro de las Frutas Agrias. Programs 1-4 Tizapán el Alto, sobre Tierra Blanca, sobre Piedra de Tizate. Programs 1-8

"Pescadores de la Ribera de Chapala". Elaboration: Instituto Mexicano de Tecnología del Agua. Production: Oficina de Comunicación del Lago. Mismaloya, un Pueblo de Pescadores. Programs 1 and 2. La Isla de Los Alacranes. Program 1: "Usos Distintos, Problemas Comunes". Program 2: "Entorno y Expectativas". Program 3: "Acuerdos y Desacuerdos de Grupos e Instituciones."

Chapter 12

Social Characterization of the Lerma-Chapala River Basin

SONIA DÁVILA-POBLETE[1] AND ANA HELENA TREVIÑO-CARRILLO[2]
Instituto Mexicano de Tecnologia del Agua, Paseo Cuauhnahuac 8532, Jiutepec, 62550 Mor., Mexico. email: [1]sodavila@tlaloc.imta.mx, [2]htrevino@cenca.imta.mx

Key words: Socio-economical analysis, participatory forms, organizational forms, social relationships, social structures, social interactions

Abstract: The main purpose of this chapter is to define conceptual frameworks to allow a better understanding of the social dimensions involved in the studies related to an integrated water management approach. The chapter has been divided into three main sections: first, the conceptual framework was identified, to understand the river basin as a unit for social analysis and the different ways in which it has been applied in the social sciences. The second section, called social structures and interactions, deals with the main social characteristics of the Lerma-Chapala river basin. Consequently data related to the overall population growth, economic fluctuations and water use by small and large-scale irrigation producers, as well as industrial, urban and rural household water users are presented. The last section deals with participation and organization forms conceived by the social relationships between water users.

1. INTRODUCTION

The environmental conditions and the quality of life in a region depend on solid management and appropriate use of natural resources including air, soil, water and forests. Therefore, an integrated resource management is fundamental for the social and economic development of any country.

This study will show the relationship that exists between access to water and social issues such as public policies and social forms of organization. We are aware that there is a need for a broader study to have an overall picture of the relationship between human behavior and natural resources, yet this chapter will focus only on the organizational aspects related to water issues.

We begin this chapter by laying out some conceptual frameworks to have a better understanding of the social dimensions involved in the integrated water management approach.

To fulfill the above, we used the Lerma-Chapala river basin as an example. We started by considering a river basin as a two-sided issue that cannot be separated: on one hand, the hydrological unit whose regional limits are given by its physical characteristics such as its surface and groundwater sources, as well as its local and regional environmental aspects. On the other hand, the different social groups that exist with its various socio-demographic, administrative and eco-productive processes, cultural characteristics and individual forms of management, all linked to each other through the hydrological cycle.

This chapter has been divided into three main sections: first, the identification of the conceptual framework that leads to understand the river basin as a unit for social analysis and the different ways in which it has been applied in social sciences. Here we explain why the "river basin" concept cannot only be considered as a space defined by its geographical characteristics, but also by its social, economic, political and cultural relationships, since all these parts are integrated by the hydrological cycle as a whole.

Therefore, this section's main concern is to highlight the social content of the river basin's definition and to show how the socio-organizational forms are generated by water use. For this purpose, it was necessary to sort out different frames of reference assigned to the river basin concept such as: an independent unit of analysis; a regional unit; a socio-economic unit and an administrative and management unit.

The second section, called social structures and interactions, deals with the main social characteristics of the Lerma-Chapala river basin. Consequently it points out some data related with the overall population's growth, its economical fluctuations and the water use among the small and large-scale irrigation producers, as well as by industrial, urban and rural household water users.

These factors, coupled with the river basin's conceptual framework, leads to demonstrate that a river basin is a complicated network of different physical, social, economic and organizational aspects. Through our research activities, we found that the Lerma-Chapala river basin, is not only overpopulated with growing water scarcity, but it is also overexploited and polluted, and regional natural resources – particularly water – have undergone severe and noticeable degradation in recent decades.

These data leads us to the last – but not least important – section, which deals with participation and organization forms. Here we see how, although most inhabitants coexist and use water from the same river basin, generally they conceive different social and organizational forms according to their access, the way they use this resource, their location in relation to the river basin (upper, middle or lower), their position in the social structure or their relationships.

The above bring us the need to establish a form of social control over water use, and to review how government designed and implemented public policies geared to improve water management for all users as well as to create organizational forms at the national, state and local levels.

This chapter shows some of the main characteristics of the Lerma-Chapala River Basin Council (*Consejo de Cuenca Lerma-Chapala*) and its structure and organizational processes implemented by the National Water Commission (*Comisión Nacional del Agua-CNA*). Based on these processes, we analyze how each group of water users understands and foresees its participation in planning and regulating water management, as well as how government institutions project their participation in the decision-making process.

2. THE RIVER BASIN AS A SOCIAL ANALYSIS UNIT

Although the river basin concept has different meanings, the most well known has been used to refer to it as an area in which water is considered the main component for the integration of all the elements that constitute a geographical space. This concept suggests a large region with similar climate, biology and geology.

If we restrain our conceptual framework to this definition, we will only see the importance of the biophysical impact on a river basin, such as soil erosion, deforestation, sedimentation, chemical pollution, flooding and changes in the water level without acknowledging how people relate to them and to nature itself.

Therefore, any definition given to the "river basin" concept must consider the social aspects as crucial for an integrated resource management in the use, protection and conservation of water and all other natural resources. To accomplish such a task, a comprehensive understanding of the systems and subsystems that make up the natural and social structures is required.

The above statements lead us to analyze how the river basin concept can be defined in different ways depending on the type of research or on the particular resource management required. Nevertheless, any definition of the term will have to consider natural and social factors involved. This includes economic activities, social groups living in the area, demographic and ethnic characteristics. If we consider the river basin as a resource management and administration unit, we will have to look into different aspects of regional development, and also at the investment and organizational structure assigned to the management of natural resources.

Hogan (1993) proposed that the river basin should be considered the "natural" unit of analysis, in which there is a need to understand the relationships between natural and social processes. He also defined the river basin as a strategic ecosystem for the observation and analysis of socio-demographic and environmental relationships.

In this chapter we will use the term river basin as referred to a land-mark or territory characterized by a social, economical and political system of relationships that occur while there is an ongoing process in the use of the natural resources. At the same time, this territory is a physical space bounded by variations in elevation of any geographical area, where the hydrological cycle is the integrating system[1]. To explain the content of this statement, we will also consider the river basin as a regional, socio-economic and water management unit.

2.1 The river basin as a regional unit of analysis

The river basin approach incorporated in the concept of region refers to a set of social relations bounded by a hydrological system, in which the region is used as the unit of analysis.

Previous studies that conceived the region as the unit of analysis highlighted the homogeneity that prevails in a given region, showing how this has an impact on the social relations and their structural processes.

This analytical frame of reference was used in several European countries, where a variety of theories flourished to explain the similarities and/or differences between regions. France is specially known for its contributions to this research field since the era of Napoleon's reforms when he set up a united country with a modern administration based on departments and municipalities.

These theories explained a region as something homogeneous whose boundaries were set by the similarity or difference of the environmental characteristics such as vegetation, geomorphology or hydrology. Thus, while

[1] This definition stems from the works of Comas d'Argemir (1998) on the importance of political ecology as part of nature and of social change. The literature published by UNESCO (1999) and Newson (1992) has also helped us to understand the social and natural implications of this definition.

some scholars tried to prove how each region was internally homogeneous, others were demonstrating how the geographical characteristics and inequalities in economical development caused the differences among regions.

Another point of view that provided a radical change in the definition of a region was the *landscape* (*landschaft*) approach, which included different trends of social analysis, conceived by the schools of geography and cultural ecology, historic sociology, among others. These schools defined *landscape* as a way of describing the relationships that exists between human beings and the environment, focusing -mainly- on the impact that human activities have on nature, where the analytical frame of reference relies on the physical and social aspects.

According to Bertrand (1986), the definition of a river basin from the *landscape* approach is that a territory is characterized by a dynamic combination of physical, biological and human elements that are part of a physical space in which natural and social processes occur. In our case, hydrological resources bound all these processes.

Another perspective within this theory focuses on the *development poles* where an environmental approach was added, not only related to water, but also to forestry and agriculture.

Walter Christaller began his central location theory by using a theoretically homogeneous space model in which he recognized a relation between locations according to their size and specialization levels, setting a hierarchical criteria for urban centers with respect to their *hinterland* or *area of influence* (Gutiérrez, 1993).

Many other approaches were added to this proposal, beginning with the inclusion of factors such as the theories of *development* or *growth poles*,[2] which considered theoretical perspectives similar to those of economical equilibrium, giving rise to theoretic postulations that dealt with territorial location and equilibrium.

Years later Cristaller, Lösch and Izaard's influence, as advocates of regional location theories, gave rise to different approaches to understand disproportionate or unequal regional development. For example, François Perroux (1993), with his *growth pole* theory, defined the varying intensities of concentrated development areas where the central pole exerted influence that "...lead to global growth, modification of productive structures and transformation of organization" (Ávila, 1993:16).[3]

[2] See Lewis, Rosenstein and Gunnar, among others. Walter Isard, theorist with the largest influence on the developmental approach, who proposed a theory of partial equilibrium in the sixties (Ávila, 1993:15).

[3] This model may be used to explain the processes of change in advanced societies, but in less developed regions it is limited by the degree of socioeconomic wealth.

On the other hand, Albert Hirschmann and later John Friedmann considered that the dynamic factor of the economy is the initial concentration of investment in industrial expansion centers. However, expansion through the *development poles* depended not only on industrial productivity but also on the capacity of social development in countries where *growth poles* were formed (Ávila, 1993).

Coraggio, Boisier and De Mattos[4], from the regional model's critical trend, pointed out that any regional development process – including management of its own resources – must have a large regional negotiation capacity as well as the ability to increase participation in the region. This implies the possibility of decentralizing decisions, with greater emphasis on regional action and management, and paying special attention to shared responsibilities by different social agents: the central government on one hand and the regional community on the other.

In summary, the approaches described above – homogeneous space, landscape, development poles and regional development processes – allow us to come closer to the definition of the river basin as an analytical unit circumscribed by a specific space defined by hydrological resources. Here the social and natural characteristics are part of the systematic unit in which different social processes (productive, socio-demographic, and political-administrative) are decisive for its configuration as a region.

2.2 The river basin as a socio-economic unit of analysis

The "river basin" as a socio-economic unit of analysis implies the study of the social relations that exists among human beings, with the main purpose of identifying the economical activities and the impact of actions upon the water resource.

The analysis of water usage by a social group in a river basin provides an infinite number of approaches. For example, the different definitions of power and social control, as well as studies related to the theory of conflicts, allow us to understand disputes, negotiations and alliances that emerge from the various and complex social interactions generated by the use of water in a river basin.

Today, many social researchers are focusing their analytical procedures to understand the relations between relatively scarce resources and unequal social opportunities to access them. This has been viewed from different perspectives such as power relations among social groups; forms of control of government agencies in water management, and social or community organizations that originate from the use of water.

[4] For a more complete description, see their articles presented in *Lecturas de análisis regional en Mexico y América Latina* by Hector Ávila (Editor).

Among the scholars that belong to this group, we can mention Karl Wittfogel (1966) who analyzed the interactions between hydrological societies, and Arthur Maass, less well known, who opposed Wittfogel observations about water distribution rules.[5]

2.3 The river basin as a management unit

According to Maynard M. Hufschmidt (1991), the river basin as a management unit should be understood as the process that incorporates the planning and implementation stage carried out through a series of institutional and organizational agreements from which specific tasks arise.

Other authors conceive it as a process consisting of a series of sequential steps that take into account the design, planning, construction and operation of the infrastructure, and the conservation of the resource, for which the essential instruments in the preliminary phase are monitoring and feedback of information.

In many Latin American countries – long before overall national planning models were proposed – water management was based on the implementation of regional development models within the spatial limits given by a river basin, where the process was selected and planned differently for rural and urban areas. Later these development and management models changed from regional to national levels, as known nowadays.

A new approach for water management begins with a full understanding of environmental characteristics of the river basin. This implies identifying the management principles within a hydrological region rather than within a geographical administrative division such as the state, municipality or county. It also means that management policies are no longer seen as planning tools for the rural or urban development – as in the past – but as the basis for the administration of water resources.

Another aspect that is considered in the policies for water management is the mechanisms used by the State to control the participation of the private sector. All this is with the purpose of consolidating a modern State with the capacity to regulate, intervene, plan, arrange and stimulate economical activities to sustain them. This planning and programming considers the river basin as the basic reference unit.

The river basin as a management unit is better understood as a comprehensive space that allows a spiral feedback between environmental impact and the changes that society must make to prevent the rapid

[5] Maass and Anderson (1978). Using a sociological approach applied to water administration, we find Hufschmidt (1991); Aguilera (1992); Barking and King (1979); Dourojeanni, (1994), among many others.

deterioration of the hydrological resources and overexploitation of the aquifers. Thus, the river basin provides the framework for policy making, based on an analysis of the effects of human interaction with the environment.

The foregoing explanation allows a better understanding of its definition as a management unit[6], in which the space defined by a river basin is a regional planning unit. This implies that efficient water management must cover specific areas based on related or similar problems within certain critical areas.

Therefore, the river basin as a management unit has two action levels: on one level, it is a management unit where the river basin is the base for planning and implementing policies. On the other, it functions as an intervention unit, to enforce the implementation of the management and administrative policies. At this level, it deals with sub-units or parts of the former, such as water sub-basins or smaller micro-basins, aquifers or other individual applications of the resource that intercept the main river basin.

An example of this is the integrated water resource management approach given on a regional or river basin level that has become widespread in recent decades. This approach emphasizes two aspects. First is the recognition and incorporation of the regional hydrological characteristics to solve complex distribution and water administration problems and the enforcement of defined rules for the design and implementation of a financial compensatory system for environmental sustainability. Second, it is the incorporation of mechanisms by which different water use interests are negotiated and conciliated. For this to happen there is recognition of the socio-political factors and the diverse social, economic and political organizational forms that intervene in water problems.

Consequently, water management is currently conceived as a decision-making process that must consider **technical and social control.** Both exist side-by-side affecting the administrative processes used in planning, organization and social division of labor. It also tends to seek for mechanisms needed to perceive the political tendencies and identify the demands and conflicts that arise with the use of water.

Management policies have become the bylaws to regulate access to water and this is foreseen through the implementation of actions aimed at social interactions. It must be pointed out that these bylaws are not simply administrative changes, but means of regulating conflicts between groups

[6] It is important to underline that in this case the river basin is not considered a unit of analysis but a unit *per se*. In Mexico, the National Water Law, Article 3, Section IV; establishes that: "The river basin, including its aquifers, are the management unit of the water resource." (CNA/SARH, 1992:8).

and/or organizations, channeling social demands and proposing the creation of spaces for joint negotiation.

The foregoing lead to understand management forms as an institutional-social relationship in which social actions have a double meaning. On one hand, are the public institutional actions[7] that create policies and mechanisms to carry out the administration of public services. On the other are the community management patterns for the supply and consumption of these services, and whose main benefit lies in the community's ability to manage natural resources and the **social power**[8] that these goods represent.

In summary, the river basin as a management unit refers to the actions directed to attain balance between people's needs and availability of resources, considering the geographical, economical, political, institutional, social and cultural aspects within a river basin.

Based on the above mentioned units of analysis and referring once again to our initial definition of a *river basin*, as "a land-mark or territory characterized by a social, economical and political system of relationships that occur while there is an ongoing process for utilizing the natural resources. At the same time, this territory is a physical space bounded by variations in elevation of any geographical area, where the hydrological cycle is the integrator system"[9]. We can infer that in a river basin there are social groups, which access in different ways the water resources (surface, groundwater or both). Water users are organized diversely (through socio-demographic patterns and processes) and are situated at different altitudes (upper or lower side of the river), yielding a set of different and hierarchical social interactions, through (direct or indirect) political or economic relationships. Although many social groups may not know about or are unaware of their relationships with others, all of them have a common bond provided by the use of the same hydrological resource: the river basin.

With this conceptual framework in mind we will now present the Lerma-Chapala river basin, as an example where the social dimension is a key aspect in its characterization.

[7] The public institutional actions includes in their managerial elements plans and control forms for the use of resource. These are part of a broader spaces for social action, such as politics and centralization of decision-making related to a territorial setup.

[8] Social power management takes into consideration the interests of diverse groups or sectors that compete for these goods or services. This is a political space for mediation in resource management.

[9] The complete definition is given in footnote 1.

3. SOCIAL CHARACTERIZATION OF THE LERMA-CHAPALA RIVER BASIN

Using the above-described concepts, we will characterize the Lerma Chapala river basin as a regional unit of analysis whose boundaries lie between the arid North, with low population density and recent migrations, as well as little water available. To the South, the basin neighbors with the Mesoamerican region, which has greater access to water and also has a higher population density. The Lerma-Chapala river basin covers partially the states of Querétaro, Guanajuato, Jalisco, Michoacán, and Mexico State.

The Lerma-Chapala river basin as a socio-economic unit of analysis, can be characterized as one of the most important hydrological systems in Mexico, not only because of its size, economic activity, and population density, but also because of its outstanding political, social and historical importance. Regarding its agricultural productivity, it has a large plain surface known as *El Bajío* or the Granary of Mexico for several centuries, and which comprises large regions of the states of Querétaro, Guanajuato, Jalisco and Michoacán.

This river basin has a wide range of economic activities from small-scale rain-fed agriculture to large producers of exportation goods or agro-industries. There is also an industrial corridor between the cities of Querétaro, Celaya, Irapuato and León, and other economic activities that result in urban concentrations in medium-sized cities and twenty other locations with more than 50,000 inhabitants.

The increase of the river basin's various economic activities is reflected on the population growth in the entire river basin, estimated to be 26.2% over the last decade, with an annual average growth rate of 2.32% (Table 1).

Comparing the three sub-regions, the Upper Lerma has the highest average growth rate (2.55%) and the Lower Lerma, the lowest (1.35%). In 1990, the Lower Lerma had the highest percentage of urban population (70.8%) and the Middle Lerma the lowest (32.8%).

This demographic structure may be the result of industrial growth drawing farmers out of the fields, causing important changes in labor structure. Another possible explanation is related to the high migratory rate. According to the National Institute for Geography, Statistics and Information (INEGI, 1997:18), between 1987 and 1992, "...at a national level, the states contributing with the greatest numbers of international migrants[10] were, in order of importance, Michoacán, Jalisco, Guanajuato, Mexico State, Zacatecas, Durango, Guerrero, Chihuahua, Mexico City and Nuevo León, with a total of 67.1% of all international migrants".

[10] The flow of international migrants is mainly toward the USA.

Table 1 Demographic growth in the Lerma-Chapala river basin [11]

Hydrological Sub-region	River basin population			Average annual growth rate 1990-1995	Urban population 1990
	1990	1995	2000*		
Upper Lerma	5,185,617	5,881,913	6,643,736	2.55%	3,254,087 (62.8%)
Middle Lerma	2,497,176	2,800,788	3,158,422	2.32%	1,702,390 (32.8%)
Lower Lerma	1,229,853	1,315,013	1,445,554	1.35%	870,458 (70.8%)
Total	8,912,646	9,997,714	11,247,712	2.32%	5,826,935 (65.4%)

Source: INEGI. Censo Nacional de Población y Vivienda 1990, Conteo de Población y Vivienda 1995 CD version

*Estimated data.

Among the five states in the Lerma-Chapala river basin, four are leading contributors to international immigration. The fact that Querétaro is not in this group may be related to its growth rate (the lowest) and the high percentage of urban dwellers related to short-term migration to the state's capital. INEGI's data show that this state "characteristically receives a large migrant population. Between 1970 and 1992, the percentage of the population migrating from other states rose from 8.3 to 21.9%" (INEGI, 1997:221).

This socio-demographic data will be better understood by defining the labor characteristics and the income in the three sub-regions according to the economically active population to which they belong (Table 2).

Table 2 shows that in the three sub-regions, the service sector is the largest, as is true in most of the country. According to the INEGI, "as a result of changes in the Mexican economy ... [between 1970 and 1996] the most marked [differences] are the decrease in the number of agricultural and livestock workers from 40.3 to 22.1% and the increase in merchants and street sellers from 7.9 to 15.8%" (INEGI, 1998:31).

In the Lower Lerma sub-region located in Querétaro, the primary sector is the second most important, after the service sector, in contrast with the other two sub-regions. This confirms that the creation of the industrial corridor in this region has stimulated its industrial activity.

[11] All the Tables presented in this study were designed and elaborated by Sergio Vargas, M.Sc.

With respect to income in the region, 9.04% are unpaid[12] workers, while 55.41% of the economically active population earns up to two times the minimum wage and 35.55% earns more.

Table 2 Economically active population and income level in the Lerma-Chapala river basin

Sub-region	Economically active population by sector				Income levels		
	Total	Primary	Secon-dary	Service	Without pay	≤ 2 minimum wages	> 2 minimum wages
Upper Lerma	1,289,619 100%	318,159 24.67%	403,222 31.27%	568,238 44.06%	118,470	722,787	423,502
Middle Lerma	653,170 100%	136,383 20.88%	247,002 37.82%	269,785 41.30%	51,725	338,091	246,659
Lower Lerma	298,824 100%	107,861 36.10%	79,409 26.57%	111,554 37.33%	27,959	152,902	108,546
Total	2,241,613 100%	562,403 25.09%	729,633 32.55%	949,577 42.36%	198,154 9.04%	1,213,780 55.41%	778,707 35.55%

Source: INEGI. Censo Nacional de Población y Vivienda (1990). CD version

These differences in income and population distribution among the three sectors in the basin lead us to take a closer look into the Lerma-Chapala river basin as a social unit of analysis. This includes the study of the cause and effect of an unequal relationship among water users as well as the creation of conflicts among them in the three sub-regions.

Some preliminary results indicate that one of the main problems in the river basin is water shortage for all users and, specially, for household and productive users. With more problems in the aspects of supply and distribution of surface water than that of groundwater, this problem increases when we take into consideration the transfer of large volumes of superficial water to the Mexico City Valley[13] and to the city of Guadalajara (Table 3).

Table 3 shows that of the total of 1,337 $Mm^3\ yr^{-1}$ of water extracted for urban and rural use, 524 $Mm^3\ yr^{-1}$ (39%) are transferred to the Valley of Mexico and the city of Guadalajara. The remaining 813 $Mm^3\ yr^{-1}$ (61%) is distributed among 8,969,000 inhabitants living in 6,314 localities.

This Table also refers to the coverage of drinking water and sewerage. It shows that the five states[14] in the Lerma-Chapala river basin have a good coverage of infrastructure in the medium and large-sized cities (with more than 10,000 inhabitants). Fresh water is distributed to 75-88% of the

[12] According to the INEGI, these workers have activities that aid in the maintenance of their family but do not receive compensation, monetary or in kind, that may be exchanged in the market (INEGI,1998:140).

[13] Mexico City is located in this valley and is the main water receiver.

[14] The variation in access to water services and availability is notable between the five states.

households, and 65 to 78% of these households have access to sewerage. In the smaller communities, fresh water is available for 70% of the homes and sewerage for 60%, while in the rural areas only 60% of homes receive drinking water and 28% have access to sewerage.

Table 3 Water extracted for urban and rural use

Localities size (inhabitants)	Number of localities	Population Thousands	Extraction $Mm^3\ yr^{-1}$	Dis-charge (lps)	Service coverage (%)	
					Fresh water	Sewerage
Below 2,500 (Rural)	5,991	2,867	105	1,196	60	28
2,500 - 10,000	239	1,095	50	721	70	60
10,000 - 50,000	66	1,368	81	1,334	75	65
More than 50,000	18	3,539	577	7,296	88	78
SUBTOTAL	6,314	8,869	813	10,547		
Transferred to Mexico City Valley			284			
Transferred to Guadalajara			240			
TOTAL			1,337			

Source: Plan Maestro de la Cuenca Lerma Chapala, 1993.

The distribution of water in municipalities,[15] farms, livestock and industry, and for the generation of electricity is presented in Table 4. These data show that while the farming and livestock sectors are still the main water users, consuming 6,920 Mm^3, or almost 77% of the total, there is an ongoing urbanization in the region. With the increasing urban population, municipal users are now in second place, with a consumption of 1,264 Mm^3 or 14% of the total.

During the last decade, agriculture has undergone many changes based on modifications in the plans and programs for this sector. Among the changes we can mention removal of government support for farm credits, government-funded technical assistance, price guarantees, and the liberalization of marketing transactions for foreign goods under the North American Free Trade Agreement (NAFTA). All of them have had an important impact in the sector, such as the tendency to move from agricultural to industrial activities, which has stimulated the formation of companies within the large urban centers and the creation of in-home workshops in rural regions.

[15] Mainly urban households, commerce and small industries use water distributed to municipalities. Some locations contemplate these water only for rural household usage.

This productive transformation in the river basin has forced many farmers to change their crop patterns from the traditional products to more profitable items and which also are more easily exported, however, they require more irrigation water, such as broccoli, asparagus and certain types of peppers. This change is clearly perceived in the demand for water, as illustrated in Table 5.

Table 4 Water usages in the Lerma-Chapala river basin (Mm^3)

Sub-Region	Muni-cipal	Agri-culture	Lives-tock	Indus-try	Electri-city	Total	Water supply	
							Surface	Ground water
Upper Lerma	563	878	60.1	117.7	551.7	2,170	1,114	1,056
Middle Lerma	387	4,380	166.0	127.2	18.5	5,080	2,587	2,492
Lower Lerma	314	1,349	87.2	32.7	0.0	1,782	1,150	633
Total	1,264 14%	6,607 73%	313.3 3.5%	277.6 3.1%	570.2 6.3%	9,032 100%	4,851	4,181

Source: Info Lerma, N° 2, May 1998. Boletín del Centro de Información de la Cuenca Lerma Chapala

Table 5 Agricultural activities (hectares)

Sub-region	Productive units	Farmed area	Area	
			Irrigation districts	Irrigation units
Upper Lerma	253,481	1,186,073	154,003	202,461
Middle Lerma	81,409	546,064	81,326	251,338
Lower Lerma	59,029	420,528	82,766	66,318
Total	393,919	2,152,665	318,095	520,117

Source: INEGI. Censo Agropecuario 1991 and Plan Maestro de la Cuenca Lerma-Chapala, 1993

Table 5 shows that of 2,152,665 farmed hectares, approximately 830,000 are irrigated with surface and groundwater. Most of this land is concentrated on irrigation units. On the other hand, the 1,300,000 hectares used for rain-fed crops yield less than one-fourth the crop volume obtained from irrigated land. Therefore, there is a constant demand to open these lands to irrigation. Ceding to the demands would result in a major disaster due to the fact that the extraction rate exceeds recharge possibilities in all the country's aquifers. For example, nowadays in Guanajuato all aquifers, with the exception of three, are overexploited, resulting in competition between water users.

Besides the above-mentioned problems, decentralization or federalization of the government has encouraged the rearrangement of the different economic or political interest groups. In a place where water is scarce, shifting from a technical problem to a political issue, this resource plays an important role within the productive sector. As a result, pressure for access to water is accelerating the organizational process where social response – although poorly coordinated at this time – implies a challenge to the regional political system.

In recent years, the model used to represent interest groups has changed from a corporate nature to political representation and, in some cases to political party interests. These new forms of representation are flustering the civil society's participation in the centralized management, where the process of decentralization between the federal and state government is forcing the former to change its relation to state authorities as well as to the users, as will be discussed in the following section.

4. PARTICIPATORY AND ORGANIZATIONAL STRUCTURE

After analyzing the Lerma-Chapala river basin as a regional, economic and social unit we need to study it as a management unit, with its existing forms of participation and organization such as the Lerma-Chapala River Basin Council.

The Lerma-Chapala River Basin Council is the first and main one of its type in Mexico. The National Water Commission (*Comisión Nacional del Agua, CNA*) regulates its organization, structure, and composition.

The National Water Commission is a governmental institution that is a *desconcentrada*[16] division of the Secretary of Environment and Natural Resources (*Secretaría de Medio Ambiente y Recursos Naturales, Semarnat*).

The legal framework -still in use- establishes that the Nation owns all the water, and designates the federal government as its **only** administrator. Former president Carlos Salinas de Gortari at the inauguration of the National Water Commission's board of directors initiated with the following speech:

[16] Among the different administrative structures, all Mexico's governmental institutions belong to a main institutional framework but have two types of administrations that differ in the management of their budget. Agencies that are *Descentralizadas* have an independent budget, so they can obtain grants and loans; those that are *Desconcentradas* have to obtain and report their budget and economic activities through the Secretary of Treasure and Public Credit.

"The National Water Commission is an instrument to carry out an integrated management of water, finance and income, in such a way that it will efficiently link it to the national efforts in the creation of hydraulic works with the systems for supply, distribution and water use" (CNA, 1993:43).

On April 13, 1989, the first River Basin Council was created in the Lerma-Chapala region, which began a program to structure the works of hydraulic and sanitation in the river basin, which includes more states than any other in Mexico. The Mexican President and the governors of the five states, signed a coordination agreement, which provided the following general objectives to be observed by the members of the council: 1. The conservation and recovery of water quality. 2. The allotment and regulation of water usage. 3. Efficient water use. 4. Management and conservation of the river basin and its tributaries (Colección Lerma-Chapala, 1994:102).

Five months later, in September 1990, an Evaluation Council was organized to follow up on their achievements and commitments. The Council added to the already existing participation of the five state governors, the presence of what was then the Secretary of Agriculture and Hydraulic Resources; Fisheries; Health; Treasure and Public Credit; Social Development; Public Expenditure; and Energy, Mines and Industry; as well as the General Directors of the Mexican Fertilizers (*Fertilizantes Mexicanos)*; Pémex (*Petroleos Mexicanos*); the Federal Electricity Commission and the National Water Commission.

In January 28, 1993 this Evaluation and Consulting Council became known as the Lerma-Chapala River Basin Council based on the National Water Law that came into effect in December 1992. Later, this River Basin Council, with the same activities and participants, agreed to include among their activities, the guidelines of the 1995-2000 Hydraulic Program (Poder Ejecutivo Federal, 1996). In addition Chapter 1, Article 4, Section 2 of the Administrative Code of the 1992 National Water Law states that: "the authority and administration of national waters and of its inherent public goods corresponds to the President, who will exercise this power directly through the National Water Commission (CNA)". Chapter 3, Article 9 of the same law states that among the attributes of this Commission is "preparing, updating and enforcing a national water program" (CAN-SARH, 1992).

Based on this law and its regulations, the federal government, through the CNA, considered that to identify the hydrological imbalances and settle actions with local water users, projects addressing more than one state were needed. With this in mind, the hydraulic plans were redrafted seeking regional integration through river basin management.

Therefore, in 1995, the Mexican government, in which the composition of the River Basin Councils was defined, introduced the Hydraulic Program

1995-2000 and new regions were created. This program contemplates 314 river basins grouped in 72 sub-regions, 37 regions, 13 administrative regions that will have to be transformed into river basins and 23 River Basin Councils[17].

Since the Lerma-Chapala River Basin Council had been formed earlier and was already in operation, it only had to change its name to Lerma-Santiago River Basin Council, and add new members to the Council to reflect the area of influence that now includes the Santiago Valley (Table 6)[18].

Table 6 Sub-regions in the Lerma-Santiago Hydrological Region

Hydrological Region	Hydrological Sub-region	River Basin
12 Lerma-Santiago	12A Alto Lerma	1 Cuitzeo (Closed watershed) 2 Alzate 3 Ramírez 4 Tepetitlán 5 Tepuxtepec 6 Solis 7 Salamanca
	12B La Laja	8 Begoña 9 Pericos
	12C Medio Lerma	10 Pátzcuaro (Closed watershed) 11 Alto Turbio 12 Angulo 13 Rio Lerma
	12D Bajo Lerma	14 Sayula (Closed watershed) 15 Duero 16 Chapala

Source: CNA. Regionalización de Cuencas Hidrológicas, 1997

In addition to the definition of the regional unit or the River Basin Council's boundaries, the CNA had to design a management unit for the administration of the Council. In other words, an organizational structure

17 The boundaries of the basin councils have been redefined but official recognition has not yet been given.

18 The criteria used to classify and determine the boundaries of the river basin councils are to be used as a management unit that allows the implementation of certain programs and projects. For this reason, each public institution creates its own classification according to its specific programs and plans. These classifications make it difficult to find water management data that coincide for all. This occurs often, even within an institution such as the CNA.

had to be designed to allow the participation of each state, and to attain fair water distribution approved by consensus, according to surface area, and type and number of users. Table 7 shows this current organizational structure.

Table 7 Current Structure of the River Basin Council

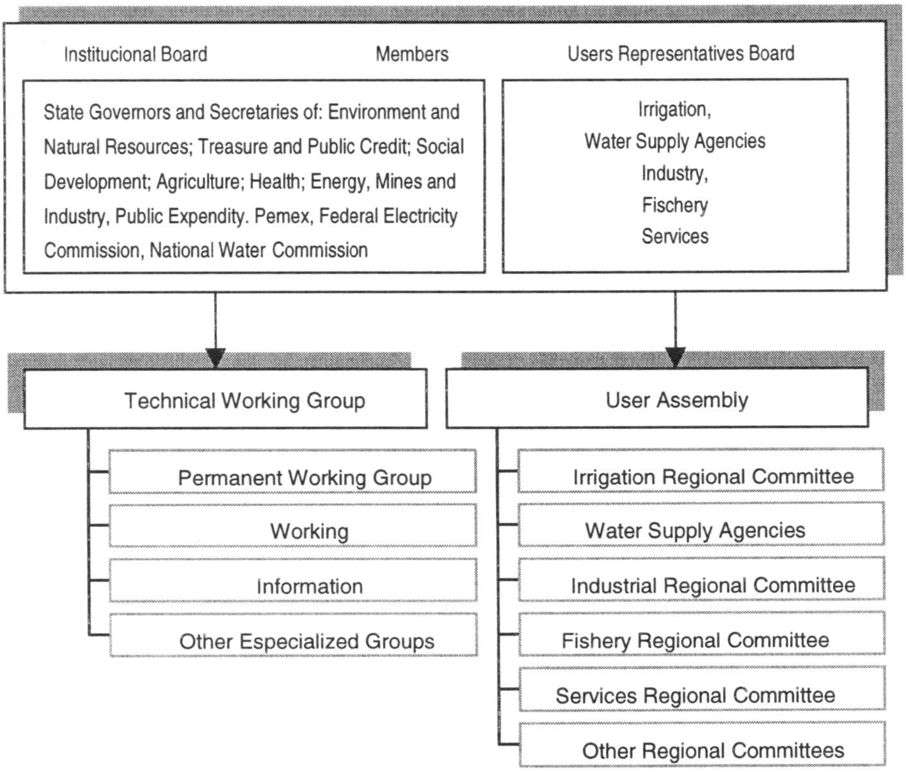

Source: River Basin Council Department

Although this organizational structure has worked many years for the members of the Lerma-Chapala River Basin Council, the CNA authorities and the users are questioning it. The authorities state that the activities of the Technical Working Group should be clearly specified. Since its creation, this group has been in charge of the main activities carried out by the Council, such as water distribution among users in the five states, technical and maintenance aspects of system operations and the legal-administrative mechanisms of the basin councils.

Another argument is that representatives of the institutional sector and state governments are their **"appointed directors or elected heads"**[19]. The numerous commitments of these authorities make it extremely difficult to schedule meetings with the minimum required quorum.

In addition, with the country's new river basin regionalization, many river basin councils include the participation of more than one state, and -at the same time- some states participate in more than one council. The **"appointed directors or elected heads"** that are members of more than one council would have to spend a lot of time in meetings.

On the other hand, users do not agree with the current structure because they feel that not all the water users are represented. They consider that the water user representatives to the User's Assembly, elected by the regional committees for irrigation, water supply, industries, fisheries and service sectors were, in some cases, appointed by a small group of privileged users in that sector. Most users do not know these representatives nor do they have any way of communicating with them to discuss their problems, requests or proposals to be presented in the River Basin Council.

The CNA has been encouraged to propose a new structure for the basin councils, as shown in Table 8.

The main difference between the existing structure and the proposal reside in substituting the **"appointed directors or elected heads"** currently representing government and institutional authorities with other members or representatives assigned by these authorities.

Another modification is that the number of user representatives would increase from six to twelve. This does not imply representation of a larger range of users because the representatives will be selected from the same regional committees. Selection of candidates is still the prerogative of small groups of the privileged water users. The inclusion of a regional committee that has NGO, academic and municipal representatives do not broaden the range of water users because they cannot participate in the decision-making process, since they are only guests.

There is a substantial difference in the structure as related to the Technical Working Group. The four commissions have become the Evaluation and Follow-up Committee, under which there would be two main groups: the Council Administrator with six areas; and the Technical Working Group with two, resulting in a total of eight sub-commissions or working groups (Tables 7 and 8).

[19] Authors' use of bold characters and quotation marks.

Table 8 Proposed structure of the River Basin Council

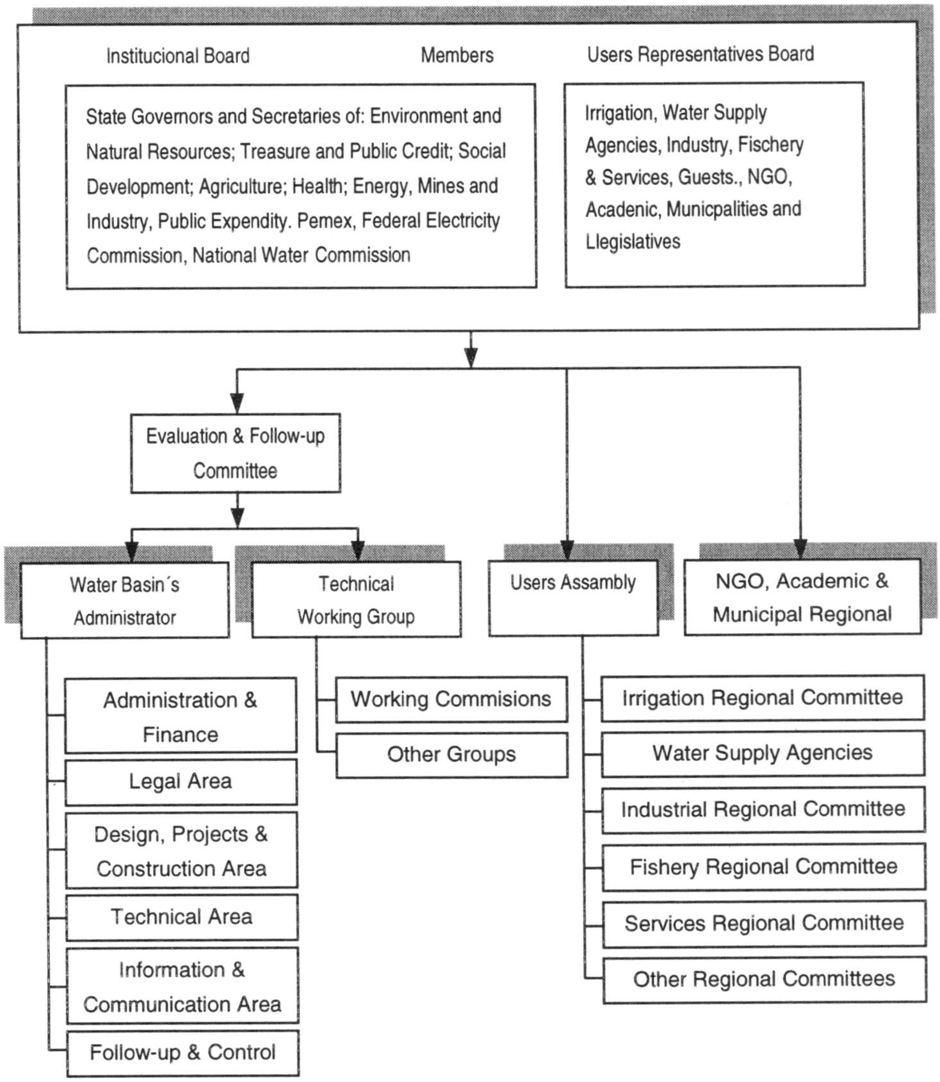

Source: River Basin Council Department

The proposal suggests a structure that would be in a better position to provide supervision and follow-up of the technical and administrative activities of the River Basin Councils. However, user participation is not easier because it does not open the way for real representation. That is, user participation in a social organization like the Lerma-Chapala River Basin Council implies the opening of spaces where different social groups and institutions can interact, complement each other and discuss conflicts. This can happen among users that belong to the same socio-economic unit because they are engaged in the same activities or between those that share the same resource or physical space as part of a regional unit. It could also take place between users that have to form alliances to obtain some benefits and compete for others. This is the case of agricultural producers, who join efforts to obtain certain public services or infrastructure for irrigation, but at the same time they have to compete for the access to these services or for better prices and markets for their products.

Some authorities in the CNA are aware that much work is needed to complete the proposed structure, because they know that the only way in which the management unit functions is as an open space for institutional and social participation. They know that the longer they postpone water user participation, the bigger the problem will be.

Therefore, the need to broaden the existing organizational spaces in the River Basin Councils to the different social groups is of extreme importance. This does not mean increasing the number of representatives, it means introducing a representative of each sector or water use through a cascade process, from bottom to top. We believe that this is the only way in which water users will consider the water problems as their own, and will begin to take real actions to participate in the conservation and preservation of this non-renewable resource.

Through this study we have seen how the Lerma-Chapala is a good example of a river basin as a regional, socio-economic and management unit of analysis. We have also proved the importance of considering all these units of analysis -with its different perspectives- linked together in order to obtain a better understanding of the technical and social relations that develop with respect to water and any other natural resources.

REFERENCES

Ávila H (1993) Lecturas de Analisis Regional en México y América Latina. Universidad Autonoma de Chapingo. Mexico.

Aguilera F (1992) Economía del Agua. Ministerio de Agricultura, Pesca y Alimentación. Madrid.

Barkin D and King T (1969) El Modelo de Desarrollo por Cuencas Hidrográficas de México. Siglo XXI. Mexico.

Bertrand G (1986) Construir la Geografía Física. 1st part. Posición. Revista de Crítica y Análisis Especial No. 3. Mexico.

CNA (1993) Plan Maestro de la Cuenca Lerma-Chapala, Documento de Referencia. Mexico.

CNA (1997) Regionalización de Cuencas Hidrológicas. Mexico.

CNA-SARH (1992) Ley de Aguas Nacionales. CNA-Gerencia de la Unidad Jurídica. Mexico. Also published in the Diario Oficial de la Federación. December 1, 1992.

Colección Lerma-Chapala (1994) Informe de los Avances de los Integrantes del Consejo. Vol. II No. 2, July, 1994. Mexico.

Comas d'Argemir D (1998) Antropología Económica. Ariel. Barcelona.

Coraggio JL (1987) Territorios en Transición, Critica a la Planificación Regional en América Latina. Revista CIUDAD. Quito, Ecuador.

Dourojeanni A (1994) La Gestión del Agua y las Cuencas en América Latina. Revista de CEPAL, No 53, August, 1994. Santiago de Chile.

Gutiérrez Javier (1993) Cristaller: La Teoría de los Lugares Centrales *In* Ávila H (ed.) Lecturas de Análisis Regional en México y América Latina. Universidad Autonoma de Chapingo. Mexico.

Hogan D (1993) Capacidad de Carga Poblacional. Rehabilitando un Concepto *In* Izazola and Lerner (Eds.) Población y Ambiente ¿Nuevas Interrogantes a Viejos Problemas?. Sociedad Mexicana de Demografía/ COLMEX/ The Population Council. Mexico.

Hufschmidt MM (1991) A Conceptual Framework for Watershed Management *In* Easter KW, Dixon J and Hufschmidt MM (eds.) Watershed Resources Management. Studies from Asia and the Pacific. ISEAS. Honolulu.

Centro de Información de la Cuenca Lerma-Chapala (1998) Boletín Info Lerma No 2, May, 1998. Mexico.

INEGI (Instituto Nacional de Estadística, Geografía e Informática) (1990) Censo Nacional de Población y Vivienda. Códice 90. INEGI. Mexico (CD version)

INEGI (Instituto Nacional de Estadística, Geografía e Informática) (1991) Censo Agropecuario. Mexico.

INEGI (Instituto Nacional de Estadística, Geografía e Informática) (1995) Conteo de Población y Vivienda Resultados Definitivos. Conteo 95. Mexico (CD version).

INEGI (Instituto Nacional de Estadística, Geografía e Informática) (1997) La Migración en México. Indicadores Estadísticos. Mexico.

INEGI (Instituto Nacional de Estadística, Geografía e Informática) (1998) Compendio Estadístico-gráfico de Empleo, 1997. INEGI. Mexico.

Maass A and Anderson RL (1978) And the Desert Shall Rejoice. Conflict. Growth and Justice in Arid Environments. The MIT Press.

Newson M (1992) Land, Water and Development: River Basin Systems and their Sustainable Management. Rotledge, London.

Perroux F (1993) Notas Sobre el Concepto de Polos de Crecimiento *In* Ávila H (comp.) Lecturas de Análisis Regional en México y América Latina. Universidad Autonoma de Chapingo. Mexico.

Poder Ejecutivo Federal (1996) Programa Hidráulico 1995-2000. Semarnap. Mexico.

UNESCO (1999) Programa Hidrológico Internacional para América Latina y el Caribe. UNESCO (Internet address)

Wittfogel K (1966) Despotismo Oriental. Estudio Comparativo del Poder Totalitario. Guadarrama. Madrid.

Chapter 13

Competition for Water in the Lerma-Chapala Basin

Economic and Policy Implications of Water Transfers from Agricultural to Urban Uses

CHRISTOPHER A. SCOTT[1], PAULA SILVA-OCHOA[2], VALENTÍN FLORENCIO-CRUZ[3], AND PHILIPPUS WESTER[4]

[1]*International Water Management Institute, email: c.scott@cgiar.org;* [2]*IHE Delft;* [3]*Colegio de Postgraduados;* [4]*Wageningen Agricultural University, email: fwester@xs4all.nl*

Key words: Water allocation, water competition, water resources management, water valuation, shadow prices

Abstract: Competition for water in the Lerma-Chapala basin in central Mexico has brought on a state of crisis. Rampant competition for groundwater has produced dramatic declines in aquifer levels (2.1 meters per year sustained over decades). In early 2000, the stored volume in the receiving waters of Lake Chapala was 28% of capacity, among the lowest levels recorded since systematic data collection began in 1934; runoff inflows have declined significantly with increased consumptive use upstream and declining groundwater levels. Surface water resources, which are subject to considerable inter-annual variability, have conventionally been used for low- to medium-valued irrigated production. Water demand is increasing for all consumptive use sectors (particularly domestic and industrial uses, but also agriculture), reinforcing the critical need for effective allocation mechanisms. The basin is subject to rapidly changing policy and economic conditions. Institutional innovations include a recent water law, basin-level water resources planning, increased user participation in water management, and nascent intra-sectoral water markets. In addition, the past few years have seen the introduction of a broad and far-reaching set of neo-liberal policies including Mexico's signing of the North American Free Trade Agreement, the lifting of price controls for agricultural commodities, and the rolling back of subsidies, particularly for electrical energy. Partly in response to – but clearly with a momentum that proceeded – the changing macro-economic context, there has been rapid urbanization accompanied by seasonal or permanent migration from rural areas to the United States. These processes have important implications for agricultural production and the associated water use. This paper presents and analyzes data to characterize current water use

The Lerma-Chapala Watershed, Edited by Hansen and van Afferden,
Kluwer Academic/Plenum Publishers, New York, 2001

and demand. Evolving water allocation mechanisms are assessed. We review the 1991 interstate treaty on the basin's surface waters and assess the tradeoffs associated with the inter-sectoral water transfers it mandates. Economic valuation techniques are applied for the urban and agricultural water use sectors, with the results indicating that the highest marginal values lie in industrial water use (US$ 0.86-1.34 m^{-3}) followed by commercial (US$ 0.26-0.94 m^{-3}) and domestic uses (US$ 0.24-0.46 m^{-3}). Lower marginal values were estimated for agricultural use, differentiated by canal water (US$ 0.06-0.24 m^{-3}) and groundwater sources (US$ 0.07-0.72 m^{-3}). The bulk transfer of 240 million cubic meters of water from agriculture to Lake Chapala for subsequent urban uses in Guadalajara city in November 1999 resulted in an estimated US$ 14.1 million in foregone agricultural benefits, equivalent to US$ 0.06 m^{-3}. Given that estimated urban water values are at least four (and up to 20) times greater than this loss, compensation of the benefits foregone by farmers appears to be economically feasible for urban water users.
An option contract arrangement in which urban users agree to compensate farmers for foregone benefits appears to be a possible approach to inter-sectoral allocation in the basin.

1. INTRODUCTION

Economic development, improved livelihoods, affordable water extraction technologies, increased variability in rainfall, and heightened awareness of the environmental requirements for water all drive growth in demand for water resources. In regions with limited water supplies, however, rising consumptive use may generate significant disparities in access to, and competition for, water across sectors and among users within sectors. As water of good quality becomes increasingly scarce, systems to allocate water in time and space must be put in place to assure continued access to this resource. Mechanisms to allocate water may take a variety of institutional forms, and must be based on prevailing and accepted water rights. Because both water demand and allocation are dynamic processes, their parallel development should result in the sustainable evolution and development of water resources in a given region.

Where allocation mechanisms are outpaced by trends in consumptive use, for instance, as driven by the rapid economic development of a particular water use sector, the resulting imbalances between demand and supply may threaten the equity and productivity of water resources. Because agriculture is the sector that traditionally receives the greatest share of water (77% of surface water in the Lerma-Chapala river basin in Mexico described in this study), the inter-sectoral allocation of water very often occurs at the expense of irrigation. Furthermore, given the invariably low level of capitalization in agriculture as compared to other sectors, specifically industry, the capacity of farmers to adapt to technological and agronomic innovations is often slow, accentuating inequitable access to water and generating significant resource deterioration.

The lack of appropriate mechanisms to allocate scarce water resources to consumptive demands generates competition that may severely degrade the resource base, both in quantity and quality. For example, static water levels of the aquifers in the Middle Lerma basin, which forms the largest area and is the predominant water user in the Lerma-Chapala basin, are declining at 2.1 meters per year (Scott and Garcés-Restrepo, 2001). In the context of river basins with a moderate to high degree of water resource development, i.e., infrastructure that allows for the capture, storage, distribution, drainage and/or disposal of water, allocation mechanisms can only be effective if backed by policy and regulatory instruments. Given that surface water invariably requires a basic set of infrastructure controls, this resource may be subject to more effective allocation mechanisms than groundwater. The fact that groundwater is usually spatially distributed tends to permit easier physical access, although capital investment costs may significantly limit the number of competing users who can actually pump groundwater.

Water market arrangements may provide an appropriate means to compensate users deprived of water through allocation mechanisms. Users with established water rights have legal recourse to compensation when other (higher priority or higher value) water uses diminish their supply. Under the assumption that inter-sectoral markets reflect the value of water for specific consumptive uses, competing uses should pay market prices for the additional water they receive. While this principle is unassailable, several very real problems may prevent market-based compensation being put in place. Chief among these is the definition of legal rights to specific volumes of water such that determining monetary compensation levels is problematic.

In Mexico, several cases highlight the difficulty in determining appropriate levels of compensation. The northern city of Monterrey extracts water from the San Juan river impounded in the El Cuchillo reservoir, significantly reducing water supplies to farmers in the downstream state of Tamaulipas (Flores-López and Scott, 1999). Despite several legal demands (including an impressive level of resource mobilization among Tamaulipas farmers to fight the case), compensation negotiations appear to break down over the lack of volumetrically defined water rights. Given that Monterrey is both the upstream riparian and the more powerful political-economic party, there is little doubt that any compensation will be negotiated on its terms, and not based on market arrangements. A second case, which will be explored in greater depth in this study, is the 1999 application of the 1991 Lerma-Chapala surface water treaty in which Lake Chapala received 240 million m^3 (Mm^3) of water stored upstream in order to meet the city of Guadalajara's annual extraction of water from the lake. Compensation claims have so far been thwarted (April 2000) by resorting to the legal basis of the treaty. This case shows the crucial importance of water allocation at the basin level, and also portrays the very difficult choices between the environment (keeping the lake full), urban water use and agricultural water use in an over-committed basin.

In general, allocation mechanisms must be in sync with actual water use trends, such that inequitable access to water as well as resource depletion and deterioration are substantively and effectively addressed. The evolving nature of allocation mechanisms in the Lerma-Chapala basin suggests that the mid-term outcomes of competition for water will be particularly severe for agriculture and the environment as domestic and industrial demands increase.

In the context just presented, the objectives of this study are to: 1) evaluate the nature, magnitude and trends of water use by sector in the Lerma-Chapala basin, 2) empirically determine water values by sector, 3) describe the institutional arrangements, both existing and potential, for water allocation in the basin, and 4) assess the economic and policy implications of inter-sectoral water transfers.

2. CONCEPTUAL APPROACH TO WATER ALLOCATION UNDER CONDITIONS OF COMPETITION

Water allocation is becoming increasingly important as river basins around the world inexorably move towards "closure," in which water use or demand significantly exceeds supply. In most instances, higher value uses (defined politically as well as economically) will tend to out-compete lower value uses, depriving them of water and leading to unregulated transfers of water out of agriculture to the urban (particularly industrial) sector. Given the major adjustments this entails, there is an urgent need to carefully consider allocation mechanisms that are just, equitable, sustainable and feasible (both technically and politically). An ideal water allocation system is characterized by flexibility in the allocation of water supplies, security of tenure for established water users, predictability of the outcome of the allocation process, equitability and fairness.

Historically, various water allocation mechanisms have been established in different parts of the world, based on either the riparian or the appropriation water rights doctrines. Under the riparian doctrine a user has the right to extract water from a river system for use on land adjacent to the river as long as the water is returned to the river undiminished in quantity or quality and in a manner that does not impair downstream use. From this stringent definition it is clear that in practice the pure riparian doctrine does not exist, as almost any use will diminish the quantity of water. Nonetheless, this doctrine is used in various forms throughout the world, especially in countries with humid climates, and generally operates without any form of permit or regulatory administration.

Under the appropriation doctrine, water use concessions or licenses are issued to users by the state, which ensures their right to divert or store and use a certain quantity of water. Many variations on the appropriation doctrine exist throughout the world, the most common two being prior appropriation and proportional appropriation. The most significant difference between these two is how they treat water shortages. Under prior appropriation, the first rights that were issued on the river have priority or seniority based on their date of origin implying that rights issued later are the first to be curtailed in times of shortage. Under the proportional appropriation system concessions are issued for the use of a fixed quantity of water, although the quantity of water that may be used in any year is adjusted to reflect water availability within the river basin. Thus, all concession holders share in any shortages or surpluses of water proportionally.

Lee and Jouravlev (1998) postulate that one of the principal benefits of water markets are the almost unlimited possibilities for water reallocation, with the only constraint being the ingenuity of the trading parties. Water markets entail the allocation of water through the free exchange of water rights (be it use or ownership rights) for money. Water rights transfers can take a variety of forms, depending on factors such as the structure of the market, legal and third party considerations, the characteristics of the water rights, transaction costs and above all the needs of the parties to the transaction. Three types of transactions can be identified, namely sales, lease contracts and option contracts (Lee and Jouravlev, 1998). Sales consist of the permanent transfer of title, including all benefits, costs, risks and obligations associated with the water right. Sales are most typical of inter-sectoral transfers, with irrigated agriculture being the dominant water seller and urban users the principal buyers. The leasing of water rights involves the sale of water, but not of the water right. Leases tend to be short term, consisting of the temporary exchange of a quantity of water for monetary or other remuneration, and frequently are used for water transfers within irrigation systems. Water trades, whereby the turns are simply exchanged in time, are not considered part of water markets, as supply and demand are not regulated by prices. Option contracts are long-term agreements to lease, but not to sell, a water right when a given contingency arises, for example a drought. Under option contracts, the receiving party pays a negotiated fee to the granting party even if the option is not exercised, and also pays for income lost when the option is exercised. This paper will assess whether a flexible short to medium-term option contract arrangement may be suitable to compensate farmers for agriculture-to-urban water transfer in the context of the Lerma-Chapala case to be presented.

Crucial to the design of appropriate option contracts are:

- Definition of the contingency conditions under which water will be transferred as well as the process by which the contract is exercised
- Duration of the contract and/or conditions for its renegotiation prior to expiration
- Volumes to be ceded (by whom in the case of multiple parties or states)
- Specification of compensation (both monetary amount and process) for lost income resulting from the water transfer
- Mechanisms to redress grievances on the part of ceding, receiving or third parties

The combination of proportional rights and administrative allocation[1] prevailing in the Lerma-Chapala basin appears to meet all but the final two of these conditions. The analysis below seeks to substantively address the compensation issue by assessing marginal values for water across sectors.

3. WATER CRISIS IN THE LERMA-CHAPALA BASIN: JUSTIFICATION OF THE STUDY

The Lerma-Chapala basin in Mexico has total consumptive water demands that significantly exceed average water supply (Aparicio, 2001; Consejo Consultivo de Evaluación y Seguimiento del Programa de Ordenación y Saneamiento de la Cuenca Lerma-Chapala, 1991). Situated in the very center of the country (see Figure 1), the 54,400 km^2 basin (including two adjacent closed basins) covers five states and is the source of water for 15 million people living within the basin as well as in Mexico City and Guadalajara.

The Lerma river rises on the slopes of the Toluca volcano and flows over 500 km through forest, rich farmland, urban areas, and semi-arid scrubland before emptying into Lake Chapala, Mexico's largest freshwater lake. When storage volumes in the lake permit, controlled outflows are passed to the Santiago river that empties in the Pacific Ocean. Surface and groundwater resources in the Lerma-Chapala basin are subject to rampant competition and rapid depletion, while water quality is a major concern (Mestre, 1997). Lake Chapala, the receiving waters of the basin, demonstrates the

[1] See Dinar *et al.*, 1997 for further discussion of allocation mechanisms.

cumulative impacts of upstream activities (natural and anthropogenic) and serves as an indicator of the sustainability of water resources development.

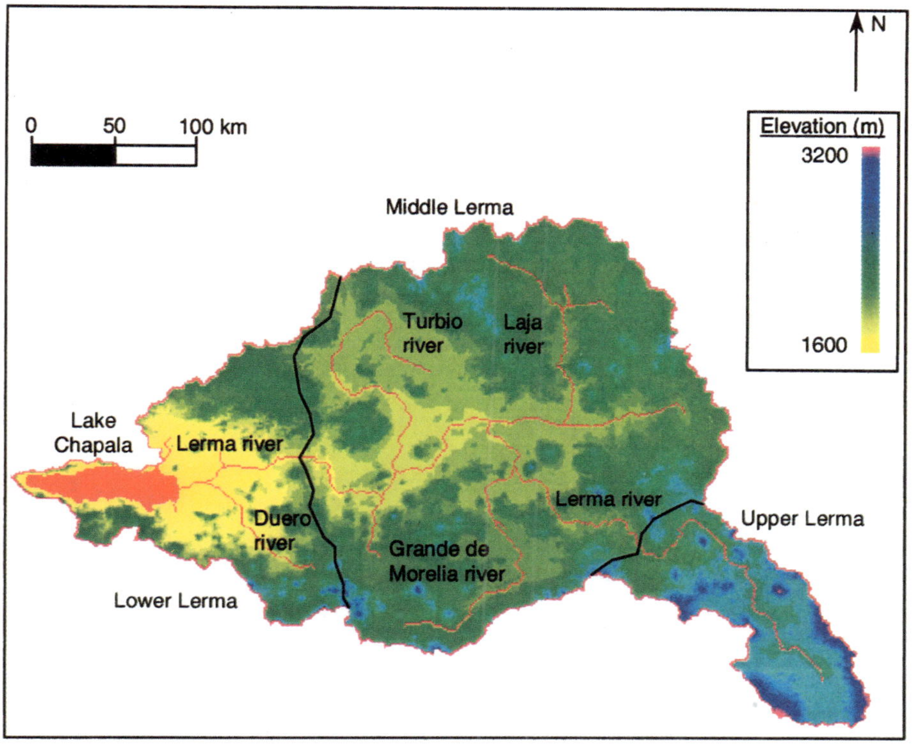

Figure 1 The Lerma-Chapala basin with Upper, Middle, and Lower Lerma sub-basins (administrative units)

Rapidly increasing upstream extractions since at least the 1960's, coupled with a declining trend in rainfall, have resulted in decreased Lerma river inflows to the lake with dramatic decreases in the stored volume (Aparicio, 2001). Two low storage periods characterize the lake's behavior since 1934 when systematic data collection began – one in the mid-1950s and the present 1999-2000 period. Average inflows of the Lerma river were significantly higher in the 1950s than they are now (Aparicio, 2001) and the lake recuperated to full storage above 8,000 Mm^3 in 1959. The irrigated area in the basin grew from around 200,000 ha in the 1950s to 800,000 ha in the 1980s (Wester *et al.,* 1999). Given the present over-exploitation of groundwater throughout the basin (Marañón-Pimientel, 1999) and the development of surface water in the upper sub-basins, it is unlikely that the lake will recover without exceptional runoff as generated through a major hurricane.

The quality of surface and groundwater in the basin is acknowledged to be a significant problem (Mestre, 1997; Hansen *et al.,* 1995), clearly driven by economic development and the accompanying heavy over-exploitation of basin water resources. Related IWMI studies have shown the distribution and extent of aquifer depletion (1.00-2.58 m yr^{-1}) and agricultural water demand (Scott and Garcés-Restrepo, 2001; Wester *et al.*, 1999). The basin authorities (CNA) do not calculate water balances annually, and for lack of accurate cropped area data particularly in farmer-managed irrigation, it is difficult to quantify the extent of water resource depletion at the basin level. Nevertheless, official data do indicate the distribution of water sources (Figure 2). We focus our analyses on the Middle Lerma "sub-basin" which is actually an administrative unit that encompasses several hydrologic sub-basins. It should be emphasized that the Middle Lerma comprises the bulk of water use, both from surface and groundwater sources. This part of the basin essentially corresponds to the state of Guanajuato (with parts of Michoacán and Querétaro), covers over half the area and accounts for the largest share (56.2%) of water use in the basin.

Although the Lerma-Chapala basin has a large urban population (estimated at 11 million inhabitants), consumptive water use remains predominantly agricultural as indicated by Middle Lerma data in Figure 3.

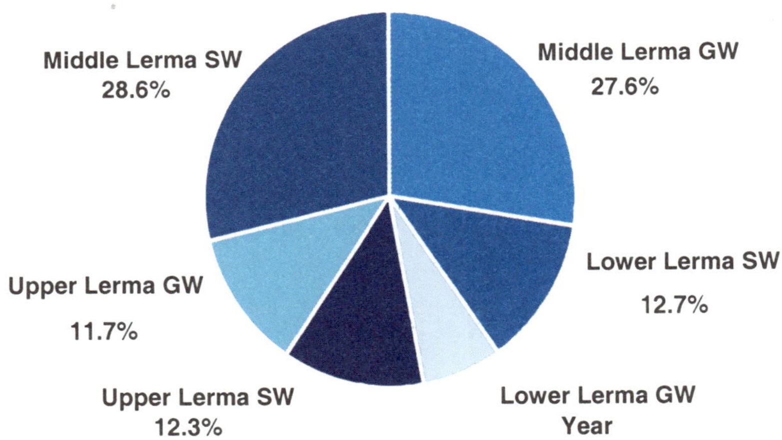

Source: Centro de Información de la Cuenca Lerma-Chapala, 1998

Figure 2 Sources of water supply, Lerma-Chapala basin, 1996-97 total volume = 9,032 Mm^3 (SW = surface water; GW = groundwater)

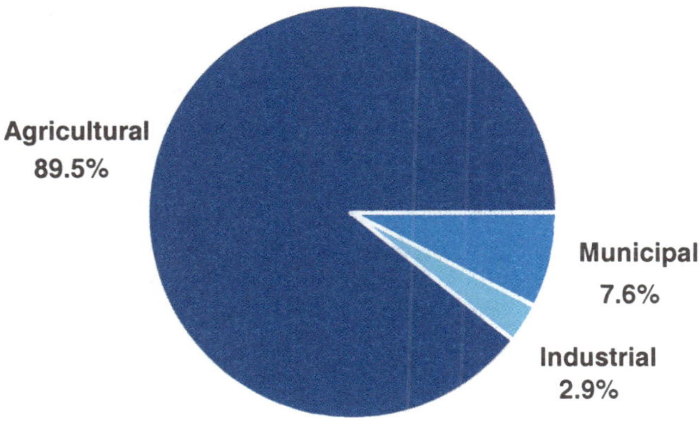

Source: Centro de Informaci6n de la Cuenca Lerma-Chapala, 1998

Figure 3 Water use in the Middle Lerma region, 1996-97 total volume = 5,080 Mm^3

The principal economic sectors consuming water in this basin – agriculture, domestic, commercial and industrial – all rely on groundwater. The surface water resource, with some exceptions, is consumed for agriculture (an important distinction here lies between water consumed, i.e., depleted or evaporated, and water used in-stream, as for hydroelectric power generation). This observation raises an important issue, to which we can only make reference in passing, that other water uses particularly environmental quality have intrinsic value although their economic value as conventionally defined relegates them lowest priority among competing uses. In practice, environmental impacts become externalities, although in an ideal sense these should be internalized in water allocation mechanisms.

Current trends (1995-98) for the state of Guanajuato, which represents the largest water user in all sectors in the Middle Lerma, indicate that the number of urban users, supplied almost exclusively from groundwater, is rising at 4.1% annually (CEASG, 1999). This is a combination of a 4.0% annual increase in domestic and commercial users (together, the bulk of users), but also notably a 9.2% annual increase in industrial users (Figure 4).

Under these trends of demand growth coupled with aquifer decline, we estimate that total urban pumping costs will increase by over 200% between 1999 and 2010. Recent trends in pumping costs (Figure 5) corroborate this estimate. Parenthetically, we should note that most urban wastewater return flows are reused in agriculture – this is the "raw" urban-agricultural interface (Scott *et al.*, 2000). Throughout the Lerma-Chapala basin, urban centers were estimated to have generated 17.0 m^3 s^{-1} of wastewater in the mid-1990s

(Mestre, 1997). At current growth rates, this figure is likely to have increased to 20 $m^3 s^{-1}$ in 2000, virtually all of which is reused in agriculture.

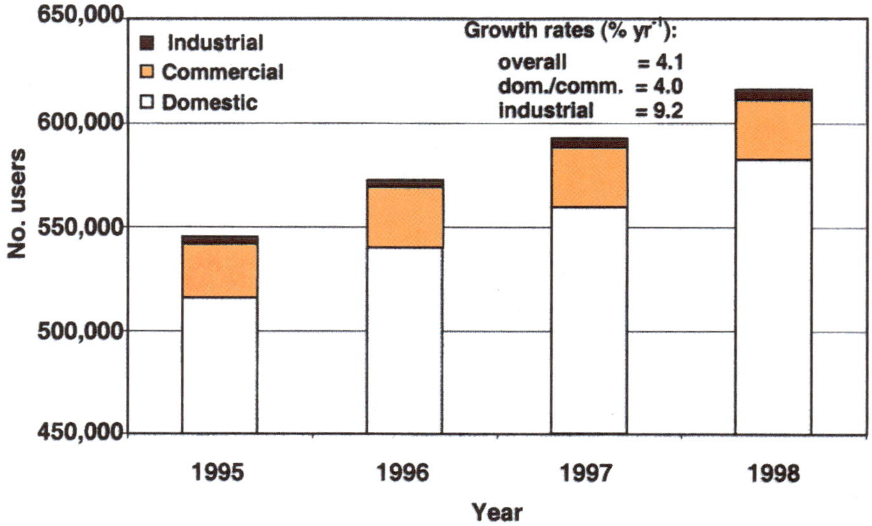

Source: CEASG, 1999

Figure 4 Urban water users by sector, state of Guanajuato, 1995-98

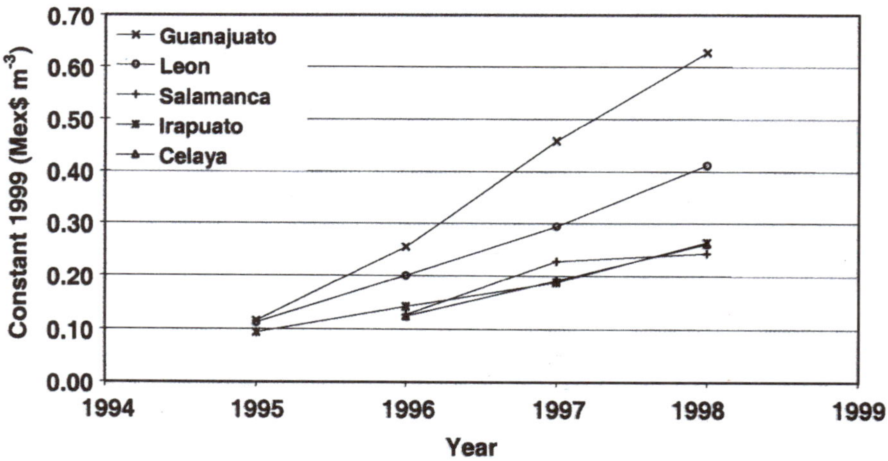

Source: CEASG, 1999

Figure 5 Increase in pumping costs, selected Middle Lerma cities, 1995-98

The precipitous declines in lake levels in the 1980s, discussed above, mobilized public opinion in Guadalajara whose inhabitants depend on the lake for 240 Mm^3 annually of municipal water supply. Importantly, the vacation and weekend homes of well-off, influential Mexicans were left high and dry as a result of the receding lakeshore. Lost tourism revenues and the threat to Chapala's symbolic value as Mexico's largest lake exerted further pressure to take action. Thus, a combination of factors provided strong impetus for coordinated action to preserve the lake.

Cognizant of the unsustainable water resources management in the basin, the five states that comprise the Lerma-Chapala basin (Mexico State, Querétaro, Guanajuato, Michoacán, and Jalisco) and the federal government signed a treaty in 1989 to regulate water use with the objective of stabilizing the falling lake levels. As a result of the 1989 creation of the National Water Commission (Comisión Nacional del Agua, CNA) and the 1992 Law of the Nation's Waters, the treaty was institutionalized as the Lerma-Chapala Basin Council. For the purposes of this paper, it is illustrative to briefly assess the institutional and legal context in which the basin council operates, and the process by which water allocations are made. More in-depth analyses of the institutional arrangements for basin management of the Lerma-Chapala are provided in Wester *et al.*, 2001.

Water rights in Mexico are quasi-proportional with administrative allocation mechanisms in place. River water is the property of the nation and the federal government has the right to concession water-use rights to users. The concession titles set out the quantity of water a user is entitled to, although in practice the actual quantity is adjusted annually to reflect water availability, in the case of surface water. The situation surrounding groundwater is more complex, as the Constitution does not define it as national property, but rather states that overlying landowners may bring groundwater to the surface as long as this does not affect other users. In 1946 the Constitution was amended to the effect that the federal government may intervene in aquifers in overdraft, by issuing pump permits or declaring that new pumps may not be installed. Based on a ruling of the Supreme Court in 1983, groundwater is now considered national property, although this is not reflected in the Constitution or the 1992 Water Law (Palacios-Vélez and Martínez, 1999). Groundwater concessions in the Lerma-Chapala basin are granted on a volumetric basis with a maximum extraction or pumping rate specified (and limited by electrical power transformer capacity).

The 1992 water law provides a modern regulatory framework for water management and stipulates that the participation of water users and interested third parties is mandatory in water management at the river basin level. To this end, responsibilities for a variety of water resource management activities have been accorded to the CNA Regional Offices

(*Gerencias Regionales*). Also, River Basin Councils (*Consejos de Cuenca*) are to be established under the auspices of CNA, in which the federal, state and municipal governments as well as water users of different sectors and other interested groups are represented. These parties share the responsibility for allocating water resources and fostering integrated water management at the basin level and formulating and sanctioning regional water plans. The combination of these changes has major implications for the equity of distribution of this critical resource, for its resulting productivity and efficiency, and for its protection.

Nevertheless, the 1992 law clearly identifies the CNA as the country's sole water authority, charged with managing water resources, both in terms of quality and quantity. The law defines the nature of water allocations to users, with priority given to domestic use, but also contains provisions that have impacted markedly on the agricultural sector. These changes provided for the transfer of government irrigation systems to water users and formalized the allocation of water to water user associations in the form of time-limited rights. IWMI's focus on river basins in Mexico has revealed that water allocations are increasingly important assets to communities, industries, irrigation user organizations, and the environment. These allocations, particularly those related to irrigation, have come under intense pressure during the past five years. As we have demonstrated above, this is a result of the expanding industrial sector, increased domestic needs in both the urban and rural areas, and periods of extended drought throughout the Lerma-Chapala basin.

The Lerma-Chapala *River Basin Council*, officially constituted in January 1993, is the oldest river basin council in Mexico. It evolved in the following manner. On April 13, 1989 the federal government and the governments of the five states in the river basin signed a historic agreement that adopted four main objectives to improve water resources management in the basin:

- Control and regulate surface water use and distribute water fairly among users according to a new water allocation policy
- Improve water quality by treating municipal and industrial effluents (clean up the basin)
- Increase water-use efficiency
- Conserve the river basin ecosystem

On September 1st, 1989 Consultative Council was formed to follow up on the four objectives. This council in turn became the Lerma-Chapala River Basin Council on January 28, 1993. On the council are represented the five states that comprise the basin, various federal and state agencies, and users' representatives of various sectors. However, user representation has not yet been effectively incorporated in the decision-making process (Scott *et al.*, 1998). Mechanisms for allocation of surface water were agreed on by the five states that constitute the Council, through signing of a surface water treaty in 1991. The Council also launched an investment program to clean the Lerma river, apparently with good results; pollutant discharges into Lake Chapala are reported to have fallen by 50 percent since 1991 (Mestre, 1997).

As noted, water allocation in the Lerma-Chapala basin is of crucial importance to the livelihoods of 15 million people living within the basin as well as in Guadalajara and Mexico City that are provided with domestic water through inter-basin transfers. The 1991 agreement sets out how much water each irrigation district, farmer-managed irrigation system, and specific urban demands (i.e., Guadalajara's annual 240 Mm^3, equivalent to 200 liters per person per day for 3.3 million people) in the basin should receive, depending on water availability in Lake Chapala and upstream reservoirs. Based on an average water availability of 4,740 Mm^3 per year from 1950 to 1979, the agreement authorizes the following maximum withdrawals (if availability is above average):

- 240 Mm^3 from Lake Chapala for Guadalajara,
- 90 Mm^3 from Lake Chapala for agricultural use,
- 3,240 Mm^3 for upstream agricultural uses.

The treaty recognizes that maximum withdrawals, do not allow sufficient volume to satisfy the lake's annual evaporative loss of 1,440 Mm^3. Thus, if lake storage is "abundant" (> 6,000 Mm^3), maximum total allocation is 3,570 Mm^3 as indicated above. For "medium" (3,300-6,000 Mm^3) and "critical" (< 3,300 Mm^3) lake storage levels, water is allocated to different agricultural demands using formulas based on a combination of runoff and stored volume in upstream reservoirs capable of serving the particular agricultural user. Apart from urban demand for Guadalajara, the treaty does not address the issue of specific volume allocations for non-agricultural uses – clearly a shortcoming given the data presented above on the basin's rapidly rising domestic and industrial demands.

Figure 6 shows the total actual allocations as a function of the surface water available since the 1991 agreement was implemented. The difference between available and allocated surface water must be large enough to account for lake evaporation (the 1,440 Mm^3 magnitude of this loss is obviously reduced as lake storage and surface area drop). Rainfall in 1999-2000 was the lowest in forty years, although carry-over volume in the reservoirs from 1999 put availability for 2000 slightly higher than it had been in 1998. The Alto Río Lerma Irrigation District (ARLID), the largest in the basin for which we estimate water values below, was allocated significantly less water than needed for 1999-2000. Its maximum allocation is 955 Mm^3, but for 1999-2000 it was allocated only 650 Mm^3.

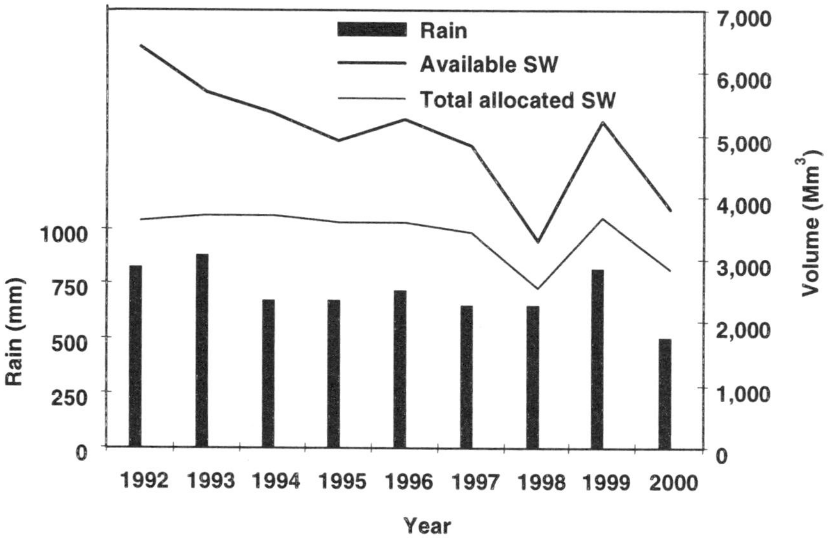

Source: Consejo de la Cuenca Lerma-Chapala, 1991-99

Figure 6 Lerma-Chapala basin surface water allocation, 1992-2000

Although its two main dams, Solís and Tepuxtepec, contained around 1,000 Mm^3 in October 1999, the basin council decided to pass 240 Mm^3 to Lake Chapala, to ensure that the city of Guadalajara received its full allocation. As a result, the area irrigated by surface water in ARLID was reduced to less than 20,000 ha, compared to a maximum of 77,000 ha. We have noted that farmers' low levels of capitalization and access to credit impede rapid adjustment to such dramatic changes. Nevertheless, our analysis of the economics of agricultural production below suggests that a switch to vegetables could mitigate some of the loss that this reallocation entailed.

This particular case of water transfer (upstream agriculture to the city of Guadalajara) is the only inter-sectoral re-allocation of water explicitly contemplated in the 1991 agreement. No doubt, other inter-sectoral transfers, particularly of groundwater, take place in an *ad hoc* fashion with no empirical data on their magnitude, e.g., cities increasingly pump from the aquifers that have historically been used for agriculture. The combination of agricultural and urban extraction represents an over-exploitation of this resource. The trends in water use indicate that municipal (domestic and commercial) and industrial extractions (and as a result, depletions) are increasing rapidly in the present economic environment, reinforcing the need for effective allocation mechanisms at the basin level. A combination of institutional innovations including aquifer management councils, and a revised surface water allocation agreement (under discussion in April 2001) may help in reducing total extractions to sustainable levels; in addition, hydrologic and economic factors make current extraction levels exceedingly difficult to maintain in the mid- to long-term future.

An essential component of any attempt to stabilize extractions is the compensation of farmers who will increasingly cede irrigation water to urban water users. In the Lerma-Chapala case, never have farmers been compensated for water passed to Lake Chapala and Guadalajara. At present, farmers in ARLID are starting to take part in the river basin council meetings, and demanding that Guadalajara pay them for the 240 Mm^3 that has been taken out of "their" dam and passed on to the lake. However, the basin council or government has not even started considering such compensation. Water is national property in Mexico, and the irrigation districts are given concession titles by the government (CNA), in which they are allocated a percentage of the available water in the basin in direct correlation with the amount of land in the irrigation district. Hence, the 1991 agreement is fully legal, and the CNA can reallocate water each year, based on availability. In order to mitigate the economic impact on farmers, future water allocation agreements must address the compensation issue directly. This paper estimates water values by sector to assess the foregone agricultural benefits (and those accrued by urban water use) in order to inform the debate over appropriate compensation levels.

4. WATER VALUATION METHODOLOGY AND RESULTS

The significant real or foreseen inter-sectoral water transfers in the basin have major social and economic implications. However, the dynamic policy and economic context makes any meaningful long-term water use projections of total water demand by sector unreliable. In the three sections that follow, we assess incremental changes in water available, or allocated, to the principal use sectors (urban and agriculture) and make projections of mid-term (10-year) commodity prices. While estimates of water demand and commodity prices for the year 2010 are more reliable than longer-term projections 25 or more years into the future, we recognize that complex forces influence these trends.

On a methodological note, we estimate marginal values instead of total values, i.e., those resulting from incremental changes in supply. Urban water values (domestic, commercial and industrial uses) are estimated using price elasticity methods based on minor changes (5-15%) in water availability under existing water price structures. Similarly, for agricultural water use, we estimate value using shadow prices as a proxy for marginal value. Given our interest in assessing accrued and foregone benefits resulting from water allocations from agricultural to urban uses, we are more interested in estimated values than in net benefits (value less price). It should be noted that the results of the economic analysis are presented in inflation-adjusted 1999 Mexican pesos (Mex$), unless otherwise noted. A dollar exchange rate of US$ 1.00 = Mex$ 9.50 was used for 1999.

4.1 Urban water use, demand and price elasticities

Data on water use and fees for the city of Guanajuato, capital of the state with the same name, were provided in digital form by the urban water board, SIMAPAG (Sistema de Agua Potable y Alcantarillado de Guanajuato). The following analyses are based on extensive data cross-checking, followed by the application of valuation techniques to be described. It should be noted that Guanajuato's water fees were approximately 23% higher than those of other municipalities in the Middle Lerma in 1999, and 17% higher than other state capitals in Mexico in 1996 (CEASG, 1999). It was decided to use Guanajuato as a case study for reasons of access to water board officials as well as the quality and form of the data available, despite the somewhat higher than average fees.

Water supply for municipal purposes in the city of Guanajuato (total population 110,000) is provided by SIMAPAG, a financially autonomous public utility with an independent administration[2]. At present there are 21,374 users (combined domestic, commercial and industrial sectors) with coverage of over 95% for supply and 80% for sewerage. Due to the absence of manufacturing or heavy industries in the city, the industrial use sector is comprised of hotels, restaurants and car-wash establishments, together with small industries such as pottery workshops and water bottling plants. The industrial sector actually has more the characteristics of a services sector as conventionally defined. Government offices and schools are categorized as commercial users along with retail stores, other offices and non-manufacturing businesses in general. A relatively small number of users who are not connected to the distribution network are supplied directly from

[2] Starting in 1983, the provision of urban water and sanitation services in Mexico was decentralized to the municipalities (Saade, 1997). Nevertheless, the federal government (since 1989 through the CNA) must grant water use concessions. In principle, concessions must be paid for; however, with the prevailing sense of water as a free good and the resulting low levels of fee collection coupled with ongoing friction among federal, state and municipal authorities over revenue collection and distribution, most municipal water boards are unable to cover their operating costs. As a result, in the state of Guanajuato in 1997, for example, municipalities owed the CNA a 1990-97 cumulative average of Mex$ 547 (Mex$ 416 in 1999 constant pesos, equivalent to US$ 44) per connection for water use rights (CEASG, 1999).

SIMAPAG wells or tanker trucks (the latter at approximately Mex\$ 5.83 m^{-3}, or US\$ 0.61 m^{-3}).

The distributions in percent and absolute numbers of users, volume consumed, billing and fee collection are shown in Figure 7. Although the domestic sector consumes 77% of the total volume, it accounts for only 57% of collected fees. For all three sectors, the fee collection efficiency for 1998 was 69% (i.e., actual collection as a percent of billing).

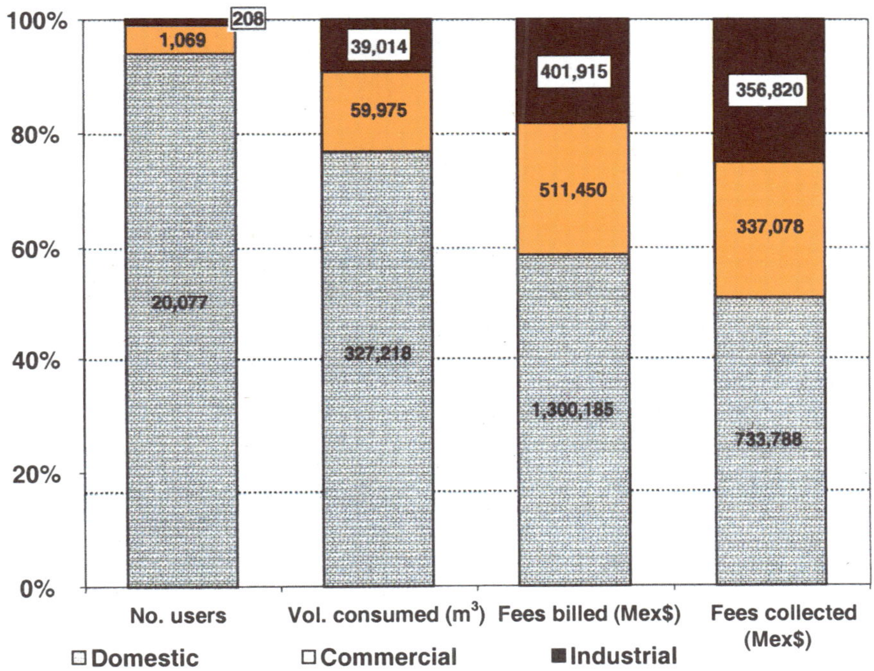

Source: unpublished SIMAPAG data, 1999

Figure 7 Water use and collection of fees, city of Guanajuato, July 1999

Water prices vary by sector (Table 1). Domestic and commercial price structures are quite similar, both with a minimum charge (underlined) followed by eleven increasing rates by volume blocks (0-13, 14-20 m^3 $month^{-1}$ etc. shown in *italics* to distinguish these from volumetric prices shown in **bold**). On the other hand, the industrial price structure has a fixed or base fee and only two increasing block rates. For example, a domestic user who consumes 2 m^3 $month^{-1}$ is billed a flat Mex\$ 26.55, the same as the user who consumes 12 m^3 $month^{-1}$. The domestic user who consumes 22 m^3 $month^{-1}$, on the other hand, is billed Mex\$ 55.44 (= 22x2.52), and so on.

The industrial sector has a fixed connection fee of Mex$ 48.89 month^{-1} with volumetric pricing applied from the first unit of water consumed.

Table 1 Urban water price structure by sector, January 1999

Sector	Fixed fee per connection per month	Fee (Mex$ m^{-3}) by *volume block*												
		0-13 (flat fee)	*14-20*	*21-30*	*31-40*	*41-50*	*51-70*	*71-90*	*91-100*	*101-110*	*111-130*	*131-160*	*161-190*	*191-9999*
Domestic	0	26.55	**2.30**	**2.52**	**3.18**	**3.57**	**3.98**	**4.60**	**4.95**		**5.27**	**5.70**	**5.96**	**6.34**
Commercial	0	36.36	**2.48**	**2.98**	**3.77**	**4.40**	**4.95**	**5.60**	**6.70**		**7.02**	**7.37**	**7.69**	**8.07**
Industrial	49.89	**7.06**								**8.53**				

Source: unpublished SIMAPAG data, 1999

With this fee structure and the collection efficiencies indicated above, SIMAPAG has had financial surpluses over the past years. Fee increases are authorized annually, in principle with a standard 1.5% increase per month (in nominal pesos) to allow for estimated annual inflation of 18%. Nevertheless, as indicated in Table 2, the actual increases have been marginally higher, as the increases may vary by sector and volume block. Note that prior to 1996, industrial use was considered part of commercial with the same annual increase. In 1997, these sectors were separated, producing the large increase in industrial fees in 1997.

Table 2 Annual fee increases by sector (%)

Year	Domestic	Commercial	Industrial
1996	18.73	18.78	18.78
1997	19.16	19.36	63.70
1998	19.29	19.36	19.40
1999	19.27	19.37	19.38

Source: unpublished SIMAPAG data, 1999

To understand the implicit value that consumers assign to water, sufficient data are required to estimate the price elasticities of demand, which vary by type of consumer (income class), seasonal and climatic factors, location, etc. Calculating price elasticities for water is a complex and demanding econometric task in its own right. The data required to estimate the elasticity of a certain curve of water demand are a set of observations for which real (inflation-adjusted) prices vary sufficiently to yield a set of price-quantity points that trace out a demand curve. The data can be observed from a cross-sectional sample of a number of cities for the same year or from a time series sample, encompassing real variations in price, for one city. The data would be typically fit with regression techniques to find the demand function. Subsequently, the price and income elasticities would be derived form the resulting demand function. During the data collection phase of our research, we were not able to get access to adequate cross-sectional data. Additionally, Table 2 shows that real increases in water fees are very low. Under these conditions, time series analysis was not possible; therefore, price elasticities could not be calculated explicitly. For the present purposes, we tested instead several reasonable *a priori* estimates that are within the range of data usually found for price elasticity for urban water demand (Schneider and Whitlach, 1991; Gibbons, 1986) and that, based on an analysis of the water consumption histograms and imputed demand behavior, are likely to encompass the real elasticity values.

As shown in Figures 8-10, for all three sectors the highest numbers of users fall in the first volume use block. However, specific water consumption behavior varies by sector. The histograms for September 1999 are presented for illustrative purposes; similar distributions were found for the rest of the year (the standard deviations over 18 months of the first, second and third domestic volume use block, which together represented 58% of the total domestic volume consumed, were only 1%, 1%, and 3% respectively). Clearly a minimum level of demand represents basic consumption for which there is no substitute or behavior compensation, while other uses are of lesser value and have ready substitutes. The marginal value of water in municipal use obviously depends on how much supply one already has; at the extremes of scarcity and abundance, the marginal value can be infinite or negligible.

For the domestic sector, water consumption is concentrated in a range between 0 and 30 m^3 $connection^{-1}$ $month^{-1}$, which is not likely wide enough to accommodate appreciable price effects, given that the uses for this volume consumed have no substitute. As a result, demand is inelastic.

In the commercial and industrial sectors, demand is better distributed across volume blocks suggesting that demand is more elastic. In both these sectors, single consumers account for the highest volume use block, and are not considered representative. Given that the city of Guanajuato does not have large-scale manufacturing or heavy industries, the industrial sector demand behavior demonstrated may be more illustrative of the service sector (hotels, etc.). Under these water demand conditions, the commercial and industrial sector price elasticity values will probably be lower (in absolute value) than domestic elasticities.

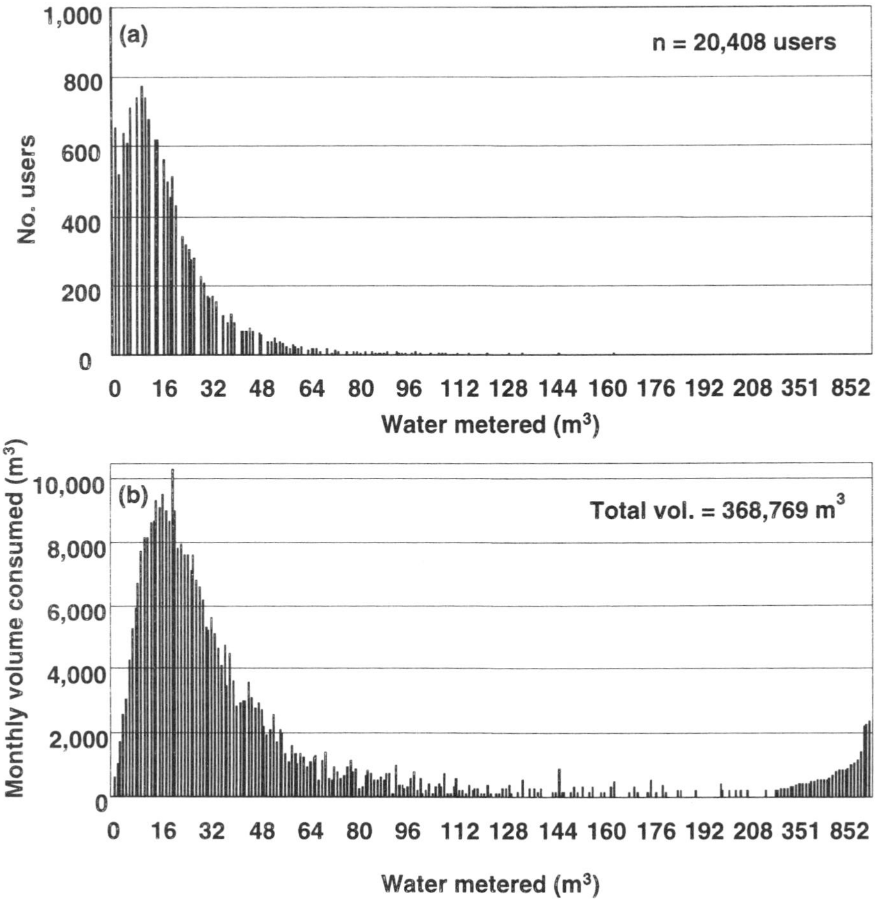

Source: unpublished SIMAPAG data, 1999

Figure 8 Histograms of metered volume per user per month (a) by number of users (1864 users consume 0 m^3) and (b) by consumed volume, city of Guanajuato domestic sector, September 1999

4.2 Guanajuato urban water valuation

Given the emphasis of this study on the economic implications of intersectoral water transfers resulting from competition or allocation mechanisms, we are interested in establishing the marginal value of water resulting from changes in supply or availability. Similar to the analysis subsequently presented for agricultural water use, marginal values are estimated based on incremental changes assumed to be realistic in the medium term.

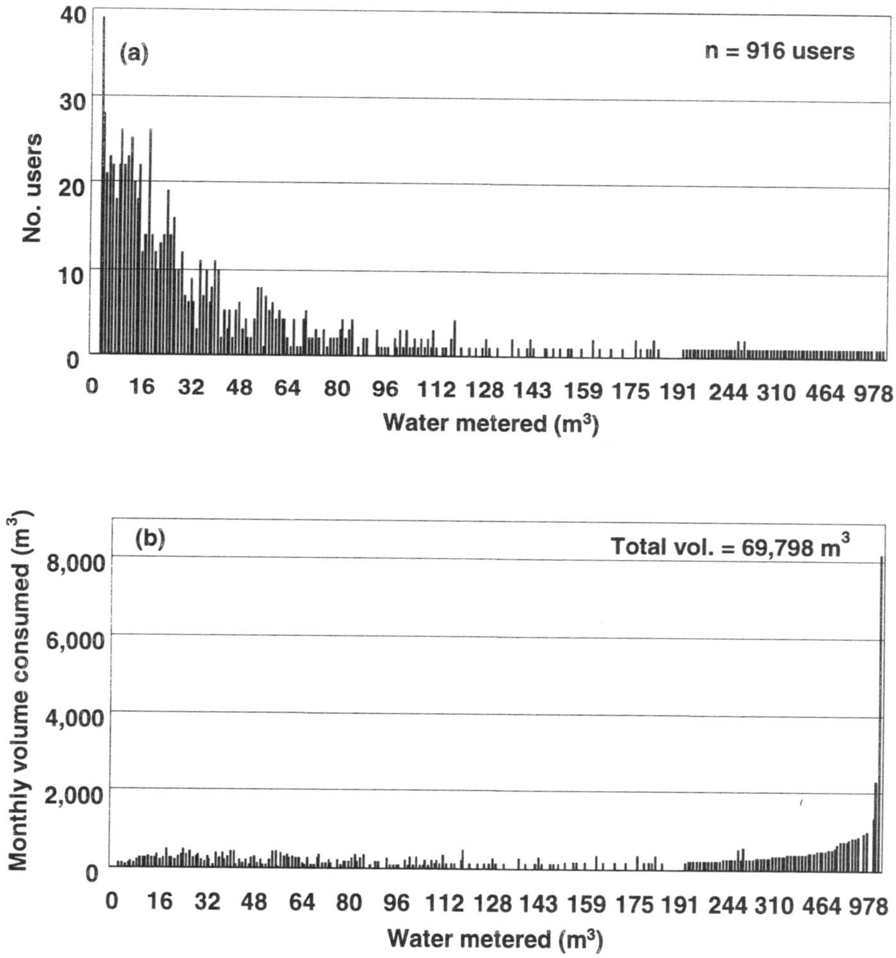

Source: unpublished SIMAPAG data, 1999

Figure 9 Histograms of commercial metered volume (a) per no. of users and (b) by consumed volume, city of Guanajuato commercial sector, September, 1999

As described in the discussion above on elasticities, the most appropriate measure of the marginal value of water to the individual urban consumer is through the use of water demand functions. The corresponding area under the demand curve indicates the willingness to pay for (or value of) an increment in supply of pre-treated, pressurized water delivered to the household faucet.

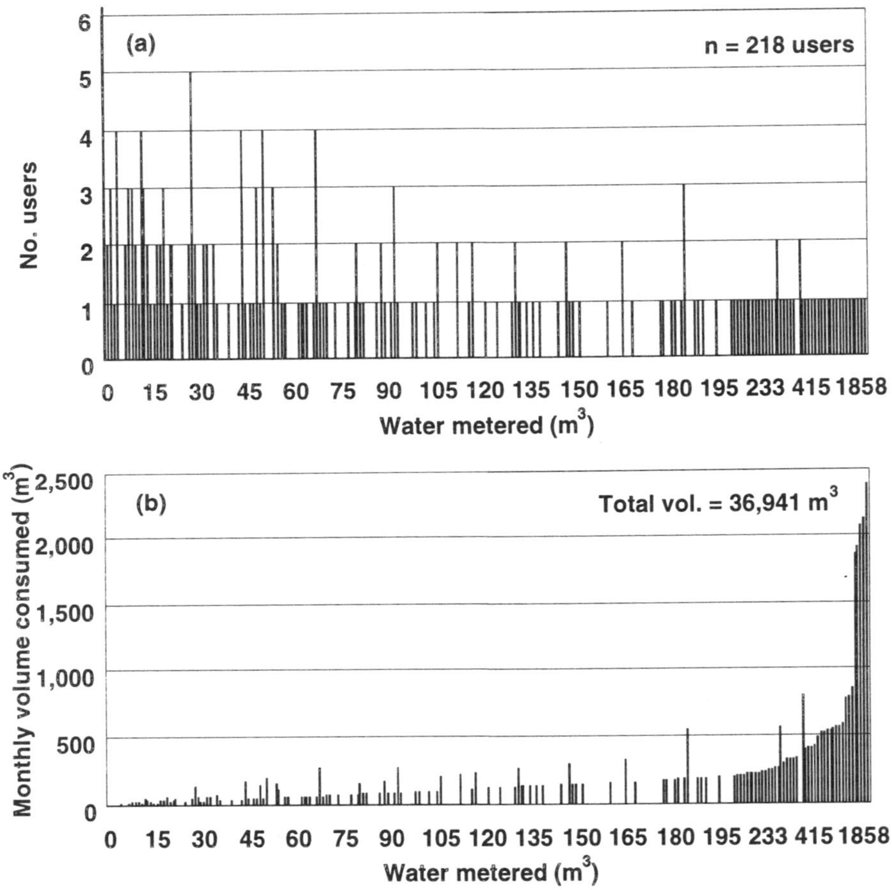

Source: unpublished SIMAPAG data, 1999.

Figure 10 Histograms of metered volume per user per month by no. of users (a) and (b) by consumed volume, city of Guanajuato industrial sector, September 1999

The area under the demand curve is calculated from the following formula[3]:

$$V = [(P_2 * Q_2^{1/E})/ (1- 1/E)] * [(Q_2/ Q_2^{1/E}) - (Q_1/ Q_1^{1/E})]$$

Where V is the total marginal value (Mex\$), $Q_2 - Q_1$ is the incremental water consumed (m^3), P_2 is the price (Mex\$ m^{-3}) currently paid for Q_2, and E is the constant absolute value of elasticity over the increment Q_1 to Q_2 and is not equal to -1.0. The unit value is simply:

$$V_u = \frac{V}{Q_2 - Q_1}$$

To estimate the consumer surplus (S in Mex\$ m^{-3}) of the incremental volume, P_2 is subtracted from V_u under the assumption that this fully covers the cost of service to the user. As a result:

$$S = V_u - P_2$$

The calculated values of surplus were found to be exceedingly low given the Guanajuato price structure. SIMAPAG uses a combination of surface and groundwater sources. In 1998, which was an extremely dry year for runoff, the volume from surface water was 2.9 Mm^3 while the groundwater volume was 4.3 Mm^3. The operation and treatment costs (excluding maintenance) vary by source, as would be expected. Surface water costs Mex\$ 0.24 m^{-3} while groundwater costs Mex\$ 1.36 m^{-3} to pump from the aquifer and through the 10 km pipeline to the city. Based on the actual mix of surface and groundwater supplied during the first six months of 1999, production costs varied from Mex\$ 0.62-1.27 m^{-3}. As a result, of the total Jan.-Sep. 1999 SIMAPAG expenditure, an alarming 72% was spent on operations, with 21% spent on administration and 7% on debt interest payments.

In view of the elevated production costs, and based on the subsequent section which estimates agricultural shadow prices for water (methodologically equivalent to the marginal values calculated here), we restrict our analysis to marginal values. For each use sector, several calculations of unit value were made in order to test its sensitivity to the magnitude of the change in quantity ($Q_2 - Q_1$) and to price elasticity E. The unit marginal value V_u results (expressed in Mex\$ m^{-3}) for the first three volume blocks in each of the three sectors are shown in Tables 3 - 5. Q_1 was taken as the annual average volume consumed in the corresponding use block (to better represent actual volume consumed in each use block), and

[3] This methodology follows Young (1996) and Gibbons (1986).

was increased by 5%, 10% and 15%. Unit marginal values were calculated using fixed unit prices P_2 for the reference month of January 1998. For presentation in Tables 3-5 and subsequent analyses in this paper, the results were inflation-adjusted to real (1999) Mex$. As a result, prices P_2 for the increased quantity Q_2 may differ from the values presented in Table 1, which are in 1999 Mex$. The elasticity values tested were -0.5, -0.7, and -0.9. The lowest volume blocks account for the bulk of demand in each sector, with the exception of commercial demand where a few high-volume users skew the histogram in the upper 464-978 m^3 $month^{-1}$ range (Figure 9); as noted, these are not considered representative.

Table 3 Domestic sector unit marginal values (Mex$ m^{-3})

Quantity Q_1 (m^3)	Quantity increment (%)	Price P_2 (1998 Mex$ m^{-3})	Unit marginal value V_u (Mex$ m^{-3})		
			E = |-0.5|	E = |-0.7|	E = |-0.9|
5.75	5	3.74	4.40	4.34	4.31
	10	3.57	4.40	4.28	4.22
	15	3.42	4.40	4.23	4.13
16.76	5	1.95	2.29	2.26	2.24
	10	1.95	2.40	2.34	2.30
	15	2.16	2.78	2.67	2.61
26.87	5	2.16	2.54	2.50	2.49
	10	2.16	2.79	2.68	2.62
	15	2.70	4.02	3.69	3.53

Table 4 Commercial sector unit marginal values (Mex$ m^{-3})

Quantity Q_1 (m^3)	Quantity increment (%)	Price P_2 (1998 Mex$ m^{-3})	Unit marginal value V_u (Mex$ m^{-3})		
			E = |-0.5|	E = |-0.7|	E = |-0.9|
3.83	5	7.58	8.92	8.79	8.72
	10	7.24	8.92	8.67	8.54
	15	6.92	8.92	8.56	8.37
17.19	5	2.12	2.49	2.46	2.44
	10	2.12	2.61	2.54	2.50
	15	2.54	3.27	3.14	3.07
25.38	5	2.54	2.99	2.90	2.89
	10	2.54	3.29	3.11	3.05
	15	3.20	4.76	4.32	4.14

These results illustrate that the marginal value of urban water is more sensitive to quantity available (Q_1) and price (P_2) than to increment (Q_2 - Q_1) or elasticity (E). Nevertheless, under present assumptions of price, quantity and expected increments over the mid-term (ten years to 2010), there is surprisingly little variation in marginal values within each sector. Domestic use marginal value was found to vary in the range of Mex$ 2.24-4.40 m^{-3}; commercial in the range of Mex$ 2.44-8.92 m^{-3}; and industrial in the range of Mex$ 8.15-12.77 m^{-3}. The more significant variation, obviously, was among sectors. This has major implications for the present inter-sectoral allocation study.

Table 5 Industrial sector unit marginal values (Mex$ m^{-3})

Quantity Q_1 (m^3)	Quantity increment (%)	Price P_2 (1998 Mex$ m^{-3})	Unit marginal value V_u (Mex$ m^{-3})		
			E = |-0.5|	E = |-0.7|	E = |-0.9|
35.03	5	7.08	8.33	8.22	8.15
	10	7.03	8.67	8.43	8.30
	15	6.99	8.99	8.64	8.44
143.51	5	7.47	8.78	8.66	8.59
	10	7.45	9.18	9.18	8.80
	15	7.44	9.59	9.59	9.00
244.28	5	7.35	8.65	10.40	11.42
	10	7.34	9.50	11.09	11.95
	15	7.32	10.89	12.16	12.77

Given our discussion above on rising urban demands receiving water transfers from agriculture, the data presented here demonstrate the accrued value this increment in urban supply is likely to generate. We recognize that in order to determine the actual value that transferred water would produce, it would be necessary to carry out water valuation studies specific to both the city and the use that would receive this water. Nevertheless, our recommendation above to consider option contract arrangements suggests that compensation should be based on foregone benefits (in this case agricultural); hence, it is imperative to value agricultural water in a comparable manner, i.e., using marginal valuation techniques. The analyses and results of urban water use just presented, when compared with agricultural water values as demonstrated below, do indicate that urban users have the ability to pay for compensation.

4.3 Agricultural water use

For agricultural sector water use, we analyzed the Alto Río Lerma Irrigation District (ARLID) based on 1999 production cost data collected from the Fideicomisos Instituidos en Relación con la Agricultura (FIRA; Institutional Trusts in Relation to Agriculture, in English), Banco Rural (BANRURAL; the Rural Bank, in English), and Secretaría de Agricultura Ganadería y Desarrollo Rural (SDAYDR; Ministry for Agriculture, Livestock and Rural Development, in English). These data are updated seasonally based on interviews of a cross-section of farmers, and are considered to be reliable and representative. Farm-gate crop prices were taken from the Consejo Nacional Agropecuario (National Agricultural Council, in English) and are considered representative of ARLID.

Part of the Middle Lerma, ARLID is the largest irrigation district in the entire Lerma-Chapala basin (Kloezen *et al.*, 1997). It was constructed in the 1930s and has a maximum irrigable area of 112,772 ha. Within the entire Lerma-Chapala basin, surface water availability averages 4,740 Mm^3 of which 43% (2,020 Mm^3) is allocated to the government-run districts (which are now transferred to users) and 31% (1,450 Mm^3) to the farmer-managed irrigation systems (Consejo Consultivo de Evaluación y Seguimiento del Programa de Ordenación y Saneamiento de la Cuenca Lerma-Chapala, 1991). The balance flows to Lake Chapala. Of the nine irrigation districts in the basin, ARLID receives approximately 47% (955 Mm^3) of the water allocated to irrigation districts (Consejo de la Cuenca Lerma-Chapala, 1991-99) in addition to groundwater (CNA, 1999a and 1999b). Water markets are nascent in the district although stable water prices have not emerged; hence, we calculate shadow prices for water and land.

In this section, we describe the linear programming model developed for ARLID to determine the optimal cropping pattern to maximize net benefits, as well as the methodology used to calculate the shadow price of water used for irrigation based on constraints in water, land, labor and machinery. The principal crops in terms of area planted and total production (maize, beans, sorghum, wheat, barley, broccoli, alfalfa, and strawberries) were modeled with area constraints set on the basis of historical data (1990-99). We considered surface canals (type I) and groundwater (type II) sources of irrigation which can physically command 76,677 ha and 32,461 ha in ARLID, respectively. The model within the constraints set allocated water and land, while labor and machinery were based on production costs as described below, under the assumption that contracted labor may be required in excess of household labor. A monthly time-step was used. The historical variations in prices are shown in Figures 11 (grains) and 12 (vegetables and forage), which also indicate projected 2010 prices based on linear regression.

The very real crisis of Mexican agriculture is evident from the alarming declines in prices. Table 6 summarizes the model parameters and constraints, and shows the calculated values (and percentage variation from the baseline model 0 for prevailing 1999 conditions) for net income, cropped area, and irrigation volumes applied.

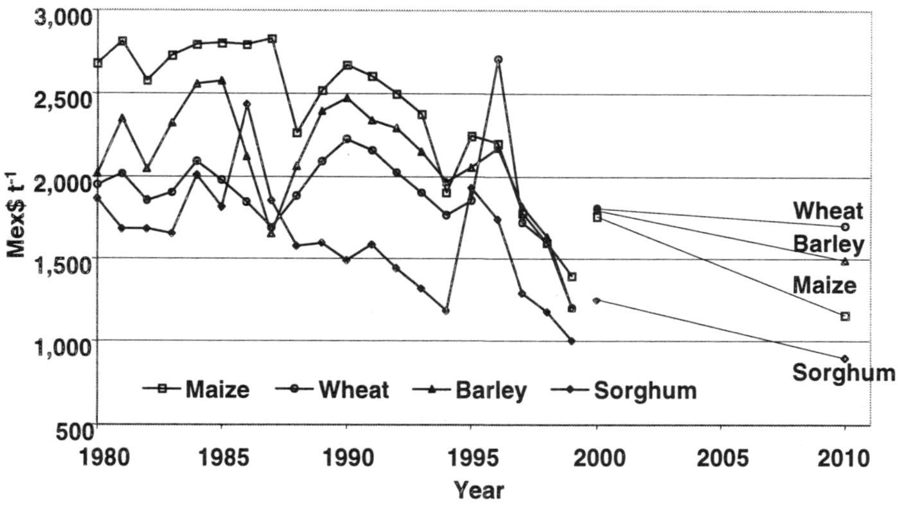

Source: Consejo Nacional Agropecuario, 1980-89 and 1990-99

Figure 11 Grain prices at farm-gate, 1980-99 and 2000-10 trend (1999 Mex$ t^{-1})

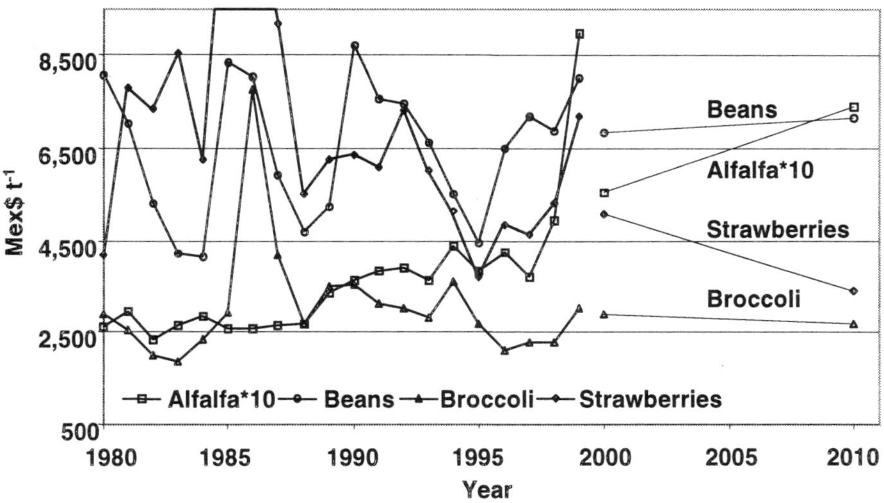

Source: Consejo Nacional Agropecuario, 1980-89 and 1990-99

Figure 12 Vegetable and forage prices at farm-gate, 1980-99 and 2000-10 trend (1999 Mex$ t^{-1})

A few clarifications are in order prior to discussing the marginal value of water calculated and presented as shadow prices in Tables 7 and 8. The 872 and 330 Mm^3 for canal water and groundwater respectively are the actual volumes applied in 1999. Prevailing 1999 prices were used for models 1-6, with variations forced in irrigation and cropped areas to observe impacts on net income. Irrigation volume reductions of 18% (both canal and groundwater) are equivalent to 2% annual reductions compounded over 10 years. As we will discuss subsequently, the 2% figure was calculated from the trend in actual canal irrigation diversions in the Middle Lerma for the 1993-97 period (in 1998, canal irrigation was totally restricted). Additionally, groundwater volume must be reduced by 2% annually to account for current levels of growth in industrial and municipal consumption. The 28% reduction in canal water (model 5) is equivalent to the 240 Mm^3 transfer of water from the Middle Lerma to Lake Chapala, as resulted from the 1999 implementation of the surface water allocation treaty. Minimum cropped area was constrained by placing either a bound equal to the 1989-1999 average area planted or the actual 10-year minimum. Maximum bounds were either the 10-year maximum for grains or twice the 10-year maximum for vegetables (the latter under the assumption that vegetable production will increase in response to market forces). Models 7-10 were based on projected 2010 prices as indicated in Figures 11 and 12 (or break-even prices, whichever were higher), with similar variations in irrigation and cropped area restrictions.

Table 6 Parameter values and results of linear programming

#	Water available		Price	Area constraint		Net income		Cropped area		Irrigation volume applied	
	I. Canal (Mm^3)	II. Ground (Mm^3)	(year)	Grains	Veg. & forage	(million Mex$)	Variation (%)	(1000 ha)	Variation (%)	(Mm^3)	Variation (%)
0	872	330	1999	Min. (10 yr avg.)	Max.	1,027	0	148	0	1,022	0
1	872	330	1999	Min.	Max.	1,084	6	149	1	1,022	0
2	872	330-18%	1999	Min.	Max.	1,030	0	140	-5	975	-5
3	872-18%	330-18%	1999	Min.	Max.	946	-8	122	-18	839	-18
4	872-18%	330-18%	1999	Max.	Max.	937	-9	120	-19	839	-18
5	872-28%	330	1999	Min.	Max.	950	-7	121	-18	815	-20
6	872	330	1999	Min.	Max.*2	1,205	17	146	-1	1,023	0
7	872	330	2010	Max.	Max.	838	-18	145	-2	1,022	0
8	872	330-18%	2010	Max.	Max.	803	-22	136	-8	975	-5
9	872	330	2010	Max.	Max.*2	889	-13	144	-2	1,023	0
10	872	330-18%	2010	Max.	Max.*2	842	-18	135	-8	975	-5

The results of the linear programming estimates of shadow values are presented in Tables 7 and 8, and may be interpreted as the (marginal) value obtained if one additional unit of water was available. Models 0-6 are based on current 1999 farm-gate prices and should be distinguished from models 7-10, which use projected 2010 prices. Maximum and minimum (non-zero) shadow prices are shown, with the month and crop that gave the particular shadow price.

Table 7 Shadow prices for canal water (Mex\$ m^{-3})

Model	Max.	Month	Crop	Min.	Month	Crop
0	2.16	Apr.	Maize	1.23	Jan.	Barley
1	2.16	Apr.	Maize	0.54	Jan.	Barley
2	2.16	Apr.	Maize	0.54	Jan.	Barley
3	2.28	Apr.	Maize	0.54	Jan.	Barley
4	1.66	Apr.	Sorghum	0.76	Jan.	Wheat
5	2.28	Apr.	Maize	0.82	Jan.	Barley
6	2.16	Apr.	Maize	0.54	Jan.	Barley
7	1.58	Jan.	Wheat	0.78	Jun.	Sorghum
8	1.58	Jan.	Wheat	0.78	Jun.	Sorghum
9	1.58	Jan.	Wheat	0.78	Jun.	Sorghum
10	1.58	Jan.	Wheat	0.78	Jun.	Sorghum

Table 8 Shadow prices for groundwater (Mex\$ m^{-3})

Model	Max.	Month	Crop	Min.	Month	Crop
0	2.40	Apr.	Maize	1.67	Jun.	Maize
1	2.40	Apr.	Maize	0.74	Feb.	Wheat
2	150.03	Sep.	Strawberry	0.77	Feb.	Wheat
3	150.03	Sep.	Strawberry	0.77	Feb.	Wheat
4	152.32	Sep.	Strawberry	0.75	Feb.	Wheat
5	2.40	Apr.	Maize	0.80	Feb.	Wheat
6	146.62	Sep.	Strawberry	0.74	Feb.	Wheat
7	2.43	Apr.	Maize	0.62	Jun.	Maize
8	6.85	Jan.	Strawberry	0.66	Jun.	Maize
9	3.44	Sep.	Strawberry	0.66	Jun.	Maize
10	6.21	Nov.	Broccoli	0.64	Sep.	Strawb.

The linear programming and marginal value results for agricultural water use clearly indicate that water places significant limits on agricultural productivity, even without considering inter-sectoral allocation to meet growing urban demand. We discuss the implications of these results in the conclusions below.

5. DISCUSSION AND CONCLUSIONS

The 240 Mm^3 transfer to Lake Chapala (model 5) resulted in a Mex$ 134 million loss (12%) compared to model 1 with similar constraints. This foregone benefit for Middle Lerma farmers, roughly equivalent to Mex$ 0.56 m^{-3}, is a fraction of the imputed benefit if all this water were to reach urban users in Guadalajara. However, much of the volume actually provided environmental benefits in the lake and the Lerma river riparian areas – a value we clearly recognize, but are unable to estimate. It is clear that canal water, the source of agricultural water transferred to the lake and Guadalajara, at Mex$ 0.54-2.28 m^{-3} has roughly the same (or slightly lower) estimated value as domestic water as determined in the previous sections. Nevertheless, given that increases in domestic water will be accorded priority by law, the real opportunity for water-market based compensation systems lies in transfers of agricultural water to commercial and industrial uses, where estimated values are significantly higher.

Significant reductions in net income are foreseen under the projected 2010 crop price structure. While these projections are based on simple linear extrapolations, not world market and production forecasting, and hence should only be taken as indicative, it is evident that major declines in farm incomes will result from falling prices. The combination of trends in price, particularly for grains, and water availability in the basin paint a bleak picture for agriculture in the medium term. Rising pumping costs in the case of groundwater will compound these effects.

IWMI research has shown the important hydrologic linkages between the canal system and groundwater recharge (Scott and Garcés-Restrepo, 2001) such that approximately 50% of canal water applied to crops recharges the underlying aquifers and subsequently becomes available for pumping. As a result of this recycling, half of each unit of canal water provides subsequent additional benefit as groundwater. By way of example, assuming marginal values of Mex$ 1.80 m^{-3} for canal water and Mex$ 2.40 m^{-3} for groundwater with 50% recycling, the aggregate value of surface water is Mex$ 3.00 m^{-3}. This may not be the most economically efficient way to manage surface water; nevertheless, given the prevalent irrigation practice of high water applications relative to crop water demand, aggregate water valuation studies must account for this process.

The results obtained for groundwater are perhaps more complex than for canal water. The astronomical shadow prices of Mex$ 146-152 m^{-3} calculated in models 2-4 and 6 are not realistic or probable, given that crops would die or farmers would pump before conditions reached such extremes. The values indicated for groundwater under the other models (Mex$ 0.62-2.43 m^{-3} for grains and Mex$ 0.64-6.85 m^{-3} for vegetables), are revealing in that they equal or exceed values estimated for domestic consumption, and compete with commercial and industrial values. This reinforces our contention that groundwater is the critical water resource whose sustainable management defines the future of most economic activity in the region. It also indicates that surface water, which has both lower marginal value and existing infrastructure to allow inter-sectoral water transfers, should be used to meet downstream urban demands under the condition that adequate compensation is paid for foregone agricultural benefits.

The minimum shadow prices for grains irrigated with groundwater, particularly wheat in the range Mex$ 0.74-0.80 m^{-3} at 1999 prices (models 1-6) raise serious questions for the profitability and sustainability of this mode of production. Our field data to be published separately demonstrate that present agricultural pumping costs are in the range of Mex$ 0.06-0.14 m^{-3}. It is possible that subsidies on agricultural energy may be lifted or severely reduced after the July 2000 presidential elections in Mexico. This would push groundwater costs for agriculture into the range of urban pumping costs for the cities located in agriculture areas (Mex$ 0.25-0.40 m^{-3} as shown in Figure 5, excluding the city of Guanajuato which has to pump through a 10-km long pipeline). We have noted above that the combination of rising energy prices and falling water tables may increase pumping costs by 200% by 2010, indicating that groundwater irrigation of grains may no longer be feasible.

To conclude, water allocation mechanisms are crucial in water-short "closed" basins. The specific arrangements must be tailored to the prevailing patterns of demand, existing water rights and allocation systems, and must account for accrued and foregone benefits resulting from water transfers. In the case of the Lerma-Chapala basin in Mexico, we have demonstrated that water allocation mechanisms have not kept up with rapidly changing water demands. Municipal growth accompanied by rapid industrialization in the basin has generated a significant demand that will need to be met through the reallocation of water currently used for irrigation. Our analysis has focused on estimating marginal water values in different sectors in order to assess economic implications of inevitable reallocation of water from agriculture.

The prevailing legal and institutional arrangements in the basin appear to make option contract water allocation appealing. Under this approach, urban centers would sign contracts to purchase water from existing agricultural users in the event of future water shortage. The contract would not need to be exercised every year, under the assumption that surface water use is declining and agricultural groundwater use will remain stable over the next ten years. In order to exercise the option contract, however, compensation would need to be paid to parties who cede water. The analyses above clearly demonstrate that the accrued value in urban water use would allow for compensation of agricultural users.

REFERENCES

Aparicio J (2001) Hydrology of the Lerma-Chapala watershed *In: The Lerma-Chapala Watershed: Evaluation and Management* (Hansen AM and van Afferden M eds) Kluwer-Academic/Plenum Publishers, London.

Banrural (1999) Costos de producción. Irapuato, Gto. Mexico.

Centro de Información de la Cuenca Lerma-Chapala. (1998) http://ciateq.mx/~lermaham/lerma.htm

CEASG (Comisión Estatal de Agua y Saneamiento de Guanajuato) (1999) Diagnóstico de Organismos Operadores de Agua Potable, Alcantarillado y Saneamiento 1995-1998. Guanajuato, Mexico: CEASG.

CNA (Comisión Nacional del Agua) (1999 a) Diagnóstico Regional: Proyecto Lineamientos Estratégicos para el Desarrollo Hidráulico del la Región Lerma-Santiago-Pacífico. Mexico City, Mexico: CNA.

CNA (Comisión Nacional del Agua) (1999 b) Informe de Distribución de Aguas. Celaya. Mexico: CNA.

Consejo Consultivo de Evaluación y Seguimiento del Programa de Ordenación y Saneamiento de la Cuenca Lerma-Chapala (1991) Acuerdo de Coordinación de Aguas Superficiales. Colección Lerma-Chapala. Vol. I, No. 5. Querétaro.

Consejo de la Cuenca Lerma-Chapala, Sub-comité Técnico (CNA), Boletines 1 (1991-92) to 8 (1998-99). Guadalajara.

Consejo Nacional Agropecuario (1990) Estadísticas Básicas del Sector Agropecuario (1980-1989) Mexico, D.F.

Consejo Nacional Agropecuario (1999) Estadísticas Básicas del Sector Agropecuario (1990-1999) Mexico, D.F.

Dinar A, Rosegrant MW and Meinzen-Dick R (1997) Water Allocation Mechanisms. Principles and Examples. Policy Research Working Paper 1779. Washington D.C.: World Bank and International Food Policy Research Institute.

Fideicomisos Instituidos con Relación a la Agricultura (FIRA) (1999) Costos de producción. Irapuato, Gto. Mexico.

Flores-López F.J and Scott CA (1999) Simulación de Alternativas del Manejo del Agua en la Cuenca del Río San Juan, Mexico. Serie Latinoamericana No. 9. International Water Management Institute, Mexico, DF Mexico.

Gibbons DC (1986) The Economic Value of Water. Washington, DC: Resources for the Future.

Hansen AM, León A and Bravo L (1995) Fuentes de contaminación y enriquecimiento de metales en sedimentos de la cuenca Lerma-Chapala. Ingeniería Hidráulica en Mexico X(3):55-69.

Kloezen WH, Garcés-Restrepo and S. Johnson SH (1997) Impact assessment of irrigation management transfer in the Alto Río Lerma Irrigation District, Mexico. Research Report 15, Colombo, Sri Lanka: International Irrigation Management Institute.

Lee TR and Jouravlev AS (1998) Prices, Property and Markets in Water Allocation. Santiago, Chile: United Nations Economic Commission for Latin America and the Caribbean.

Marañón-Pimientel B (1999) La gestión del agua subterránea en Guanajuato. La experiencia de los COTAS. Estudios Agrarios 5 (12): 153-173.

Mestre JE (1997) Integrated approach to river basin management: Lerma-Chapala case study – attributions and experiences in water management in Mexico. Water International 22(3): 140-152.

Palacios-Vélez E and Martínez R (1999) Aspectos constitucionales, legales e institucionales. In Políticas opcionales para el manejo de la sobreexplotación de acuíferos en Mexico. Estudio sectoral. World Bank/CNA, Mexico City.

Saade HL (1997) Toward more efficient urban water management in Mexico. Water International 22(3): 153-158.

Schneider ML and Whitlach EE (1991) User-specific water demand elasticities. Journal of Water Resources Planning and Management 117(1): 52-73.

Scott CA and Garcés-Restrepo C, 2001, Conjunctive Management of Surface Water and Groundwater in the Middle Río Lerma Basin, Mexico. In Asit K. Biswas and Cecilia Tortajada (eds.)_Integrated River Basin Management. Oxford University Press, New Delhi.

Scott CA, Zarazúa JA and Levine G (2000) Urban Wastewater Reuse for Crop Production in the Water-Short Guanajuato River Basin, Mexico, Research Report 41. Colombo, Sri Lanka: International Water Management Institute.

Scott CA, Wester P and Garcés-Restrepo C (1998) Integración vertical de funciones para el manejo sustenible de las cuencas: observaciones conceptuales. 3rd International Seminar on Irrigation Management Transfer. Torreón, Coah. Mexico. Sep. 2-5.

Secretaría de Agricultura Ganadería y Desarrollo Rural (SDAYR) (1999) Dirección de Estadísticas Agropecuarias. Costo de producción. Celaya, Gto. Mexico.

Wester P, Marañón-Pimentel B and Scott SA (1999) Institutional responses to groundwater depletion: The aquifer management councils in the State of Guanajuato, Mexico. International Symposium on Integrated Water Management in Agriculture. 16-18 June 1999. Gómez Palacio, Mexico.

Wester P, Melville R and Ramos-Osorio S (2001) Institutional Arrangements for Water Management in the Lerma-Chapala Basin *In The Lerma-Chapala Watershed: Evaluation and Management* (Hansen AM and van Afferden M eds) Kluwer Academic/Plenum Publishers, London.

Young RA (1996) Measuring Economic Benefits for Water Investments and Policies. World Bank Technical Paper No. 338. Washington, DC: The World Bank.

Chapter 14

ProEstado–MAUA®

A Systems Dynamics Approach for the Design of a Sustainable Water Use Strategy in the Middle Lerma-Chapala Basin

JUAN MANUEL HUERTA[1], MONTSERRAT SERRA[2], AND RICARDO SANDOVAL[2]

[1]*PDM de México, S.A. de C.V.; email: prospect@mail.internet.com.mx* , [2]*Comisión Estatal de Agua y Saneamiento de Guanajuato, Mexico*

Key words: Systems dynamics, systemic thinking, systems approach, dynamic simulation, policy analysis, scenario formulation and testing, water resources management

Abstract: A statewide model has been built to implement an integrated, long-term water resources management approach. The model breaks down the state geography into major watersheds. For each watershed, a holistic view that integrates the water cycle and man social and productive activities on the watershed is presented. Two examples of policy analysis utilizing the model are also presented.

1. BACKGROUND

Thirty years ago, Guanajuato was considered Mexico's "bread basket" supplying most of its grain needs. Endowed with a valley that crosses the state from the east to the west, measuring almost 100 miles long and between 20 and 30 miles across, the "Bajío", as this low valley is known, had the ideal conditions for a diversified agricultural production.

The Bajío, crossed by the Lerma river, endowed with a benign climate and provided with large number of freshwater aquifers continues to be one of Mexico's prime agricultural producing regions. So what has happened in the last 30 years? Why is it more necessary today than in the past to carefully watch the water resources of Guanajuato? There are three factors that, over time, have contributed to the current water resources problems:

1. Population. Guanajuato's population has been growing at a fairly high rate. This growth is affecting water resources since human settlements modify the natural paths of water. Also, the volumes of water withdrawn from rivers and aquifers and the returned wastewater have dramatically increased. Population is also the cause of soil cover modifications by deforestation and urbanization that have changed rain patterns.

2. Accelerated Agricultural Production. Driven by Mexico's population growth and agricultural product exportation, the pressure to increase agricultural production in the state of Guanajuato has been growing. Because of this pressure, Guanajuato's agricultural producers have struggled to improve the yield of their land. Outdated technologies, limited amounts of surface water and a marked variability in the precipitation patterns have forced the producers to supplement the surface water which is used in irrigation with well water. This practice has resulted in over-pumping the state's aquifers.

The pernicious problem of aquifer over-pumping was identified more than 40 years ago when a drilling ban for new wells went into effect for the Bajío Valley and other regions of Guanajuato. Nevertheless, the illegal drilling of wells has continued relentlessly. The tangible effect of this practice has been a fast and worrisome decrease of the aquifers' static levels.

3. Surface Water Supply. The supply of surface water from the Lerma river to Guanajuato has also been decreasing over time. The reasons for this are various. First, at the source of the river, water is being extracted and pumped over the mountains to supply drinking water to Mexico City. Second, population growth along the Lerma river basin is demanding more drinking water, and also, water for agricultural production and industry. Last, Lake Chapala at the end of the Lerma river basin, has been gradually drying up for over two decades, requiring additional volumes that are taken away from the water that in the past was distributed among the states on the basin including Guanajuato.

It is evident that the Lerma river carries smaller volumes of water to Lake Chapala than in the past, but a more obvious cause for the lake's drying up is the near explosive growth of Guadalajara, the second largest city in Mexico. Guadalajara is geographically adjacent to Lake Chapala and draws its drinking and industrial waters from the lake. The national water authorities adamantly refuse to accept this fact as the main source of trouble for the lake. They believe that a reasonable level in the lake can be achieved by enforcing a 20 year old treaty that adjusts the distribution of the Lerma river waters among the states on its basin by using as the deciding criteria the following year's long term precipitation forecast.

1.1 The planning process in Guanajuato in the decade of the 1990s

At the beginning of the decade of the 1990s, the Government of the state of Guanajuato set as one of its highest priorities the need to change the direction of the state development. To do so, an effort to identify the strengths and weaknesses of the state – that also took into account the changes that Mexico and the World – was undertaken.

Between 1991 and 1994 the project Guanajuato XXI Century (Flores *et al.*, 1994), was developed, with the objective to "find a new mechanism to direct the state development", and as one of its short-term goals to "generate possible long term futures that will indicate what actions to take in the present to reach the desirable future". An integral part of this project was the State Foresight Study (Huerta and Cerón, 1997), whose purpose was to "generate expected futures that can be used as guidelines for the state development". To prepare the Study, a foresight dynamic simulation model of the state of Guanajuato, ProEstado-Guanajuato was built.

Supported by Gaston Berger's (1964) idea that the forces that shape the future can be found in today's reality, ProEstado-Guanajuato was built to: (a) portray Guanajuato's today "reality", (b) identify the forces that are immerse in this reality and (c) make the reality and the forces interact to create the future change.

ProEstado-Guanajuato is a systems dynamics simulation model, built as a set of differential equations that are numerically solved over time by an algorithm that generates time dependent behaviors for the model's variables.

Systems dynamics simulation models, like ProEstado, have the inherent property that the policies, which control the behavior of the model (also known as policy variables), are separate from the mathematical structure of system that represents the phenomenon under analysis. This is very much like an automobile, where the driver applies his/her policies through control mechanisms (steering wheel, throttle, and brakes) so that the "transportation system" (the engine, the transmission, the brakes, the suspension) performs the way the driver wants it.

ProEstado-Guanajuato was originally built to examine Governor Vicente Fox's (1995) statewide policies, in search of sustainable ones, to be able to ensure that "present and future generations have the resources they need for their development". This examination identified important weaknesses in several of Guanajuato's sectors, one of them water resources.

1.2 Selecting water resources long-term planning tools for the state of Guanajuato

Confronted with an uncertain future for the state water resources, the State Water Commission designed a planning approach based on two ideas:

1. To view the water resources planning and management problems as part of a much broader system, made up of technical, social, economic, political, environmental and institutional "realities", that had all to be taken into account concurrently in order to understand the true roots of problems.
2. To take a systemic, long view (20 to 30 years) of water resources, because only by looking at the distant future will it be possible to evaluate what policy will make the resources sustainable.

To implement the necessary planning approach, the State Water Commission needed a method that combined: (a) an integrated water resources management framework as the platform for the implementation of its planning approach and (b) system thinking as the paradigm for the effort to be undertaken.

Searching into the water resources literature for an approach that was close to what the State Water Commission needed brought many articles on integrated water resources management, or IWRM as it is commonly known, but none matched what Guanajuato required.

The same was true when the search was directed to long-term analysis. A plethora of articles about applications of the system thinking for long-term analysis can be found, but non-directly related to water resources.

One of the very few articles that address both IWRM and system thinking emphasizes the need to employ the systems approach to: "integrate planning and management of water resources, since the systems approach provides a very efficient solution to complex water resources problems"(Simonovic, 2000).

The current literature on IWRM focuses primarily on descriptions of the "real world" needs for integration. For example, an article on IWRM from the Southern Africa region (Mtetwa, 1998) defines IWRM as: (1) the hydrological cycle, that is, the downstream and upstream uses of water, the surface and groundwater supply and, most important, rainfall, (2) the full range of sectoral interest, where all relevant objectives and constraint of society are considered and (3) the future needs as legitimate claims to the water resources. This chapter does not outline a mechanism to accomplish the integration for planning or management purposes. "System thinking" is a relatively new concept (Senge, 1994) and is being used for business applications while the "systems approach" is a time tested control and aerospace engineering "way of thinking and solving" problems. Both "system thinking" and the "systems approach" are based on dynamic theory and the feedback principle.

"System thinking" methods tend to be more qualitative while the "systems approach" tends to be more quantitative. Systems dynamic simulation is a most useful tool that the engineer has available when it is required to understand the time related behavior of a complex system, such as the state of Guanajuato.

1.3 Using ProEstado-Guanajuato as the base model for MAUA®

ProEstado-Guanajuato had already been deployed by the state of Guanajuato planning authority when the State Water Commission initiated the process of looking for planning tools that would support the planning approach discussed above. The examination of ProEstado – Guanajuato by these authorities indicated that the model had the structure to represent water demands of people for life support, food production, and industrial development. ProEstado-Guanajuato even included an aggregate water supply and demand model that had been employed to pinpoint the precarious condition of this sector in the state. The capabilities of ProEstado-Guanajuato soon convinced the State Water Commission that this model was a good starting point for designing and building MAUA (in Spanish, Modelo de Abasto y Uso del Agua, that translates into Water Supply and Use Model). However, a drastic adaptation of ProEstado-MAUA® was required to convert it from an aggregate state level model to a watershed specific model. The development and construction of ProEstado-MAUA® started in June of 1998. It was concluded in January of the following year.

The systems dynamics approach of ProEstado-MAUA® breaks the single-water use analysis paradigm and focuses on a more important subject, the final use and control of the resource, regardless of who uses it. Once directed towards this central objective, ProEstado-MAUA® pinpoints flaws in policies that are being designed to "solve the problems" and that, contrary to their intended purpose, may even reinforce the existence of the problem.

ProEstado-MAUA® is also useful as a tool for helping to reflect on ways to avoid future problems instead of, as it is being typically done, reacting to problems after they have occurred. In the systems dynamics paradigm this is possible, since the methodology itself enables us to find the causal loops through which we can ascertain the future levels of performance of the state water resources systems.

One of the functions of ProEstado-MAUA® is to increase the overall knowledge of a situation, since knowing about its possible futures tells us what we should be doing today. In short, ProEstado-MAUA® provides a solid platform for water resources planning since:

1. It has the capability of integrating in a single model the visions of the various water uses that go beyond the traditional single-water use approach.
2. It provides the means to visualize what happens when we simulate the implementation of a policy. This capability, when shared with others, serves the purpose of improving our communications.
3. It handles physical as well as social variables in the same model, helping us to determine the impacts on both sides that result from the implementation of policies over an extended period of time.

2. MODEL DESCRIPTION

ProEstado-MAUA® is based on theory, information and data. It is also based on experts' knowledge and field visits to establish the "Ground Truth"; that is, what actually exists on the terrain, as far as natural and man-made water bodies and structures, as well as the way things really are done.

ProEstado-MAUA® breaks down the state of Guanajuato into its major river basins (Figure 1), that are: Turbio, Guanajuato, Temascatío, Laja-Peñuelitas, Laguna Seca, Laja-Allende, Lerma, Río Verde, and Pánuco.

For each watershed, ProEstado-MAUA® can generate time related behaviors at three levels of the physical environment (Figure 2):

Atmosphere, where precipitation is modeled as part of the Water Cycle

Surface, where the following components are modeled:

Socioeconomic. This sub-model has three related components: Population, Education, Labor Force and Employment.

Industry. The industries that this sub-model considers are: leather, footwear, agro-industry, automotive, auto parts, clothing manufacturing, bakery, chemical and petroleum industries.

Agricultural production. The agricultural production sub-model represents crops for the spring/summer production cycle and also, for the fall/winter production cycle. It also considers the production of perennial crops. The primary objective of this module is to generate production inputs with emphasis on land, water and labor.

Watershed. The objective of this sub-model is to concentrate population, agricultural production and industrial production into one geographic entity, in order to compute water demands per use type, as well as wastewater.

Land Portion of the Water Cycle. This sub-model has the following components: runoff, infiltration, evaporation, water storage in rivers, lakes, dams, impoundments as well as water extractions by type of use: agricultural, domestic, industrial.

<u>*Subsurface,*</u> This sub-model handles extractions and aquifer recharges for the more than twenty aquifers of Guanajuato.

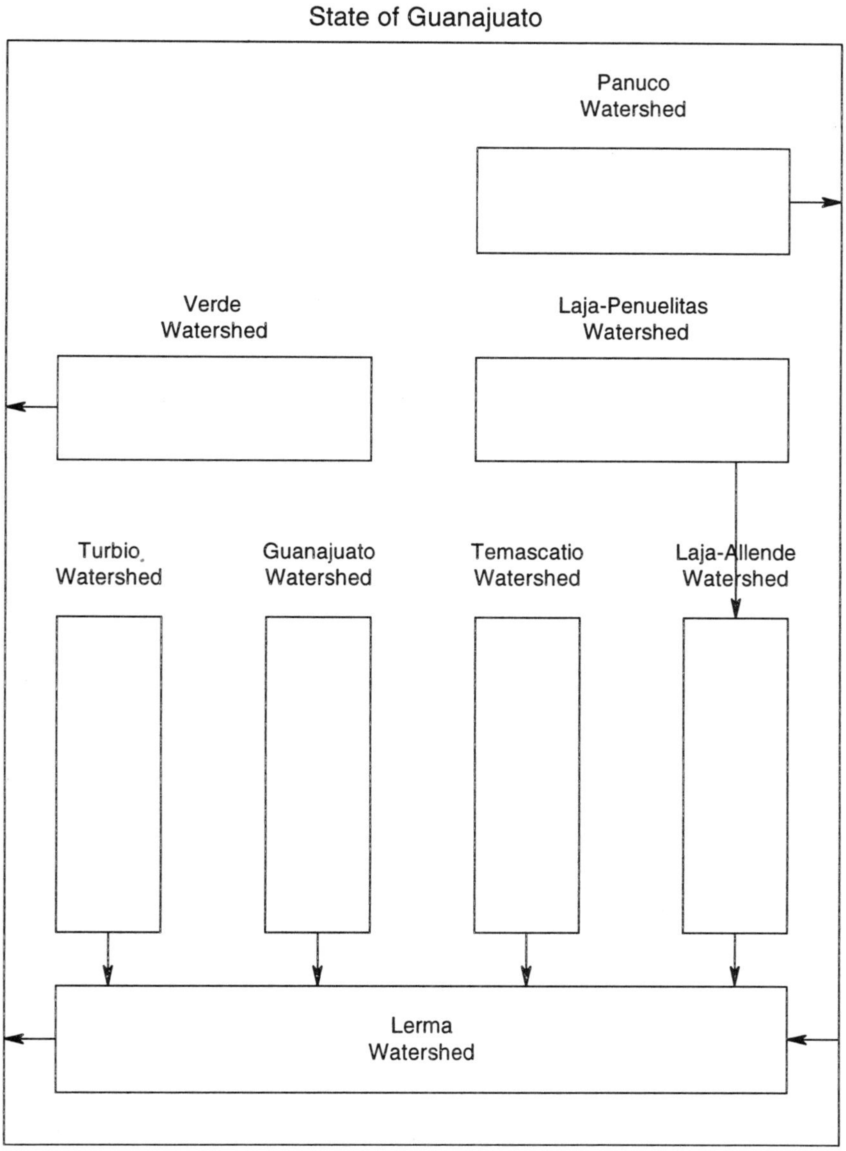

Figure 1 Principal watersheds in the state of Guanajuato

Since ProEstado-MAUA® has about 2,500 equations and 800 variables, no attempt is made in this chapter to describe it in detail. The model is based on dynamic systems theory that considers the behaviors of individual systems components as well as the dynamic interactions among them. These two effects are simultaneous. Under this paradigm of analysis, time is considered the independent variable and it is also the axis along which the simulated phenomenon is being observed.

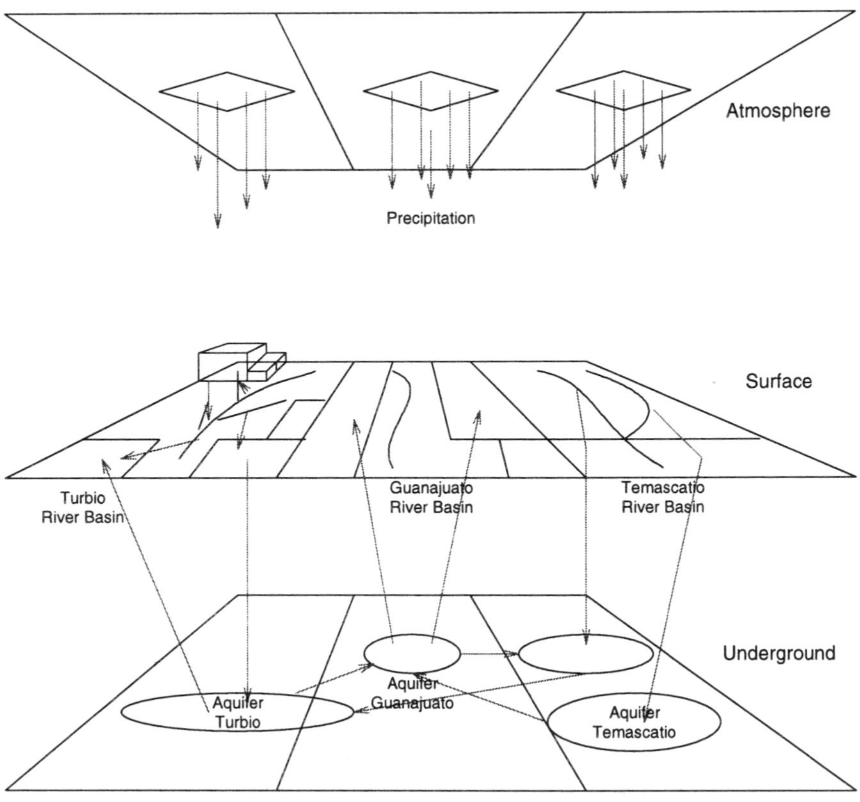

Figure 2 An example of a three level view of watersheds

To simulate time-related behaviors, the model utilizes a system of differential equations of the form:

$$d\mathbf{X}(t)/dt = \mathbf{A}(t)\mathbf{X}(t) + \mathbf{B}(t)\mathbf{U}(t)$$

where:

t = time, is the independent variable,

X(t), the "state" vector variable that defines the status of the system at time t,

d**X**(t)/dt, the rate of change of the "status" in the time interval t +Δt,

A(t), a vector variable of parameters that is used to specify the "reality" of the modeled system,

B(t), a vector variable of coupling parameters for the policies that are implemented in the modeled system, and

U(t), a vector variable that contains the policy being tested.

The main components of an individual watershed model are described below.

2.1 Water cycle

The water cycle model of ProEstado-MAUA® is based on WatBal (Yates, 1996), a differential equation model constructed to compute the balance of a watershed through the simulation of the water cycle. The components of the water cycle that were implemented in ProEstado-MAUA® are the following:

Atmosphere. The water cycle begins with precipitation. ProEstado-MAUA® has two forms to generate precipitation: (1) with an empirical model that employs evaporation and cloud formation to generate precipitation, and, (2) through the use of exogenously input precipitation time series.

Surface. At the ground level, ProEstado-MAUA® represents in the first place, the "geometry" of river, lakes and other water bodies that exist on the watershed, including their interconnections. Second, for a given level of precipitation, the model computes surface runoff, evaporation, and infiltration. Finally, it simulates the operations of rivers and damns as close to reality as the interviewed experts who manage the waterworks could describe them.

Subsurface. Understanding the dynamics of the more than 20 aquifers in the state of Guanajuato is of crucial importance since many of them are over-exploited and their static levels are decreasing several meters each year. The aquifer's model structure consists of the aquifer itself, the inputs by infiltration and the natural flows from neighboring aquifers. The aquifer's outputs consist of water extractions for the various uses, and the outflows to neighboring aquifers. An aquifer's inputs, outputs and level are all time functions.

2.2 Socioeconomic

This model is composed of four sub-models: Population, Education and Labor Force, Industry and Labor Demand, and Agriculture.

2.2.1 Population

The population pyramid database for this sub-model stores information for 18 five-year age female groups, and the same for 18 five-year age male groups. Each group contains five one-year population groups (e.g. the first group goes from ages 0 to 4, the second, from 5 to 9 years, etc.).

For every year of simulation, the population change for each group, with the exception of the first one, that is computed as follows: The input to i^{th} group, are the individuals that leave the $(i - 1)^{th}$ group (roughly, one fifth of the population of the $(i - 1)^{th}$ group). The outputs of the ith group are the ones that leave this group to be included in the $(i + 1)^{th}$ group, plus the ones that died. Death rates are functions of social services such as health, nutrition, etc., and are age group specific.

For the first group (0 to 4 years), the inputs are the newly born (males and females). For each simulated year, the model computes the number of births as a function of the females in each fertile group and the specific group fertility rate (fertility rates are function of environmental conditions, such as level of income, level of education, regional habits, etc.).

For each simulated year, the population module database contains the watershed population by group and sex that can easily be added to compute the total watershed population.

2.2.2 Education and labor force

Before a simulation run starts, ProEstado-MAUA® has to associate the initial Guanajuato population with its corresponding level of education. This computation is performed by age group. For example, for the group 9 (40 to 44 years), a group that is beyond school years, the model computes how many have (1) no education, (2) primary education, (3) secondary education, etc. This process is performed for each group that is beyond school years.

The education model, is a "throughput model", that is, for each simulated model all groups are "passed" through the education model, that acts as a filter to yield a vector containing the number of individuals that finished basic education, secondary education or higher education. Adding the vectors of all groups by level of education, the model computes the labor force available by level of education.

2.2.3 Industry and labor demand

From the various industrial sectors in the state of Guanajuato, only the nine highest ranked in terms of capital, labor and total output was modeled in detail. These sectors are: leather, footwear, agro-industry, automotive, auto-parts, clothing manufacturing, bakery, chemical and petroleum.

Industrial data was not available at the watershed level. Therefore, the models for the industries listed above were built at the state level where data were available. For each sector, a time variable Cobb-Douglas model was constructed. Given a level of output for a specific industry, the main purpose of these models was to compute yearly labor demands in number of individuals. Matching labor supply with labor demand, the model can compute unemployment or shortages of labor availability.

2.2.4 Agricultural production

The use of water for irrigation amount to 87% of the total water use in Guanajuato. For this reason it was decided to build a very detailed model for agricultural production. The first attempt was to build a model based on "direct" water use volume. This effort was soon stopped since the available data was incomplete and inaccurate. The second attempt was to look at computing water requirements for agriculture using an "indirect" method. This model proved to be far more accurate then any of the approaches that utilize direct methods

Since the type of crop and cultivated area were known for each watershed, using FAO´s Penman–Monteith Method, time-dependent water requirement profiles for each crop were estimated. At any point in time, the model can compare the water required for the plant against the humidity of the soil. If the water required for the plant is greater than the soil humidity then irrigation is started. Irrigation is stopped when the plant and the soil humidity are equal.

This method computes the net volume of water for irrigation at the plant itself. To compute gross volumes of the liquid, the net water requirements are divided by the efficiency of the irrigation system, as well as by the efficiency of the distribution systems.

The agricultural production model has the mathematical structure to simulate separately fall/winter and spring/summer production cycles. It also has the structure to simulate production of perennials.

2.2.5 Watershed

Because the industry and education models of ProEstado-MAUA® are state aggregate models and not watershed specific (this is because data exists only at the state level), it was necessary to introduce the concept of *watershed.* A *watershed* concentrates all information available for a particular watershed (i.e., population, labor force, industrial production, agricultural production, commerce and transportation).

The determination of how much of an industrial sector corresponds to each watershed, is accomplished by a matrix multiplication.

3. MODEL COMPUTER IMPLEMENTATION

For the computer implementation of ProEstado-MAUA® the dynamic simulation package Stella Research ®, Version 5.0, for Windows (1999), was employed.

Stella has an input interface and editor that enables the user to build models through the use of "icons" that represent state variables, connectors or auxiliary variables. These last ones, can be constants or function of various state variables.

Stella provides the user with a graphical interface to set model runs and show model results. The data provided are directly input to the model equations.

Stella also has numerical integrators to solve the system of differential equations of a model. It utilizes Euler and Runge-Kutta (second and fourth algorithms) to integrate the systems of differential equations. Figure 3, above shows a block diagram of ProEstado-MAUA® implemented in Stella Research, Version 5.1.

4. USE OF THE MODEL

ProEstado-MAUA® was designed and built to examine long-term, complex water and water related issues such as: *What is the outlook of Guanajuato's agriculture over the next 20 years?*

To examine this question, a set of policy variables from ProEstado-MAUA® can be selected to formulate the issue in terms of the model structure. For example, if we are interested in decreasing water demand for agriculture and, at the same time, decrease demand of drinking water, the program will indicate which policy variables we can select (Table 1).

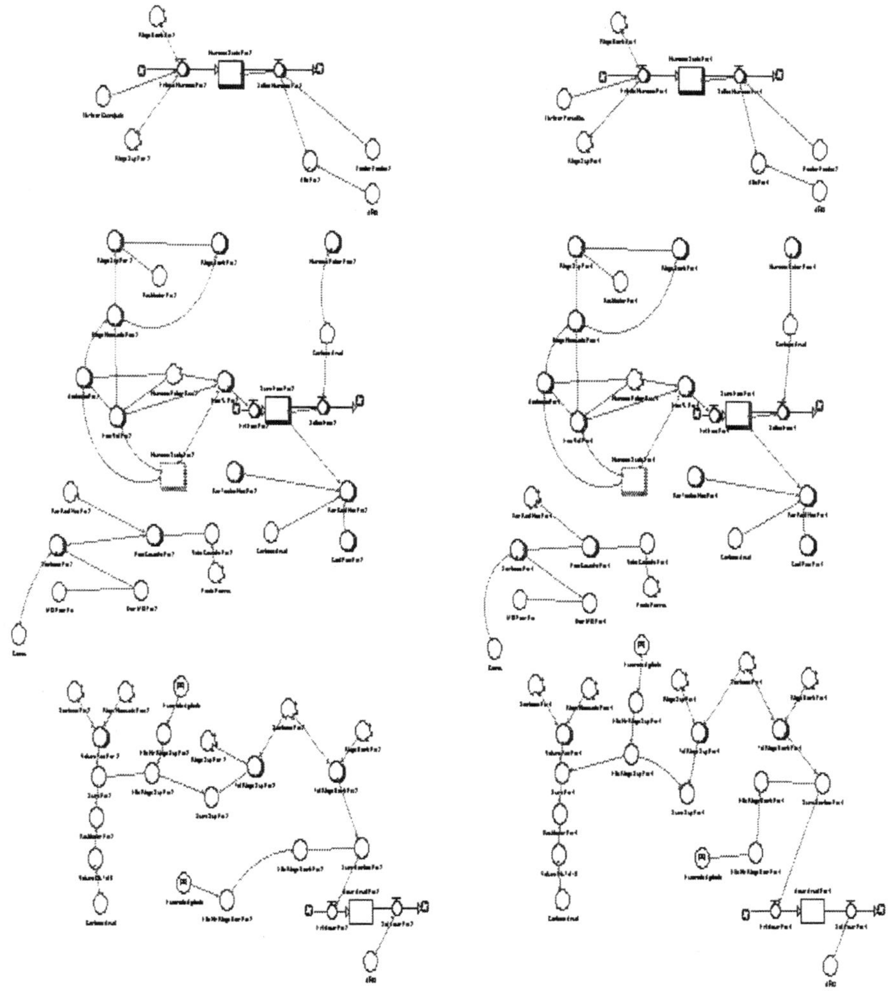

Figure 3 Block diagram of ProEstado-MAUA® implemented in Stella

Table 1 Sub-models and their policy variables

Sub-model	Policy variable
Agricultural production	-Cultivated area -Irrigation technology -Change of crop
Socioeconomic	-Use of water savings devices -Number of drinking water wells

Options include: (1) reduction of *Cultivated area*, (2) improvement of *Irrigation technology*, or (3) *Change of crop*. Combinations of these options can be applied simultaneously.
To decrease demand of drinking water, we can select either or both of the available policy variables options: (1) *Use of water saving devices*, or, (2) reduction of the *Number of drinking water wells*.

It should be clear to the reader that although one can reach the objective of the "decreasing water demand for agriculture" with any of the policies shown in Table 1, it is not the same thing to reduce *Cultivated area*, as to *Improve Irrigation Technolog*y, because these two policy variables will have different impacts on the watershed.

By *reducing cultivated areas*, unemployment among agricultural workers will result. If, on the other hand *irrigation technology* is improved, unemployment will not occur, to the contrary, more labor is needed to handle the improved technology.

Through simulation runs one may understand the impact that either of these options will have on variables such as *Agricultural Employment* or *Crop Production Value*. These variables are called "tracers" and are always printed or plotted during different scenario analysis, so that the user can compare one scenario with another.

5. ISSUES EXAMINED WITH PROESTADO-MAUA®

ProEstado-MAUA® has been in use for few years. In that time it has been employed for the analysis of different situations:

5.1 State water resources plan

ProEstado-MAUA® was built to support the State Water Resources Plan, a statewide 25-year plan that examines the state water resources with an integrated vision.

Two scenarios were simulated with ProEstado-MAUA® to evaluate the sustainability of the resources under each of the scenarios. They were:

Scenario 1: Inefficient water use

Scenario 2: Efficient water use

For the two scenarios, only one policy variable was tested: *Irrigation Efficiency*.

For Scenario 1, the policy is to continue present irrigation practices that employ low technology irrigation systems

For Scenario 2, the policy is to increase irrigation efficiency through the implementation of higher technology irrigation systems

The criterion to analyze the test of the two scenarios was: Which of the two scenarios can achieve long-term sustainability?

The results of the simulation of these two scenarios are presented in Figure 4. The vertical axis of this Figure, indicates the depth in meters (the numbers are negative since they are measures from the surface, which is level zero on down) of the static level of the León aquifer. The horizontal axis indicates the time measured in years.

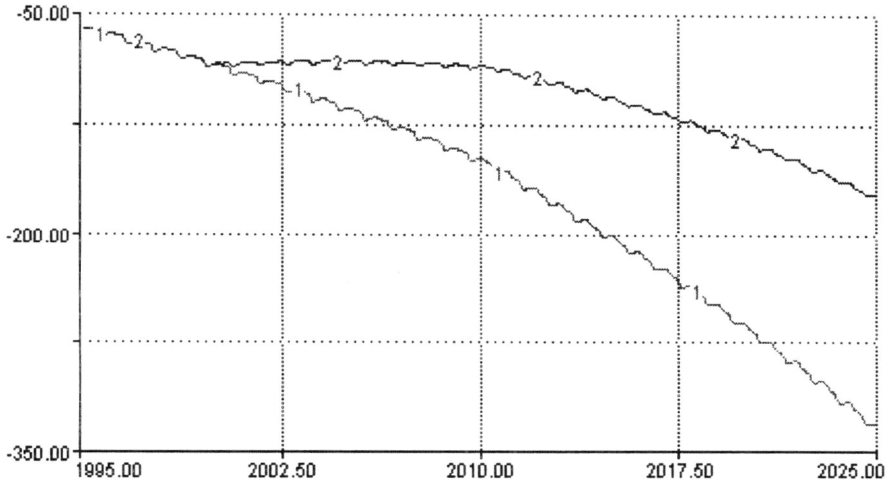

Figure 4 Static levels of the León aquifer under the two scenarios tested [1]Inefficient water use; [2]Efficient water use

The static levels of the León aquifer correspond to the two scenarios tested. Scenario 1, shows the behavior of the aquifer for the policy of continued inefficient water use, while Scenario 2 shows the results for a more efficient water use.

Comparing the two scenarios, it becomes clear that the León aquifer will significantly benefit from the improved efficiency of the irrigation systems. For Scenario 2, the decrease of the aquifer level after 2015, is due to the accelerated industrial development of the city of León that is a consequence of a marked increase in the population living in this basin.

5.2 Lerma-Chapala basin water distribution agreement

With the requirement of increasing the volume of Lake Chapala, located at the end of the Lerma river basin, national water authorities, in charge of the management of this river basin, rewrote a 20-year old agreement that distributes the waters of the Lerma river among the states in the basin.

To negotiate a fair distribution agreement with the national water authorities, the state of Guanajuato decided to undertake a study to examine the social and economic impacts that the new distribution may have on the watershed. Three basic scenarios and some variations were tested:

1. *Business-as-Usual*. The assumption is that there is no reduction from the volume delivered to Guanajuato during the last decade.

2. *Command and Control*. Within one year, the water applied on crops is reduced current values of about 1.40 m to 1.20 m for *ejidos* (government designated agricultural lands), and 0.60 m for small-scale farming.

3, *Transition*. This scenario assumes a gradual reduction of water applied to crops, over a time period of more that one year.

The results of the simulations are presented in Table 2. It is apparent, that the best scenarios are 1, 3.a and 3.b.2. Scenarios 2 and 3.b.1, produce large unemployment and substantial losses for agricultural producers.

Table 2 Summary of scenarios effects in Guanajuato

Scenario	A	B	C	D	E
1	-----	-----	-----	-----	-----
2	17%	39%	19%	9%	6%
3.a	0%	0%	0%	9%	0%
3.b.1	12%	35%	23%	4%	5%
3.b.2	0%	0%	0%	5%	0%

The percentages shown in Table 2 are calculated having as base scenario 1. Table column headings: **A** – Increase in unemployment in the agricultural sector; **B** – Reduction in crop value: **C** – Increase in unemployment in other sectors; **D** – Increase in volume of Lake Chapala; **E** – Overall economic loss. Table row headings: **1** Business–as–Usual; **2** Command and Control; **3.a** Transition, with increment in efficiency of the irrigation systems (from 33% to 55%); **3.b.1** Transition, with reduction of cultivated area for the spring/summer crop cycle; **3.b.2** Transition, with a fixed volume reduction (280 Mm^3 total), over a year period and an increment in the irrigation efficiency

In conclusion, the best scenarios with the least disturbance to the Lerma watershed in the state of Guanajuato, are those that consider an increase in the efficiency of the irrigation systems utilized throughout the state.

6. CONCLUSIONS

This chapter has presented a process defined by three stages: 1) The development of a conceptual framework where water resources planning and management problems are part of a larger system. Within this system the analysis of policies are performed for a 30 years time horizon to determine sustainability; 2) The development and construction of a systems dynamics model, ProEstado-MAUA®, that provides the conceptual framework, and, 3) The application of the model to answer different questions: a) one strategic, that has to do with the long term sustainability of water resources, and, b) one tactic, that deals with the national water authorities initiative of reducing the volume of surface water that irrigates grains. Both situations were analyzed for the state of Guanajuato.

The result of these applications of ProEstado-MAUA® are very encouraging, since they have enabled the State Water Authorities to analyze complex relationships between water, and man, social and productive activities where either of these act as restrictions or bottle necks, preventing state water resources authorities to fulfill planning objectives. ProEstado-MAUA® provides a mechanism that integrates: (1) the hydrological cycle, (the downstream and upstream uses of water), the surface and groundwater supply and, most importantly, rainfall, (2) the full range of sectoral interests, where relevant objectives and constraint of society are considered, and (3) future needs and legitimate claims of water resources (Mtetwa, 1998).

Present and future scarcity of natural resources, such as water, soil, energy and other natural resources will dictate the need of performing integrated (holistic) analysis of sustainability of geographical areas. The approach presented here demonstrates the capability to widen the space of analysis, in order to simultaneously consider the natural and socioeconomic systems of region.

REFERENCES

Berger G (1964) Phnomenologie du Temps et Prospective. PUP, Paris.

Flores C, Treviño R and Sojo E (1994) Guanajuato Siglo XXI. Fundación Guanajuato Siglo XXI, Guanajuato, Gto, Mexico.

Fox V (1995) Plan Básico de Gobierno 1995-2000, Gobernador del Estado de Guanajuato.

Huerta J and Cerón R (1997) Estudio de Prospectiva Estatal, CODEREG, Guanajuato, Gto, Mexico.

Mtetwa S (1998) Strategic Approaches to Freshwater Resources Management in Zimbabwe. Experts Group Meeting on Strategic Approaches to Fresh Water Management, UN.

Senge P (1994) The Fifth Discipline. Currency Doubleday, New York.

Simonovic S (2000) Tools for Water Management: One View of the Future, International Water Resources Association. Water International, Volume 25, Number 1, p. 76-88.

Stella Research for Windows (1999), High Performance Systems, Hanover, NH, USA.

Yates D (1996) WatBal: An Integrated Water Balance Model, International Journal of Water Resources Development, 12 (2).

Chapter 15

Institutional Arrangements for Water Management in the Lerma-Chapala Basin

PHILIPPUS WESTER[1], ROBERTO MELVILLE[2], AND SERGIO RAMOS-OSORIO[3]

[1]*International Water Management Institute (IWMI), Colombo, Sri Lanka*
email: fwester@xs4all.nl
[2]*Centro de Investigaciones y Estudios Superiores en Antropología Social (CIESAS), Mexico City, Mexico. email: melville@juarez.ciesas.edu.mx*
[3]*email: sergioramo@excite.com*

Key words: Institutional arrangements, water management, stakeholder participation, river basin councils

Abstract: In the past ten years the water sector in Mexico has undergone radical reforms, including the creation of river basin councils, in which users' representatives and federal and state governments share the responsibility to achieve integrated water management at the basin level. This article assesses the institutional arrangements for water management in the Lerma-Chapala basin and pays special attention to the origins and achievements of the Lerma-Chapala River Basin Council. Stakeholder participation in decision-making in the Council is reviewed; as is the extent to which the Council has achieved the objectives it set itself, namely equitable water allocation and water quality control. The analysis shows that, while basin level coordination mechanisms are clearly necessary to deal with increasing water scarcity and promising progress has been made, more drastic changes are needed to ensure sustainable water management and improve water-related social equity in the basin. In particular, the serious overdraft of the basin's aquifers needs to be remedied, while user representation in basin-level decision making needs to be strengthened and mechanisms for compensating farmers for the transfer of water out of the agricultural sector need to be drawn up. Lastly, more decision-making power and control over financial resources need to be decentralized to the basin and state level to enable sustainable water management.

1. INTRODUCTION

Global withdrawals of freshwater grew dramatically in the 20th century, resulting in water stress in many countries of the world, rich as well as poor. At present, some 380 million people live in countries where water scarcity is severe (Seckler *et al.*, 1998), while world-wide more than one billion people are still deprived of access to water of sufficient quantity and quality to meet minimal levels of health, income and human dignity (van Koppen, 2000). It is projected that by 2025 some 2.7 billion people will face severe water scarcity (IWMI, 2000), while already all around us are the signs of a water crisis. Droughts, salinization, groundwater depletion and water pollution are but some of the frequently mentioned problems associated with water (Falkenmark and Lundqvist, 1998). These factors, and others, are threatening river basin ecosystems, human health and the world's capacity to feed itself.

It has become conventional to cite water scarcity as the next threat to humanity, typified by the replacement of Cold War rhetoric by Water Wars rhetoric. However, as pointed out by Falkenmark and Lundqvist (1998), water scarcity is more than simply a situation where demand exceeds supply. While examples of absolute water scarcity exist, for most people the main issues are water quality, competition between incompatible uses or political and economic barriers that limit their access to water. It are the interactions between the different uses and users of water, and the conflicts that stem from these interactions that lie at the heart of the need for an integrated approach to water management.

This need is especially urgent in the Lerma-Chapala basin in central Mexico, which has reached a crisis point. Consumptive water use exceeds supply by 10% on average, making it one of the most over-committed basins in the world. Unchecked competition for groundwater has produced dramatic declines in aquifer levels, while surface water withdrawals exceed supply in all but the wettest years, as a result of which Lake Chapala, the receiving water body of the basin, is gradually drying up. Around 79% of the water diverted in the basin is used to irrigate some 800,000 ha, all of which is locally managed and shows good overall productivity, but generally low economic returns. Moreover, water pollution remains a serious problem with significant wastewater reuse for irrigation within the basin (Scott *et al.*, 2000). Most significantly, water is being transferred from the agricultural to the urban and industrial sectors, without due compensation.

As a response to the deterioration in the basin's water resource base, several institutional innovations have occurred in the basin since 1989, including the signing of a river basin coordination agreement (1989), the creation of a river basin council (1993) and the establishment of aquifer management councils (1995-onwards). Water reforms at the national level, such as the creation of a national water agency in 1989, the transfer of government irrigation districts to users (1991-present) and the promulgation of a new water law in 1992, have also significantly altered institutional arrangements for water management in the basin.

This article partially assesses the effectiveness of these changes in addressing the negative impacts of basin closure[1] in the Lerma-Chapala basin. The rationale of this assessment is to explore the larger issue of the types of institutional arrangements for water management needed at the basin level. Although a mantra of modern water management is that the management of water resources should be based on their natural boundaries, i.e. river basins, rather than on political and administrative boundaries, to date little is known empirically about the types of institutional arrangements most appropriate for achieving this. The Lerma-Chapala basin is well suited for analyzing this question, not only because Mexico's first river basin council was formed in this basin, but also because Mexico is one of the few countries in the world, together with France, England and Australia, with functioning river basin councils (Betlem, 1998; Buller, 1996; Haisman, 1999; Newson, 1997).

2. INSTITUTIONAL ARRANGEMENTS FOR RIVER BASIN MANAGEMENT

The recognition that the effective management of surface and groundwater requires a basin perspective is long-standing and widespread. However, experience with river basin authorities as action agencies, in contrast to those designed primarily to carry out studies, suggests that they are difficult to sustain (Barrows, 1998). In part, this is so because political boundaries are usually not co-terminus with basin boundaries, making accommodation between different states and government agencies difficult. In closing river basins increasing water scarcity results in a complex interplay between surface and groundwater over-exploitation, declines in water quality, inter-sectoral water transfers, threats to human health, inequitable water allocation and reduced access to water by resource-poor people. These phenomena pose profound institutional and political challenges, as succinctly pointed out by Vermillion and Merrey (1998):

1 Seckler (1992, 1996) coined this term to characterize river basins with no utilizable outflows, i.e. where the use of water that renders it unavailable for further use is approaching or equal to the level of annual renewable water. This definition differs from the hydrologic definition of a closed basin, where there are outflows but these go only to internal seas, lakes or other sinks.

> "Once basins become closed, where water is the prize in what is in essence a zero-sum game, the institutional and management issues become enormously complex, and solutions that seem optimal from a 'systems' perspective (which in practice means an economic perspective) are unsatisfactory from the perspective of particular water users. What is required are effective institutional arrangements for (...) providing the diverse water services demanded by users" (1998:177).

The question then becomes which types of institutional arrangements are best suited for water management in a basin context and whether this includes basin level organizations. Mostert *et al.* (1999) identify three organizational types for river basin management, the hydrological, the administrative and the coordination model. In the hydrological model the organizational structure for river basin management is based on hydrological boundaries, and all water development and management decisions are typically concentrated in one single agency: the river basin authority. The most famous example of this model is the Tennessee Valley Authority (TVA). The opposite of the hydrological model is the administrative model, where water management is the responsibility of states, provinces, municipalities and other bodies not based on hydrological boundaries. This model applies for most countries in the world, where water is not managed along basin lines. In the coordination model, which falls between the previous two models, water management is typically coordinated by river basin councils or coordinating bodies. Pioneering countries in applying this approach are France and England, who have had functioning coordinating bodies at the basin level for the past thirty years (Betlem, 1998; Buller, 1996). Although they make no judgment on which model is best, Mostert *et al.* (1999) tend to favor the coordination model. Likewise, based on an extensive review of river basin management, Barrow (1998) concludes that basin level organizations are necessary but that they should be coordinating bodies and not a single basin authority.

This article argues that the type of organizations needed at the basin level depends on the level of water scarcity in a basin. Turton and Ohlsson (2000) distinguish between a "first-order scarcity of natural resources" and a "second-order scarcity of social resources". They posit that water scarcity *per se* (first-order scarcity) is not the key issue, but rather whether a social entity has the ability or adaptive capacity (second-order scarcity) to cope with the challenges increasing water scarcity poses. In a similar vein, Falkenmark and Lundqvist (1998) argue that a country may, with comparative ease, utilize up to 15-20 per cent of its annual renewable water, but that withdrawals above that level require more complex institutional arrangements.

Based on historical examples and conceptual work, Turton and Ohlsson (2000) conclude that two institutional transitions (need to) occur in the water sector as water becomes more scarce: the first transition when water abundance turns to water shortage and the second transition when water shortage turns to water scarcity. The first transition, which occurs when water demand due to population growth overtakes the readily available supply of water, triggers the construction of significant hydraulic infrastructure, usually by the government, to mobilize more water. Reisner (1993) terms this transition to water supply development the birth of the hydraulic mission, embodied in a central government agency consisting of engineers. Before water was controlled locally, but after the first transition, its development and management becomes highly centralized. During this phase river basin development is important and one would typically expect to find river basin authorities.

At a given point the supply-oriented phase runs up against a barrier, namely when water demand continues to outstrip supply even though all available water sources have been developed or are prohibitively expensive to develop. This induces increased competition between water use sectors and water scarcity reaches such levels that the exploitation limits become evident and finding the best possible use of water becomes imperative. However, making the second water transition, from supply-oriented development to water demand management, requires substantial changes in institutional arrangements for water management, including the creation of river basin councils. Under favorable socio-economic and political conditions this transition can be made, resulting in a stabilization of water demand and the birth of sustainable water management. However, these transitions are not automatic and whether and how well they occur is a function of the adaptive capacity of a society.

3. WATER REFORMS AND RIVER BASIN MANAGEMENT IN MEXICO

The transition from supply to demand management is clearly underway in the Lerma-Chapala basin. Through studying the institutional arrangements for water management in the basin we can glean lessons for other basins moving towards closure. Before analyzing the current institutional arrangements for water management in the Lerma-Chapala basin, it is instructive to review the history of water management and river basin development in Mexico. Two aspects of how Mexico is structured as a country are important in this regard namely that it is a centralized federation and that water is defined as national property, making it a federal concern.

Although most works on water management in Mexico quote 1926 as the year in which the centralization of water management in Mexico started, 1888 saw the first attempts by the federal government to gain control over water[2]. Before then, although initial water rights were granted by the state, municipalities administered water rights and for all intents and purposes water was considered private property. This changed in 1888, when Congress passed the *Ley General de Vías de Comunicación* (General Law on Communication Routes) that placed lakes and navigable rivers under federal jurisdiction and specified that water concessions and the confirmation of existing water rights could only be issued by the federal government. The federalization of water management was perfected through the amendment of Article 72 of the Constitution in 1908, which placed rivers in the public domain. Based on this amendment it was concluded that surface water was no longer private property and that access to water was only possible through concessions granted by the federal government. Thus, in a space of twenty years, in legal terms surface water in Mexico passed from being private property to falling in the public domain (Aboites, 1998; Sánchez, 1998).

This trend was consolidated in Article 27 of the Constitution drawn up after the Mexican revolution in 1917, in which all surface water in Mexico is defined as belonging to the nation and falling under federal jurisdiction. Based on Article 27 the centralization of water management began in earnest in the 1920s, when President Calles launched a program for the construction of large-scale irrigation systems (called irrigation districts in Mexico). This program found its legal expression in the Irrigation Law issued in 1926, which also created the *Comisión Nacional de Irrigación* (CNI; National Irrigation Commission), the first government agency solely devoted to the design and construction of irrigation districts and their subsequent management (Oribe-Alba, 1970).

The concept of river basins as a unit of development and management, started to gain force in Mexico during the 1940s. This was related to the perception by Mexico that it needed to ensure its fair share of the rivers flowing from the USA into Mexico, such as the Colorado, Tijuana and Bravo, to make continued irrigation development in northern Mexico possible (Orive-Alba, 1960). In 1944 an international river treaty was signed between the US and Mexico clearly defining their respective water rights. Another event that strongly influenced river basin development in Mexico was the flooding of around half a million hectares in the Papaloapan river basin in September 1944. Based on a study of this disaster, CNI suggested that the problems of this region needed to be tackled in an integral manner (Poleman, 1964).

[2] See Aboites (1998) for an excellent overview of this process from 1888 to 1946.

The combination of these circumstances decisively influenced the reordering of water management in Mexico. Until then, responsibilities for water resources development and management had been spread over several federal ministries and agencies. In 1946 the then director of CNI convinced President Alemán of the need to correct this dispersion of administrative efforts and in December 1946 the *Secretaría de Recursos Hidráulicos* (SRH; Ministry of Water Resources) was created. This was a pioneering move in many respects and was the first time water resources was elevated to the level of a ministry in the western Hemisphere.

Along with the concentration of water resources development and management in the SRH, river basin authorities were set up for several of Mexico's key basins, such as the Papaloapan, Tepalcatepec, Fuerte and Grijalva, between 1947 and 1950. The objective of these authorities was river basin development, based on the TVA model. For the Lerma-Chapala basin a commission was formed, but with much less autonomy and fewer resources. Its creation was strongly related to the drought that hit the basin in 1941 and continued for several years, which resulted in sharp declines in the water levels of Lake Chapala. By 1950 the situation had become so critical that irrigation in the basin had to be restricted and several hydroelectric plants had to cease functioning. As a result, the Lerma-Chapala-Santiago Study Commission was created on 28 November 1950 to:

> "(...) achieve a complete regularization of the existing water use systems [in the basin] and a better planning of those that can be realised in the future; arrive at a full understanding of the available water resources and their potential; and effectuate a more equitable water distribution in the basin through an adequate and combined operation [of existing infrastructure]" (Vallejo-Ivens, 1963:5; our translation).

The SRH presided over the commission that consisted of representatives from the Federal District and the states of Guanajuato, Michoacán, Jalisco, Nayarit, and Mexico State. Interestingly, the state representatives were specialists appointed by the state governors and not user representatives. Although the commission's main objective was to study the water resources of the basin, from 1953 to 1959 it was also authorized to construct hydraulic infrastructure in the basin, which it did. To deal with the worsening drought, the Mexican president also charged the commission's chairman with water allocation at the basin level and the operation of hydraulic infrastructure in the basin (Sandoval, 1981; Vallejo-Ivens, 1963).

A series of studies carried out by the commission towards the end of the 1950s showed that the basin was under-developed and that there was little coordination between the government agencies working in the basin. Based on this evaluation, the commission decided to elaborate an integrated development plan for the basin. Although the scope of this undertaking was beyond the commission's mandate it was presented to the Inter-American Development Bank (IDB) in June 1962 by the commission's directors, during the finalization of a loan agreement for the construction of irrigation systems in the basin. The IDB reacted positively to this idea and suggested to the federal government that the objectives and budget of the original irrigation loan be enlarged to cover the integrated planning and development of the basin. The federal government agreed and thus a $150 million loan agreement was signed for the integrated development of the Lerma-Chapala-Santiago basin (Carrillo-Arronte, 1975).

Of the total loan package $5 million were earmarked for technical assistance. With these funds the Plan Lerma Asistencia Técnica (PLAT) was formed on September 19, 1963, to assist federal agencies in regional development planning. The chairman of the Lerma-Chapala-Santiago Study Commission was named coordinator of PLAT, and for all intents and purposes PLAT functioned as a dependency of the commission in its first years. It has been suggested that the enthusiasm of IDB, and more specifically of the United States, to finance PLAT was strongly influenced by the Cuban revolution and that it served to promote alternative regional planning methodologies.

During the first phase of PLAT, from 1963 to 1966, river basin studies were conducted and the first integrated development plan for the basin was elaborated. However, according to Carrillo-Arronte (1975:15) "A characteristic of the region [the basin] was that it was neither an entity nor a jurisdiction, hence it lacked a political and administrative spine. This was the principal constraint for the promotion of river basin development". The lack of interaction between PLAT and the states changed during the second stage of PLAT (1967-1971), when PLAT's action radius was enlarged to cover all the states in the basin. Its objective became: "the execution of studies to identify and elaborate plans, programs and projects related to the development of the natural resources and industries, principally agriculture, in the states of the Lerma-Chapala-Santiago basin" (Carrillo-Arronte, 1975:37). Thus, the river basin focus was lost. Another significant change during this period was the dissolution of the Lerma-Chapala-Santiago Study Commission in 1970, as a result of which the interactions between PLAT and SRH diminished. During the third phase of PLAT (1972-1975) more emphasis was placed on the industrial sector and PLAT's role was redefined to one of promoting regional development.

The history of the Lerma-Chapala-Santiago Study Commission shows how it oscillated between a study commission and a river basin authority. Interestingly, during the 1950s the commission had a clear mandate to manage water in the basin and stabilize water levels in Lake Chapala.

Through the addition of PLAT it was transformed into a regional planning agency that moved away from a river basin focus. In part this would seem to be related to the wet years that started in 1958, by which the lake fully recovered. On the other hand, the difficulties with including the states in river basin development planning resulted in the states becoming the focus of planning. The relationship between PLAT and the commission remains unclear, but it would seem that PLAT subsumed the commission, leading to the commission's dissolution in 1970. Unfortunately, a detailed study analyzing the commission and PLAT remains to be written.

In 1976 PLAT was disbanded together with the remaining river basin authorities in the country. At this time, President López-Portillo also merged the SRH with the Ministry of Agriculture to create the *Secretaría de Agricultura y Recursos Hidráulicos* (SARH; Ministry of Agriculture and Hydraulic Resources). Evaluating the river basin authorities, Mestre (1997) indicates that they were run by SRH with little interaction with water users and other government agencies and had evolved into powerful bodies that challenged the authority of states and federal agencies.

The creation of SARH entailed a serious demotion for the hydraulic engineers and provoked "institutional turmoil" in the water sector (Mestre, 1997). The SRH top opposed the fusion and politically expressed their demand for renewed autonomy during the election campaigns of both de la Madrid in 1982 and Salinas in 1988. Their main concern was the dispersion of water management related functions and resources over different ministries, and their lack of autonomy under SARH (IEPES, 1987), and thus they lobbied for the constitution of an autonomous water authority. This aim was achieved when Salinas became president (Rap *et al.*, 1999).

In January 1989 Salinas created the *Comisión Nacional de Agua* (CNA; National Water Commission), less than six weeks after he became president. Through the unification of all government responsibilities related to water in one agency, the hydraulic bureaucracy regained a large degree of autonomy. As part of the water reforms a new water law was promulgated in 1992. The Law calls for an integral approach for the management of surface and groundwater in the context of river basins, which it considers as the ideal geographical unit for the planning, development and management of water resources. Under the new water law, the CNA was made the country's sole water authority, charged with managing water resources, both qualitatively and quantitatively.

A key provision of the 1992 water law is the stipulation that stakeholder participation is mandatory in water management at the river basin level. To this end river basin councils, which are forums where federal, state and municipal governments as well as water user representatives share the responsibility for allocating water resources and fostering integral water management at the basin level, have been established in 26 river basins. In addition, the CNA has divided the country into 13 hydrologic regions and established an office in each region to improve river basin planning and the interaction with stakeholders.

Congruently with the establishment of new institutional arrangements for water management at the national and river basin level, the Mexican government implemented far-reaching reforms in the irrigation sector. As part of CNA's policy to modernize water management in Mexico, a program to transfer government-managed irrigation districts to Water Users' Associations (WUAs) was initiated throughout the country. Although the idea of user management of irrigation districts was not new in Mexico, this did not materialize before the nineties, largely because of the centralization of water management by the state between 1888 and 1988 (Aboites, 1998; Rap *et al.*, 1999). This changed in the nineties with Irrigation Management Transfer (IMT) in the sense that user involvement in irrigation management sharply increased, although at the same time the state's control over water was reasserted through IMT. This is apparent from the new water law, which states that the state retains ownership of the irrigation infrastructure and remains ultimately responsible for the management of the irrigation districts.

Thus, 1989 was a watershed year for water management in Mexico. Whereas the preceding one hundred years, from 1888 to 1988, were characterized by increasing federal control over water, to the point that the federal government took all decisions related to water resources development and management, after 1989 decentralization has been the norm. Currently states, municipalities and water users have a much larger say in water management decision-making, exemplified by the changes in the Lerma-Chapala basin:

- The signing of a river basin coordination agreement in 1989;
- The decentralization of domestic water supply and sanitation to states and municipalities;
- The transfer of government irrigation systems to water users, from 1991 until the present;
- The creation of state water commissions, from 1991 onwards;
- The creation of a river basin council in 1993; and
- The establishment of aquifer management councils, starting in 1995.

4. WATER RESOURCES IN THE LERMA-CHAPALA BASIN

Of the institutional innovations that have occurred in the basin since 1989 many are related to water reforms at the national level. However, several of these innovations, such as the creation of a river basin council and the establishment of aquifer management councils, are clearly related to increasing water scarcity. A brief review of the basin's water resources brings out why. The Lerma-Chapala basin covers some 54,400 km^2 and crosses five states: Querétaro (5%), Guanajuato (44%), Michoacán (28%), Mexico State (10%) and Jalisco (13%). Although the mean annual runoff in the basin only represents a little over one percent of Mexico's total runoff, the basin is the source of water for around 15 million people (11 million in the basin and 2 million each in Guadalajara and Mexico City) and contains 13 percent of the irrigated area in the country (CNA, 1999a).

From a water perspective the basin is in crisis, with demand exceeding supply in all but the wettest years (CNA, 1999a, 1999b; de Anda *et al.*, 1998; Mestre, 1997). CNA (1999b) estimates that for the 1940-1995 period average annual rainfall was 705 mm, yielding an average annual runoff of 5,757 million cubic meters (Mm^3), including the Pátzcuaro and Cuitzeo sub-basins. Annual average groundwater recharge is estimated at 3,980 Mm^3. Table 1 presents average consumptive water use for different sectors in the basin and average annual renewable water, showing a deficit of 900 Mm^3.

Table 1 Water balance of the Lerma-Chapala basin

	Surface water (Mm^3)	Groundwater (Mm^3)	Total (Mm^3)
Agriculture	3,424	3,160	6,584
Urban	40	751	791
Out of basin transfer	237	323	560
Industry	39	239	278
Other	6	148	154
Total consumptive use	3,746	4,621	8,367
Lake evaporation	2,270	-	2,270
Total depletion	6,016	4,621	10,637
Annual renewable water	5,757	3,980	9,737
Balance	-259	-641	-900

Source: CNA (1999b).

The out of basin transfers are to Guadalajara (surface water) and Mexico City (groundwater) for domestic water supply. The percentage of available water that is developed and put to use in the basin is 110%[3], showing its degree of over-commitment. To portray the effects of basin closure in the Lerma-Chapala basin it is instructive to analyze fluctuations in the water levels of Lake Chapala. Figure 1 shows these fluctuations from 1934 to 1999 and relates them to developments in the basin.

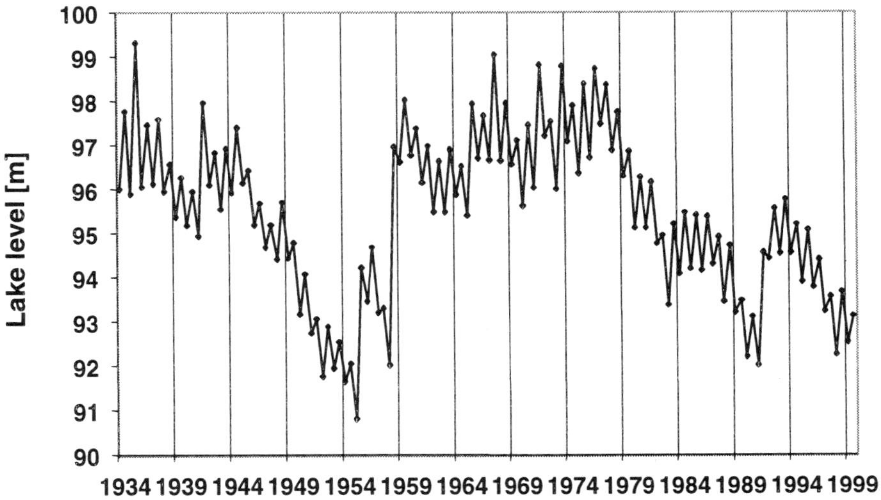

Rainfall (mm) (1)	670	648	685	757	740	668	720
Population (millions) (2)	2.5	3.0	3.6	4.5	5.9	8.7	11.0
Storage capacity in dams Mm^3)(3)	747	1,628	1,817	2,448	3,019	3,678	3,678
Irrigation (ha) (4)	n.a.	175,843	250,500	408,746	681,668	657,734	689,743
Lake inflow from the Lerma river (Mm^3) (5)	2,864	1,652	1,692	1,773	1,931	590	n.a.
Lake extractions (Mm^3) (5)	2,638	1,049	674	1,350	1,817	309	293

Figure 1 Lake Chapala water levels and basin developments

Sources for lake levels: Sandoval (1994) and CNA (1991, 1992, 1993, 1994, 1995, 1996, 1997, 1998, 1999c)

(1) Average rainfall for the decade. Sources: de P. Sandoval (1994) for 1934 to 1949; CNA (1999a) for 1950 to 1999.
(2) Population data for the end of the decade. Sources: de P. Sandoval (1994) for estimates for 1939, 1949, 1959, 1969 and 1979 and CNA (1999a) for actual figures for 1989 and 1999.
(3) Constructed storage capacity at end of decade. Source: de P. Sandoval (1994).
(4) Average irrigated area over the decade. Source: CNA unpublished data.
(5) Average over the decade. Does not include evaporation. Source: Sandoval (1994) and CNA (1991, 1992, 1993, 1994, 1995, 1996, 1997, 1998, 1999c).

[3] This water scarcity indicator is arrived at by dividing total depletion (process and non-process) by annual renewable water (see Seckler *et al.*, 1998 and Molden, 1997).

Starting in 1945, water levels in the lake declined drastically, from around 97 m [4] on average to 90.8 m in 1954, due to a drought combined with significant abstractions from the lake.

At this time around 250,000 ha were irrigated, mainly with surface water, and the constructed storage capacity in the basin was 1,817 Mm^3. This period was the first time the basin headed towards closure as far as surface water is concerned. However, thanks to abundant rainfalls towards the end of the 1950s, the lake recuperated, and levels fluctuated between 95.5 and 98.5 m from 1960 to 1979.

In the 1980s a second period of decline set in, leading to basin closure in the mid-1980s. Constructed storage capacity in the basin had increased to 3,678 Mm^3 and the average irrigated area had grown to around 650,000 ha, mainly due to an increase in groundwater irrigation. Even though abstractions from the lake for hydroelectricity had ceased, the combination of these factors resulted in declines of the lake level, from around 95 m at the start of 1980 to 92 m in 1990. After a modest recuperation in the early 1990s, lake levels are once again low, due to poor rainfall and continued over-exploitation of surface and groundwater.

5. PERFORMANCE OF THE LERMA-CHAPALA RIVER BASIN COUNCIL

The drying up of Lake Chapala in the 1980s, combined with the severe contamination of the Lerma river, gave a strong impetus to institutional change. It was clear that something had to be done to preserve Lake Chapala, which generates significant tourism revenues and foresees 2 million inhabitants of Guadalajara with domestic water. In addition, its symbolic value as Mexico's largest natural lake is high. According to Mestre:

"A wide-ranging water diagnosis existing by mid 1989 clearly presented four capital problems in the Lerma river basin: scarcity, as well as unsuitable water allocation, pollution, inefficiency of water use, and environmental depredation. To turn the tide, it became clear that it would be insufficient and imprudent to maintain that the federal government was solely responsible for this chaos and for its solution or mitigation" (1997:144).

[4] A locally defined benchmark where 100 m is defined as the high shoreline. At 97.2 m the lake holds 8,126 Mm^3 and covers 114,905 ha (de Anda *et al.*, 1998).

Hence, the federal government and the governments of the five states falling in the river basin signed an agreement in Chapala on 13 April 1989, adopting four main objectives to improve water management in the basin:

- Allocate surface and groundwater fairly among users and regulate its use;
- Improve water quality by treating municipal and industrial effluents;
- Increase water-use efficiency; and
- Conserve the river basin ecosystem and protect watersheds.

On September 1st, 1989, the Consulting Council was formed to follow up on these objectives. It created a Technical Working Group "with a straightforward agenda, to be held responsible for making things happen" (Mestre, 1997:145). Based on the 1992 water law the Consultative Council became the Lerma-Chapala River Basin Council on 28 January 1993, the first to be established in Mexico. On the Council are represented the five states in the basin, various federal and state agencies, and users' representatives of different water use sectors.

As outlined above Mexico has a long tradition in river basin development. However, the current river basin councils are very different from the river basin authorities set up in Mexico during the presidency of Alemán, which served to develop the "unlimited" potential of water resources and to channel government resources into river basin development. The river basin councils, on the other hand, are coordinating and consensus-building bodies aimed at reconciling the multiple uses of water and redistributing the costs and responsibilities for restoring environmental balance in the basin between the stakeholders. The councils have to allocate a scarce and finite resource among many competing uses and correct the over-exploitation of water.

In the final analysis, the utility of river basin councils rests in their ability to promote sustainable water management in a river basin. The following assesses the extent to which the Lerma-Chapala River Basin Council has succeeded in attaining two of the four objectives it set itself, namely water allocation and water treatment, while stakeholder participation in the Council is also reviewed. In this regard, it is helpful to keep in mind the situation in the basin in 1989: hardly any municipal and industrial wastewater was treated, domestic water fees and collection rates were very low, cost-recovery in irrigation districts stood at 18%, irrigation water users were not represented in water management decision-making, water allocations to different sectors were not made on a basin-wide basis and clandestine water use stood at around 15% of overall water use (Mestre, 1993).

5.1 Surface water

To allocate water fairly among users, the Consultative Council moved quickly. In August 1990 it formed a Technical Working Group to elaborate a surface water treaty. In August 1991, the governors of the five states in the basin and the federal government signed this treaty, which sets out surface water allocation mechanisms for all the users in the basin. The objective of the treaty is to "assure a rational, equitable and just allocation of water, adaptable to periods of shortage as well as abundance, that would sustain and reinforce the economic development of the region" (Mestre, 1993:3). A major concern behind the treaty was to maintain adequate water levels in Lake Chapala and to ensure Guadalajara's domestic water supply.

To arrive at the treaty the Technical Working Group analyzed hydrometric and dam storage data over the 1950 to 1979 period for 19 sub-basins. To protect existing water rights, actual water demands of registered irrigation users were determined and the inventory of unregistered water users was updated. Based on this assessment, average surface water availability in the basin excluding the Pátzcuaro and Cuitzeo sub-basins was determined to be 4,740 Mm^3 while the historical average for water withdrawals was placed at 3,240 Mm^3 for irrigation, 240 Mm^3 for Guadalajara from Lake Chapala and an additional 90 Mm^3 from Lake Chapala for irrigation (Consejo Consultivo, 1991).

To preserve Lake Chapala the treaty sets out three allocation policies, namely critical, average and abundant, based on the volume of water in the lake (less than 3,300 Mm^3, from 3,300 to 6,000 Mm^3 and more than 6,000 Mm^3, respectively). Each year on November 1st, the Technical Working Group verifies the volume stored in Lake Chapala and thus determines the allocation policy to be followed for the next year. For each allocation policy, formulas have been drawn up to calculate water allocations to the irrigation systems in the basin, based on the surface runoff generated, stored and used in each of the five states in the previous year. Table 2 indicates how this works for the Alto Río Lerma Irrigation District, located in the state of Guanajuato. Based on extensive modeling of these formulas the Technical Working Group came to the conclusion that the resulting water allocation would not impinge on the 1,440 Mm^3 needed by Lake Chapala for evaporation.

The reduction of volumes allocated to irrigation systems under the 1991 treaty is legally possible because surface water rights in Mexico are based on the proportional appropriation doctrine. Under this doctrine concessions are issued for the use of fixed quantities of water, although the quantity of water that may be used in any year is adjusted for all concession holders to reflect water availability[5]. Thus, all concession holders share proportionally in any shortages or surpluses of water. As Mexico follows the proportional appropriation doctrine, water rights are in the form of time-limited concessions issued to users by the federal government and the actual quantity a user receives may be adjusted annually to reflect water availability, with priority accorded to domestic water use (CNA, 1999d).

Table 2 Water allocation principles for the Alto Río Lerma Irrigation District

Lake Chapala volume	Surface runoff generated (SRG) in the state of Guanajuato (Mm^3)	Volume allocated (VA) to irrigation district (Mm^3)
Critical	if SRG between 280 and 1,260	then VA = 94.2% of SRG -262.8
	if SRG > 1,260	then VA = 924
Average	if SRG between 144 and 1,125	then VA = 94.2% of SRG–135.6
	if SRG between 1,125 and 1,400	then VA = 924
	if SRG > 1,400	then VA = 955
Abundant	if SRG between 19 and 1,000	then VA = 94.2% of SRG–17.9
	if SRG between 1,000 and 1,200	then VA = 924
	if SRG > 1,200	then VA = 955

Source: Consejo Consultivo (1991)

Since 1991, the Technical Working Group has met each year and has applied the water allocation rules set out in the treaty. Figure 2 shows the volumes of water allocated and used from 1992 to 2000 as well as the volume of water stored in Lake Chapala. This shows that the 1991 treaty has been strictly enforced, as actual use has never been higher than the allocated values. Thus, the institutional arrangements put in place to ensure a more rational, just and equitable water allocation at the basin level have worked. Strictly allocating surface water on a basin scale is no mean feat, and it is difficult to envision how this could have been achieved without appropriate institutional arrangements at the basin and national level.

Despite the good performance of the surface water allocation mechanisms at the basin level and the strict application of the 1991 treaty, Lake Chapala's volume has halved in the past eight years. Thus, the surface water treaty has failed to achieve its objective of rescuing the lake. This is so because the treaty takes the surface runoff generated in the previous year to determine water allocations. This is clearly brought out by the allocations for 1998 and 1999. In 1997 rainfall was 645 mm and dam storage on the 1st of November (used here as a proxy of surface runoff) was consequently low. Combined with a lake volume below 3,300 Mm^3 the critical allocation policy was followed for 1998, leading to the lowest allocations since the treaty was

5 This contrasts with the prior appropriation system, where first rights have seniority implying that water rights issued later are the first to be curtailed in times of shortage.

signed. However, rainfall in 1998 was exceptionally good, at 810 mm more than 100 mm above average, leading to a substantial recuperation of the volume of water stored behind dams and a slight increase in the volume of Lake Chapala to 3,361 Mm^3. As a result, the average allocation policy was followed for 1999 and 3,664 Mm^3 were allocated to water users, the highest level since the signing of the treaty. Unfortunately, rainfall in 1999 was a historic low of 494 mm. These two factors resulted in Lake Chapala's volume dropping to its lowest level since the signing of the treaty.

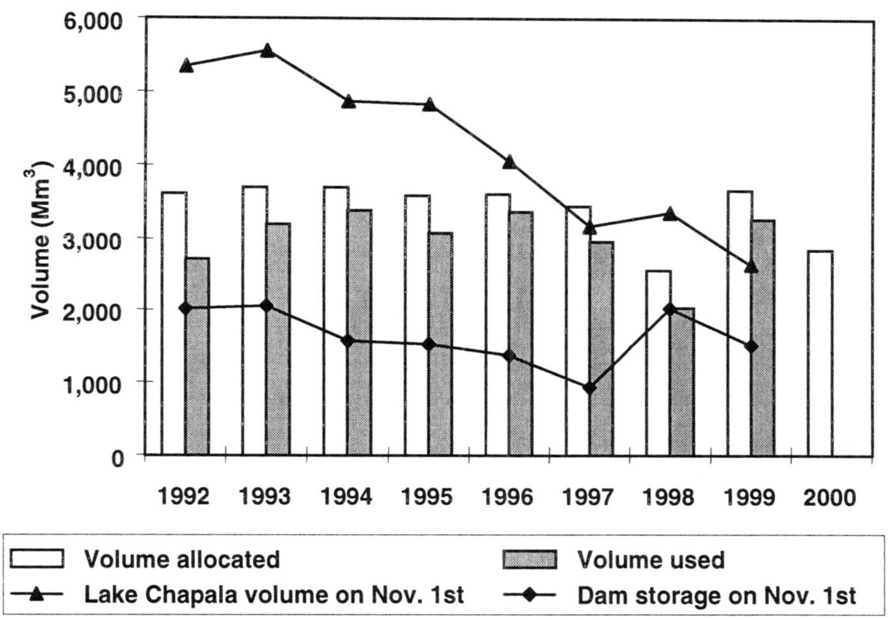

Figure 2 Surface water allocated and used in the Lerma-Chapala basin. Sources: CNA (1991, 1992, 1993, 1994, 1995, 1996, 1997, 1998, 1999c)

Although the signing of the surface water treaty in 1991 was historic, the members of the Council have recognized and discussed its shortcomings candidly. In 1999 the Council decided to revise the treaty as it was clear that it was not rescuing Lake Chapala. Over the past year detailed hydrological studies have been carried out using data from the 1945 to 1997 period (a major improvement over the 1950 to 1979 data used for the previous treaty) and numerous simulations have been run to arrive at an improved surface water allocation treaty. This shows the adaptability of the Council and the commitment of its members to construe a water allocation policy that meets urban and agricultural needs while safeguarding the environment.

An issue that the Council has not yet started to consider is how to compensate farmers for water transferred out of agriculture for urban and environmental demands. Scott *et al.* (2001) calculate that the benefits forgone for farmers in the Alto Río Lerma Irrigation District as a result of the reduced allocation to the district for 2000 amounted to US$ 14 million. Although sufficient water was available in its main dam to cover its full allocation (955 Mm^3) the district was allocated only 648.2 Mm^3under the treaty, due to the critically low volume of water in Lake Chapala and the minimal surface runoff generated in Guanajuato in 1999. To shore up water levels in Lake Chapala the Council decided to release the additional storage in the Solis Dam, the first time that surface water was physically transferred from the agricultural sector to the urban and environmental sector under the 1991 treaty.

Although the reductions in the amount of water allocated to irrigation under the 1991 treaty are fully legal, to mitigate the economic impact on farmers and to be truly just and equitable, future allocation agreements should include compensation mechanisms for the transfer of irrigation water to urban water users. A suitable mechanism would be flexible option contracts between urban centers and farmers, whereby the urban centers would enter into long-term agreements with the irrigation districts to lease water from them when a given contingency arises. For this the districts would receive a negotiated fee when the option is not exercised, and full payment of the benefits foregone when the option is exercised (see Lee and Jouravlev, 1998 for details). Scott *et al.* (2001) conclude that as marginal values for urban water use are up to eight times higher than for agricultural water use, compensation of farmers by urban water use is economically feasible. The revision of the surface water treaty now underway presents an opportunity to consider the compensation mechanisms most appropriate for the Lerma-Chapala basin.

5.2 Groundwater

According to García y García "The graveness of the fall in groundwater levels in the Bajío has been pointed out now for more than 20 years, however, the results of the actions taken to resolve this problem, both by private parties and by the government, have been disastrous" (1998:160). The current aquifer over-exploitation in the Lerma-Chapala basin of 641 Mm^3 bears out the truth of this statement. To tackle this serious problem, the five states in the basin and the federal government signed a coordination agreement to regulate the extraction and use of groundwater in the basin in 1993. In addition, a detailed action program was drawn up to establish regulations for each aquifer, which would form the basis for determining annual extractions from the aquifers. A key element of this action program was to organize users in *Grupos de Preservación del Acuífero* (Aquifer Preservation Groups), to reach agreement on reductions in extractions (Consejo Consultivo, 1993). Although it was originally planned to finish the aquifer regulations by the first half of 1994 and to legalize or disconnect the estimated 10,000 "irregular" wells in the basin (Mestre, 1993) progress has been much slower. The reasons for this are various and largely outside the control of the Council.

The key problem in groundwater management is that the Council, through the CNA, does not physically control the water extraction infrastructure (the wells), as it does in the case of surface water (the dams). Although the Constitution mandates the federal government to intervene in aquifers in overdraft by placing them under *veda*, entailing that it is prohibited to sink new wells without permission from the federal government, the experience with *vedas* has been mixed (Arreguín, 1998). According to Vázquez (1999) ten *vedas* were issued in Guanajuato between 1948 and 1964, prohibiting the perforation of new wells in the Bajío and in the North of the state. In 1983, the rest of the state was also placed under *veda*. Notwithstanding these legal restrictions, the number of wells grew incessantly, from approximately 2,000 in 1958 to 16,500 in 1997 (Guerrero, 1998). Apparently, this was the result of the lack of a coherent legal framework, an agency endowed with the authority and capacity to apply and enforce the laws and of political will to apply it. As remarked by Peña and Arreguín: "The formulation of regulations for prohibited zones [vedas] has been and is the Achilles Heel of the management and administration of groundwater in Mexico" (1999:2-6).

Based on the recognition that *vedas* have not worked and to counter the continued depletion of groundwater in the Lerma-Chapala basin, the CNA started promoting the formation of *Comités Técnicos de Aguas Subterráneas* (COTAS; Technical Committees for Groundwater) in selected aquifers in 1995, as an outflow of the 1993 agreement. Through the establishment of COTAS the CNA is seeking to stimulate the organized participation of aquifer users with the aim to establish mutual agreements for reversing groundwater depletion, in keeping with Article 76 of the water law regulations (CNA, 1999d). To date the characteristics of the COTAS being established by CNA remain unclear, specifically where it concerns their structure, tasks and autonomy in relation to the federal government. It is clear, however, that the COTAS will not be financially and administratively autonomous organizations with an elected management board and a manager, as is the case of the WUAs in the transferred irrigation districts. Instead, it will be a forum in which aquifer users, government agencies and the civil society will interact regarding groundwater management, under the auspices of CNA. In addition, they are organizations without legal status or decision-making powers, whose agreements may be taken into account or not by the CNA (Palacios and Martínez, 1999).

An interesting development in recent years has been the enthusiasm with which the state of Guanajuato is promoting the creation of COTAS, although along different lines than those proposed by CNA (Wester *et al.*, 1999). The most salient difference is that the *Comisión Estatal de Agua y Saneamiento de Guanajuato* (CEASG; Guanajuato State Water Commission) is taking the lead in organizing *Consejos Técnicos de Aguas* (COTAS). According to CEASG: "In Guanajuato, the COTAS have emerged in response to the outcry of water users and institutions, given the deplorable administration of aquifers that historically has existed in the Mexican Republic and that is worsening in areas where water is scarce" (Guerrero, 1998:35). The COTAS as envisioned by CEASG are full-fledged user organizations, whose membership consists of all the water users of an aquifer. They are to serve as mechanisms for reaching agreement on aquifer management taking into consideration the needs of the various sectors using groundwater (Guerrero, 1998). By the end of 1999 CEASG had constituted fourteen COTAS, covering all the aquifers in the state.

Groundwater irrigation in Guanajuato netted US$150 million in 1997, and probably double that in the whole of the Lerma-Chapala basin (Wester, *et al.*, 1999). If industrial and domestic uses, which depend nearly entirely on groundwater, are included it is fair to say that groundwater is the strategic resource in the basin. The long-term consequences of its continued depletion easily overshadow the consequences of Lake Chapala drying up. However, due to the complexities of groundwater management, the River Basin Council and other institutional arrangements in the water sector have been less successful in regulating groundwater extraction in the basin. In this sense, the COTAS are a timely institutional response to the pressing need for innovative approaches to managing aquifers in the basin.

As the COTAS are a recent institutional innovation, it is too early to evaluate if they will be effective in arresting the depletion of groundwater. A key issue that needs to be resolved is the doubts surrounding their attributes, structure and tasks. According to CNA the COTAS are to be consultative bodies without any decision-making powers, let alone the mandate to resolve conflicts between water users or restrict groundwater extractions. According to CEASG, the COTAS should be user-managed organizations with a legal status for groundwater management.

As it has proven very difficult to regulate individual groundwater users through *vedas* a way forward may be to bundle the water right concessions of a whole aquifer and to grant this bundled concession to the COTAS[6]. This would make the COTAS responsible for ensuring that extractions do not exceed the sustainable yield of aquifers. To do so, the groundwater users assembled in the COTAS would need to devise annual aquifer management programs indicating maximum extractions for that year. This groundwater allocation program would then need to be approved and monitored by the State Water Commissions, the River Basin Council and the CNA, and if extractions exceed the agreed allocations then the COTAS as a whole should be severely fined. The enforcement of the allocation program would be in the hands of the COTAS, who would need to devise monitoring programs to ensure that individual users do not pump more than their fair share. To accomplish this institutional change will be an extremely complicated and time-consuming process but the alternative for most groundwater users is economic ruin in the short to medium term.

5.3 Water quality

Besides major achievements in the allocation of surface water, the Council also launched an investment program to clean the Lerma river, through the planned construction of 105 municipal wastewater treatment plants. This was highly necessary, as up to 1989 hardly any of the municipal and industrial wastewater in the basin was treated. As a result, large stretches of the Lerma river were highly contaminated, as where two of its tributaries, the Turbio and Querétaro rivers, while Lake Chapala was classified as contaminated. In 1993 it was estimated that the municipal wastewater generated in the basin was 12,700 L s^{-1} and that this had risen to 17,000 L s^{-1} in 1997 (Mestre, 1993; 1997).

At the signing of the coordination agreement in April 1989 it was agreed that 48 treatment plants with a capacity of 3,700 L s^{-1} would be constructed at a cost of Mex$ 339 million. Priority was given to cities with more than 10,000 inhabitants that discharged directly into the Lerma river or Lake Chapala. On 28 January 1993 the River Basin Council agreed on a 2nd water treatment program, entailing the construction of 52 new plants and the enlargement of 5 existing ones, with a total treatment capacity of 10,835 L s^{-1}. Total costs of this second phase were estimated at Mex$ 722 million. An important innovation of the second phase was the treatment

6 Although the volume of the bundled concession would probably exceed the sustainable yield of the aquifer, in itself this should not be a problem as water rights in Mexico are based on the proportional appropriation doctrine. Hence, annual extraction volumes assigned to groundwater users can legally be reduced to reflect the sustainable yield of the aquifer as long as this is done proportionately.

plants were to be constructed, a turn-key project financed by federal, state and private funds (Mestre, 1997).

It was foreseen that the two investment programs would create a total treatment capacity of 14,561 L s^{-1}, or 85% of municipal wastewater generated in the basin. At the end of 1999, 48 plants had been constructed with a treatment capacity of 6,037 L s^{-1}, while 6 additional plants are still under construction, with a capacity of 3,513 L s^{-1} (CNA, 1999a). State wise breakdown, as well as the operational condition of existing treatment plants is shown in Table 3.

Table 3 Wastewater treatment capacity in the Lerma-Chapala basin

State	Mexico State	Querétaro	Guanajuato	Jalisco	Michoacán	Total
Capacity (L s^{-1})	2732	600	1215	643	847	6037
Number of plants	17	2	5	17	7	48
Good (capacity)	17 (2732)	0	1 (70)	13 (353)	0	3155
Average (capacity)	0	0	3 (445)	1 (40)	3 (367)	852
Bad (capacity)	0	2 (600)	1 (700)	3 (250)	4 (480)	2030

Source: CNA (1999a)

Treatment plants falling under the category bad are operating poorly, hence of the installed capacity of 6,037 L s^{-1}, only 3,155 L s^{-1}, or 18.5% of the municipal wastewater generated in the basin, are fully treated. Although the Council has not succeeded in attaining the ambitious goals it set itself, the installed capacity of 6,037 L s^{-1} up from cero L s^{-1} in 1989 is an achievement in itself. The Council readily admits that more needs to be done, although the large investments required and the lack of financial resources at the basin level makes this very difficult (CNA, 1999b).

5.4 Stakeholder participation

A challenge for the River Basin Council has been ensuring effective user representation – critical in the consensus building and coordination role envisioned in the law. Until the end of 1998, the Council was very top heavy: its president was the federal minister of agriculture (until the end of 1994) or the federal minister of the environment (from 1995 to 1998), while its members were the governors of the five states making up the basin, the federal ministers of five key ministries and the Directors General of CFE, PEMEX and CNA. It is evident from this choice of institutional design that control over water and financial resources were a driving force in the inter-agency alignments within the basin. In 1999 this changed, with user representatives from 6 different sectors being appointed to the Council. Also, the Director General of the CNA became the president of the Council while the remaining members are the five state governors. This change was based on a modification in 1997 of the water law and its regulations, to allow for larger user representation. However, the users on the Council have been nominated by the CNA, and are not known to nor necessarily reflect the interests of the water use sector they represent. To rectify this, CNA is currently working to establish a stepped form of user representation consisting of a general assembly at the basin level and water user committees at the state level. At this time it is too early to evaluate the impact of these changes on the functioning of the Council, although it is a promising sign that user representatives now hold half the votes on the Council.

6. CONCLUSIONS

The assessment of the institutional arrangements for water management in the Lerma-Chapala basin clearly brings out the need for coordinating mechanisms at the basin level in river basins facing closure. Through the Lerma-Chapala River Basin Council promising progress has been made towards improved water management in the basin. This progress is remarkable, in light of the complicated transition from a highly centralized management of water resources to one in which states, municipalities and water users have a larger say. Although it is easy to criticize the Council for its failing and it is clear that much remains to be done to achieve sustainable water management, the institutional changes in the basin in the past ten years have been fundamental and the Council's achievements substantial. Nonetheless, from a water perspective the Lerma-Chapala basin is still in crisis and time is running out. The efforts of the River Basin Council in the past ten years will need to be redoubled to tackle the significant challenges lying ahead of it.

Although surface water allocation mechanisms are working well, and the current revision of the 1991 treaty may lead to increased inflows to Lake Chapala, compensating farmers for water transferred out of agriculture needs to be considered. In closed basins inter-sectoral transfers are inevitable and it will invariably be the irrigation sector that will need to cede water. A key institutional challenge in closed river basins is how to deal with these transfers in a just and equitable manner. The Lerma-Chapala River Basin Council provides a good forum for drawing up and enforcing compensation mechanisms and for safeguarding the water rights of farmers.

A much more serious challenge that the Council and other water management stakeholders in the basin need to deal with urgently is the serious overdraft of the basin's aquifers. The COTAS are an encouraging step in the right direction, but their role in groundwater management should go beyond mere consultation. Bundling extraction rights in an aquifer and concessioning this to a COTAS is feasible under the Mexican water law and should be seriously considered. Placing aquifer management in the hands of the aquifer users, under the supervision of the River Basin Council, State Water Commissions and the CNA, shows more promise of reducing extractions than the current system of *vedas* and federal regulation.

The institutional arrangements for water allocation in the Lerma-Chapala basin revolve around who controls water. With basin closure, the competition for access to water is becoming more severe and poor people are losing their access to water, due to reductions in surface irrigation and increased costs for groundwater irrigation. Unfortunately, meeting the water needs of poor people and including poor women and men at all levels of water management decision-making is not a priority of the Council, nor of the larger set of institutional arrangements for water management in Mexico. The Council needs to seriously start considering how to safeguard and improve the access of the poor to water, and how to combat the current *de facto* concentration of water rights in the hands of the few.

ACKNOWLEDGEMENTS

The authors wish to thank Doug Merrey, Chris Perry and Randy Barker for their insightful and constructive comments. IWMI gratefully acknowledges the financial support provided by the German Government's Bundesministerium für Wirtschaftliche Zusammenarbeit und Entwicklung (BMZ) and the Deutsche Gesellschaft für Technische Zusammenarbeit (GTZ) for this study under the Research Program on Institutional Support Systems for Sustainable Local Management of Irrigation in Water-Short Basins. The staffing support provided by the Dutch Ministry of Foreign Affairs under its Associate Expert program is also gratefully acknowledged.

REFERENCES

Aboites L (1998) El agua de la nación. Una historia política de México (1888-1946). Mexico City, Mexico: CIESAS.

Arreguín J (1998) Aportes a la historia de la geohidrología en México, 1890-1995. Mexico City, Mexico: CIESAS-Asociación Geohidrológica Mexicana.

Barrows CJ (1998) River basin development planning and management: A critical review. World Development. 26 (1): 171-186.

Betlem I (1998) River basin planning and management. in F.N. Correia (ed.) Water resources management in Europe. Volume II: Selected issues in water resources management in Europe. pp. 73-104. Rotterdam: Balkema.

Buller H (1996) Towards sustainable water management. Catchment planning in France and Britain. Land Use Policy. 13(4): 289-302.

Carrillo-Arronte R (1975) Plan Lerma, organismo promotor del desarrollo regional. Cuernavaca, Mexico: PLAT.

CNA (1991) Volúmenes máximos de extracción de agua superficial para los sistemas de usuarios de la cuenca Lerma-Chapala. Ciclo noviembre 1991 - octubre 1992. Boletín No. 1. Mexico City, Mexico: CNA.

CNA (1992) Volúmenes máximos de extracción de agua superficial para los sistemas de usuarios de la cuenca Lerma-Chapala. Ciclo noviembre 1992 - octubre 1993. Boletín No. 2. Mexico City, Mexico: CNA.

CNA (1993) Volúmenes máximos de extracción de agua superficial para los sistemas de usuarios de la cuenca Lerma-Chapala. Ciclo noviembre 1993 - octubre 1994. Boletín No. 3. Mexico City, Mexico: CNA.

CNA (1994) Volúmenes máximos de extracción de agua superficial para los sistemas de usuarios de la cuenca Lerma-Chapala. Ciclo noviembre 1994 - octubre 1995. Boletín No. 4. Mexico City, Mexico: CNA.

CNA (1995) Volúmenes máximos de extracción de agua superficial para los sistemas de usuarios de la cuenca Lerma-Chapala. Ciclo noviembre 1995 - octubre 1996. Boletín No. 5. Mexico City, Mexico: CNA.

CNA (1996) Volúmenes máximos de extracción de agua superficial para los sistemas de usuarios de la cuenca Lerma-Chapala. Ciclo noviembre 1996 - octubre 1997. Boletín No. 6. Mexico City, Mexico: CNA.

CNA (1997) Volúmenes máximos de extracción de agua superficial para los sistemas de usuarios de la cuenca Lerma-Chapala. Ciclo noviembre 1997 - octubre 1998. Boletín No. 7. Mexico City, Mexico: CNA.

CNA (1998) Volúmenes máximos de extracción de agua superficial para los sistemas de usuarios de la cuenca Lerma-Chapala. Ciclo noviembre 1998 - octubre 1999. Boletín No. 8. Mexico City, Mexico: CNA.

CNA(1999a) Proyecto lineamientos estratégicos para el desarrollo hidráulico de la Región Lerma-Santiago-Pacífico. Diagnóstico regional. Guadalajara, Mexico: CNA.

CNA (1999b) El consejo de cuenca Lerma-Chapala 1989-1999. 10 años de trabajo en favor de la gestión integral y manejo sustentable del agua y de los recursos naturales de la cuenca. CNA: Guadalajara, Mexico.

CNA (1999c) Volúmenes máximos de extracción de agua superficial para los sistemas de usuarios de la cuenca Lerma-Chapala. Ciclo noviembre 1999 - octubre 2000. Boletín No. 9. Mexico City, Mexico: CNA.

CNA (1999d) Ley de aguas nacionales y su reglamento. Mexico City, Mexico: CNA.

Consejo Consultivo de Evaluación y Seguimiento del Programa de Ordenación y Saneamiento de la Cuenca Lerma-Chapala (1991) Acuerdo de coordinación de aguas superficiales. Colección Lerma-Chapala Vol. 1 No. 5. Queretaro, Mexico: CNA.

Consejo Consultivo de Evaluación y Seguimiento del Programa de Ordenación y Saneamiento de la Cuenca Lerma-Chapala (1993) Propuesta de acciones para reglementar la distribución, uso y aprovechamiento de las aguas del subsuelo en los acuíferos que se ubican en la cuenca Lerma-Chapala. Queretaro, Mexico: CNA.

de Anda J, Quiñones-Cisneros SE, French RH and Guzmán M (1998) Hydrologic balance of Lake Chapala (Mexico). Journal of the American Water Resources Association. 34(6): 1319-1331.

Sandoval F (1981) Obras, sucesos y fantasías en el Lago Chapala. Guadalajara, Mexico: Gobierno de Jalisco.

Sandoval F (1994) Pasado y futuro del Lago Chapala. Guadalajara, Mexico: Gobierno de Jalisco.

Falkenmark M and Lundqvist J (1998) Towards water security: Political determination and human adaptation crucial. Natural Resources Forum. 21(1): 37-51.

García y García E (1998) Diagnóstico electromecánico de los sistemas de bombeo de agua subterránea en el estado de Guanajuato. pp. 160-167 in Memoria del Simposio Internacional de Aguas Subterráneas, 7-9 December 1998. León, Guanajuato, Mexico.

Guerrero V (1998) Participación social en el aprovechamiento sustentable de las aguas subterráneas — El caso de Guanajuato, pp. 33-42, in Memoria del Simposio Internacional de Aguas Subterráneas, 7-9 December 1998. León, Guanajuato, Mexico.

Haisman B (1999) Sustainable water resource development and management. Notes on the Murray-Darling Basin experience. in H Manthrithilake and R White (eds.) Proceedings of the Mahaweli River Basin Management Workshop. pp. 11-21. Colombo, Sri Lanka: Mahaweli Authority of Sri Lanka.

IEPES (1987) Diálogo nacional. El agua: Recurso vital. Mexico City, Mexico: IEPES.

IWMI (2000) Water issues for 2025. A research perspective. Colombo, Sri Lanka: IWMI.

Lee TR and Jouravlev AS (1998) Prices, property and markets in water allocation. Santiago, Chile: United Nations Economic Commission for Latin America and the Caribbean.

Mestre E (1993) Avances en al gestión del agua y sus finanzas en la cuenca Lerma-Chapala. Paper presented at the 1st International Symposium on Water Economics and Environment, July 1993.

Mestre E (1997) Integrated approach to river basin management: Lerma-Chapala case study – attributions and experiences in water management in Mexico. Water International. 22(3): 140-152.

Molden D (1997) Accounting for water use and productivity. SWIM Paper 1. Colombo, Sri Lanka: IIMI.

Mostert E, van Beek E, Bouman NWM, Hey E, Savenije HHG and Thissen WAH (1999) River basin management and planning. Keynote paper for the International Workshop on River Basin Management, The Hague, 27-29 October 1999.

Newson M (1997) Land, water and development. Sustainable management of river basin systems. London and New York: Routledge.

Orive-Alba A (1960) La política de irrigación en México: historia, realizaciones, resultados agrícolas, económicas y sociales, perspectivas. Mexico City, Mexico: FCE.

Orive-Alba A (1970) La irrigación en México. Mexico City, Mexico: Editorial Grijalbo.

Palacios E and Martínez R (1999) Aspectos constitucionales, legales e institucionales. pp. 6.1-6.41 in Políticas opcionales para el manejo de la sobrexplotación de acuíferos en México. Estudio sectoral. Mexico City, Mexico: World Bank/CNA.

Peña S and Arreguín J (1999) Opciones de manejo para acuíferos sobreexplotados y para la formulación de políticas en México. pp. 2.1-2.48 in Políticas opcionales para el manejo de la sobrexplotación de acuíferos en México. Estudio sectoral. Mexico City, Mexico: World Bank/CNA.

Poleman T (1964) The Papaloapan project. Agricultural development in the Mexican tropics. Stanford, California: Stanford University Press.

Rap E, Wester P and Pérez-Prado LN (1999) The articulation of irrigation reforms and the reconstitution of the hydraulic bureaucracy in Mexico. Paper presented at the International Researcher's Conference The Long Road to Commitment: A Socio-Political Perspective on the Process of Irrigation Reform. 11-14 December 1999, Hyderabad, Andhra Pradesh, India.

Reisner M (1993) Cadillac desert. The American West and its disappearing water. New York: Penguin Books.

Sánchez M (1998) La primera transferencia. Gestión y administración federal del agua en México. in P. Avila García (ed.) Agua, medio ambiente y desarrollo en México. pp. 139-146. Zamora: Colegio de Michoacán.

Scott C, Zarazúa JA and Levine G (2000) Urban-wastewater reuse for crop production in the water-short Guanajuato river basin, Mexico. IWMI Research Report 41. Colombo, Sri Lanka: IWMI.

Scott CA, Silva-Ochoa P, Florencio-Cruz V and Wester P (2001) Competition for Water in the Lerma-Chapala basin. Economic and Policy Implications of Water Transfers from Agricultural to Urban Uses. *In: The Lerma-Chapala Watershed: Evaluation and Management* (Hansen AM and van Afferden M eds) Kluwer Academic/Plenum Publishers, London.

Seckler D (1992) Irrigation policy, management and monitoring in developing countries. In Roundtable on Egyptian Water Policy, Conference Proceedings (Abu-Zeid and Seckler D ed) Water Research Center, Ministry of Public Works and Water Resources, Cairo, Egypt and Winrock International, Arlington, Virginia, USA.

Seckler D (1996) The new era of water resources management: From "dry" to "wet" water savings. IIMI Research Report 1. Colombo, Sri Lanka: IIMI.

Seckler D, Amarasinghe U, Molden D, de Silva R and Barker R (1998) World water demand and supply, 1990 to 2025: Scenarios and issues. IWMI Research Report 19. Colombo, Sri Lanka: IWMI.

Turton AR and Ohlsson L (2000) Water scarcity and social stability: Towards a deeper understanding of the key concepts needed to manage water scarcity in developing countries. SOAS Water Issues Study Group Occasional Paper 17. London: University of London.

Vallejo-Ivens F (1963) Origen, finalidades y resultados, hasta 1963, de la Comisión del Sistema Lerma-Chapala-Santiago. Paper presented at the Third Latin American Irrigation Seminar, December 1963, Guadalajara, Mexico.

van Koppen B (2000) From bucket to basin: Managing river basins to alleviate water deprivation. Colombo, Sri Lanka: IWMI.

Vázquez ML (1999) Decretos de veda en el estado de Guanajuato. Aqua Forum. 4(15): 20-21.

Vermillion DL and Merrey DJ (1998) What the 21st Century will demand of water management institutions. Zeitschrift für Bewässerungswirtschaft 33 (2):165-187.

Wester P, Marañón-Pimentel B and Scott C (1999) Institutional responses to groundwater depletion: The aquifer management councils in the state of Guanajuato, Mexico. Paper presented at the International Symposium on Integrated Water Management in Agriculture, 16-18 June 1999, Gómez Palacio, Mexico.

SECTION V

SUMMARY AND CONCLUSIONS

Summary and Conclusions

The management of watersheds is one of the most complex tasks that scientists, users and policy makers are confronted with, especially in recent years when freshwater resources have diminished and conflicts on water allocation have increased worldwide.

The present book was written as a response to the need for systematic information on aspects related to the management of the most important and controversially discussed Mexican hydrological basin: the Lerma-Chapala watershed.

Additionally, media exposure on shrinking water levels of Lake Chapala, the largest in Mexico, underlines the urgent need of a multidisciplinary discussion, including social, ecological, economical, technical, and legal aspects.

Through the information provided in this book, future research needs and promising management options were identified. The reader will notice some different points of view among authors' discussions in the book. The editors have decided to maintain these as a reflection of the current state of dispute about the crises in the basin. For example, some worry about surface water depletion and Lake Chapala water level decrease as the main existing problems, which have to be resolved, others that groundwater depletion has to be addressed as the limiting factor in the sustainable development of the region. A hydrologist would claim that surface and groundwater are closely related. Nevertheless, institutions involved in water management frequently have influence on separate components of the water cycle.

Another argument occurs around the reasons for the low water levels in Lake Chapala during the last decades. Some authors claim that this is mainly due to cyclic climate variations, while others refer to basin transfers, excessive extractions and poor efficiency in the water use, as the main causes.

1. CURRENT SITUATION

By considering indicators, such as water in quantity and quality, ecology/health implications, social aspects and conflicts, and especially the situation of Lake Chapala, the following conclusions can be drawn on the current state of the watershed:

Water quantity: As economic activity increases, so does water demand.

Water supply in the Lerma river basin as well as in the neighboring Valley of Mexico and Guadalajara is completely allocated and regional water availability is insufficient to cope with these demands. This has led to an over-exploitation of surface and groundwater resources that endangers both the medium and long-term sustainability in the region.

It is clearly observed that the Middle Lerma watershed is the most over-exploited sub-region that strongly influences the overall groundwater deficit in the whole watershed. As a consequence, springs have disappeared and natural water bodies have shrunk significantly.

Water quality: Natural water resources in the basin have long been object of pollution from cities, industries and agriculture. Although considerable efforts have been made to increase the number and capacity of treatment facilities in the basin, today high percentages of effluents are still discharged untreated. Concentrations of some heavy metals in water samples of Lake Chapala have been occasionally above the national limits for safe drinking water sources. This situation may become worse when water levels decrease and contaminants further accumulate, thereby indicating an urgent need for monitoring.

Existing data on toxic pollutants in river and lake sediment samples were below the Severe Effect Level-criteria that have been adopted by the Canadian government for highly polluted sediments. Nevertheless, it should be pointed out that a detailed comparative analysis of the data with ecotoxicological Low Effect Level-criteria reflects a different situation. Both sediment samples of lake and rivers sampling stations exceeded this level in at least one parameter, indicating that sediments in the Lerma-Chapala basin are affected by industrial, municipal, and/or agricultural discharges.

Ecology/Health: Erosion can be considered as one of the most serious ecological problem in the basin, which affects natural resources. Soil erosion occurs moderately in 45% of the Lerma-Chapala basin and in an accelerated way in 35% of the region.

The existing nutrient concentrations in the basin water bodies, promote eutrophication associated by the growth of floating aquatic vegetation that have bloomed in the Lerma river, Lake Chapala, and the Santiago river. More recently, there have been blooms of blue-green algae, that as well could cause taste and odor problems in drinking water supply and affect ecosystem-health.

Another ecological problem is the over-exploitation of fishing resources, which are intensified by the low productivity of the system that is caused by decreasing water quality and quantity. The few data on metals in fish indicate that concentrations do not exceed limits for acute toxicity, but point to a need to assess chronic human health risks associated with fish consumption.

Social conflicts: The relatively limited water resources in the watershed and its dense population make water-related conflicts likely. Nevertheless social conflicts are difficult to quantify and even to define as purely water-related; the industry is exerting pressure to acquire water rights formerly in possession of irrigated farmland. Something similar is occurring with water supply for cities and townships.

Some users do not agree with the current structure of the River Basin Council. They consider that the water user representatives for irrigation, water supply, industries, fisheries and service sectors were, in some cases, appointed by a small group of privileged users.

Even at the local level, water-related conflicts have been reported among some families and agricultural producers. These disagreements mixed with simultaneous struggle for land, even have had fatal victims and incited rivalries.

Lake Chapala: Since Lake Chapala is the final destination of the water captured in the Lerma river basin; it reflects and to some extent amplifies the water-related problems of the whole watershed.

Above all, the current drying up of Lake Chapala, caused most probably by the sum of hydro-meteorological fluctuations and man-made over-exploitation in the up stream sub-basins, make it unlikely that the lake will recover without increased runoff.

It is clear that something has to be done to preserve Lake Chapala, which generates significant tourism revenues, represents a unique ecosystem, generates fishery, and supplies over two million inhabitants of Guadalajara with domestic water and last but not least, has an elevated symbolic value as it is still Mexico's largest lake.

2. PROBLEM ANALYSIS

Given that all the above-mentioned indicators are negatively affected, the present situation of the whole basin may be classified as seriously unbalanced. This arises the questions: What are the reasons, and is there really not enough water for all users, including Lake Chapala?

Concerning the first question we can summarize that:

- A large region of the watershed has a semi-arid climate, with temporary rainfall during few month of the year, only. This has to be considered in the integrated management of the basin.
- Fluctuations of the hydro-meteorological characteristics contribute to recurring low water availabilities in the basin.
- The largest water volume allocated to a single use in the basin is for agricultural irrigation and efficiencies in water uses are still low and improvements slow.

- Water pricing – although little by little rising – does not encourage water-efficiency measures among users, since water prices on the average are below costs.
- Important quantities of water are transferred to neighboring watersheds for water supply in Mexico City and Guadalajara. These water volumes are currently not returned to the Lerma-Chapala basin as treated wastewater.
- Concerning the social structure in the basin, restrictions in water uses are not easily implemented since they affect the existence of low-income groups as farmers that strongly depend on low-rate water availability and irrigation.
- The size of the basin and the fact that it partially includes five states make coordination actions especially difficult.
- The legal framework complicates the coordination as well as the vertical and horizontal integration and participation of user's groups and institutions in water allocation policies, both locally and in the entire region.
- Changes in the use of forest soil for agriculture and urban developments causes erosion and consequent loss of water storage capacity. Additionally, in the Upper Lerma region seasonal agriculture is practiced with low technology on steep slopes, with limited vegetation, and deficient terrace outlining.
- Water quality is suffering due to an incomplete coverage of wastewater treatment facilities and control mechanisms, such as voluntary commitments, international markets, national authorities, monitoring, public interest, and education.

The second, above-mentioned question "...is there really not enough water for all users, including Lake Chapala?" is much more difficult to answer.

For a sustainable development of the watershed, both in the productive and ecosystemic meaning, it has to be accepted that water is a limited resource. In this sense the question should be answered with a NO!

On the other hand the answer could be a YES, if the potential for optimizing water uses, which is still to be explored, is sufficient to maintain existing productivity and at the same time reduce social conflicts and guarantee the ecological intactness in the basin.

3. FINAL CONSIDERATIONS

Among considerations that should be made and actions that need to be undertaken, first of all, a review and analysis of the current situation of the watershed should be carried out with the scope to quantify the optimization potential of water uses in the basin. This analysis should quantify the amounts of water necessary to satisfy the needs of all users, including aquifers and ecosystems. The result should be a balanced allocation of the resource, taking into account the ecological needs for water of acceptable quality.

Secondly, mechanisms and instruments should be identified and applied that allow to even out this unbalance, such as adaptation of laws, reorganization of social and institutional participation, technification and modernization of irrigation infrastructure as well as improvement and construction of wastewater treatment facilities, and rehabilitation actions such as reforestation and soil conservation programs.

It is urgent to create the necessary tools for achieving sensibility, awareness and change in attitude of members of the society, by considering water as an element of life.

It is essential to estimate prices for different water uses in order to help define assignation and compensation schemes for the water resources. Indicative marginal costs and their variation according to local conditions as presented in this book, may be useful for present and future transfers among sectors, specially when water resources are scarce as they have been during the last decades.

Affordable financing instruments and an effective water pricing system may catalyze more efficient water allocation and use, provided other control mechanisms like social development, health and environmental risk assessments, water quality and quantity monitoring as well as institutional and public environmental education are included.

Through biannual evaluation of the influences of each of these actions and reporting to the public, the short and long-term improvements have to be controlled for their abilities to solve the above-mentioned problems and to approach a sustainable development of this most important basin in Mexico.

Anne M. Hansen and Manfred van Afferden
Instituto Mexicano de Tecnología del Agua. Paseo Cuauhnáhuac 8532, Jiutepec, 62550 Mor., Mexico

Index